Teaching Mathematics in Primary Schools

Teaching Mathematics in Primary Schools

Principles for effective practice

3rd edition

Robyn Jorgensen, Shelley Dole and Kevin Larkin

Routledge
Taylor & Francis Group
LONDON AND NEW YORK

First published 2004 by Allen & Unwin

Published 2020 by Routledge
2 Park Square, Milton Park, Abingdon, Oxon OX14 4RN
605 Third Avenue, New York, NY 10017

Routledge is an imprint of the Taylor & Francis Group, an informa business

Our thanks to the schools for allowing access to classrooms for photographs.

The authors acknowledge the contribution of Robert J. Wright to the first edition of this book, some of which has been retained or adapted in this new edition.

Second edition published in 2011

NATIONAL LIBRARY OF AUSTRALIA

A catalogue record for this book is available from the National Library of Australia

Index by Puddingburn
Set in 12/14.7 pt Bembo MT Pro by Bookhouse, Sydney

ISBN-13: 9781760529734 (pbk)

CONTENTS

ABOUT THIS BOOK

This book has been designed to introduce pre-service and practising teachers to current methods of teaching school mathematics to primary and middle-years students. We have drawn on a wide range of research to inform the book, but have not taken one specific approach. Rather, our intention has been to expose readers to the diversity of approaches to teaching school mathematics. Our rationale for doing this is that students learn from a range of experiences, and, with such diversity in contemporary classrooms, teachers need to have a wide repertoire of skills and knowledge if they are to provide the quality experiences needed for student success. It is recognised that the most important factor in creating success for learners is the teacher. As such, teachers in today's society have a critical role in facilitating successes for all students. By having a rich repertoire and a strong knowledge based on evidence, teachers are better able to prepare their students for a positive future.

We have opted to use the term 'school mathematics' rather than 'mathematics' because there is increasing recognition that what is taught in schools is a particular form of mathematics. It is quite different from the mathematics used in workplaces or by mathematicians, and should be recognised as such. Creating opportunities for students to learn and develop a deep appreciation for mathematical thinking and doing is essential for success in life. School mathematics plays an important role in the preparation of young people for their entry into the adult world.

Trends in research are reflected by the availability or paucity of relevant publications. In this third edition, we have updated the book with more current literature—particularly through an increased emphasis on the use of digital technologies in mathematics education. In this edition, we have also consolidated various aspects of numeracy found throughout the second edition into a dedicated chapter on numeracy (Chapter 8), in which we examine numeracy from a range of perspectives and distinguish between mathematics and numeracy.

We spent some time considering the merits of including a stand-alone chapter on the use of digital technologies in the teaching of school mathematics; however, we have opted instead to incorporate this topic into each chapter. This is so that readers will be able to see where and how technologies (digital and otherwise) can be used in the teaching of school mathematics. We argue strongly that, when used appropriately, technologies are useful tools in the enhancement of mathematical learning (Larkin 2016). They enable deeper explorations of many ideas by removing the tedium of calculations or drawings, and can extend children's mathematical understanding, particularly in geometry. However, technologies can also be used inappropriately (e.g. as a form of reward or as a time filler). We address these issues within each chapter when we propose various digital technologies to support learning and teaching.

A continued theme in this third edition is the inclusion of the 'middle years'—those years between upper primary and early secondary—since many schools maintain specific programs for learners in these years. For many students, the transition to secondary school involves revising much of the mathematics taught in primary school, creating the risk of significant repetition in learning, rather than being seen as part of the learning continuum. To help avoid this considerable overlap, we have included the extra school mathematics that is typically taught in these years, and provided new links to digital resources that can support this new mathematical content. This not only caters for teachers working in the middle years but also allows primary school teachers to see where learners will be taken in their later mathematical learning.

References

Larkin, K. (2016). Mathematics education and manipulatives: Which, when, how? *Australian Primary Mathematics Classroom, 21*(1), 12–17.

ABOUT THE AUTHORS

Robyn Jorgensen is an Emeritus Professor at the University of Canberra. Robyn has worked across the various sectors of schooling, from early years, primary, middle and secondary to vocational and post-compulsory schooling. Her research interests are centred on issues of equity and access, particularly for students from disadvantaged backgrounds—low socio-economic status, rural, remote and Indigenous students. She has received thirteen Australian Research Council grants to work in these areas. In 2009–10, she took up a position of CEO and Principal in a remote Aboriginal community. Her work is predominantly based in classrooms, where she works to better understand how teaching practices are implicated in the success (and failure) of learners. She has been Chair of the Queensland Studies Authority Curriculum Advisory Board for Mathematics; worked on the Queensland Ministerial Committee for Science, Technology, Engineering and Mathematics (STEM); and was the eminent Professor on the Australian Association of Mathematics Teachers' 'Count on Me' project, which seeks to improve mathematical learning for Indigenous Australians. Robyn has a distinguished publication record and has authored many books, journal articles and conference papers. She has been invited to be the keynote speaker at numerous state, national and international conferences. While currently living in Queensland, she has also lived in Victoria, New South Wales and the Northern Territory.

These experiences have helped to inform her breadth of knowledge of the diversity of Australian society and curriculum.

Shelley Dole is Professor and Head of School of Education at the University of the Sunshine Coast. Shelley is an educator and mathematics education researcher. Over the 25 years she has been in education, she has taught in primary, secondary and tertiary teaching institutions throughout Australia. Shelley has led two major Australian Research Council projects, focusing on numeracy across the curriculum and the development of proportional reasoning, with teachers and schools in Queensland and South Australia. She has been involved in several major research projects in Queensland, Tasmania, Victoria and South Australia, including middle-years literacy and numeracy; early-years literacy and numeracy and distance education; mental computation, number sense and invented algorithms; teaching and learning per cent in the middle school; basic facts in the early years; as well as professional development projects. In 2009, Shelley won a University of Queensland Award for Teaching Excellence, and in 2010 she was the recipient of an Australian Award for University Teaching. She is a senior fellow of the Higher Education Academy.

Kevin Larkin is a Senior Lecturer (Mathematics Education) at Griffith University, where he teaches two of the core primary undergraduate education mathematics courses at three campuses. Kevin is a mathematics educator with 37 years of educational experience. He taught in primary-school classrooms for fifteen years (both in Australia and internationally) and worked in various leadership positions for a further fourteen years. Since 2012, Kevin has worked at Griffith University. Every aspect of his teaching is informed by an overarching goal to inspire new generations of creative, motivational mathematics teachers. His teaching philosophy is based on a deep appreciation of the critical role of mutual and supportive working relationships between teachers and learners, and commitment to the provision of learning and teaching environments that are personally, socially and educationally rich. His teaching is complemented by a robust research and scholarship portfolio, which includes over 70 publications, two Office for Learning and Teaching grants, and three Griffith Grants for Learning and Teaching. Kevin has published widely in national and international publications in the areas of mathematics education, digital technologies, early-years STEM, higher education and activity theory. His particular area of expertise is the use of digital technologies in mathematics education. He is a Senior Fellow of the Higher Education Academy, and is the Inaugural Chapter Chair for the Arts Education and Law Group Learning and Teaching Academy. Kevin's expertise in learning and teaching has been recognised, among other awards, with a National Teaching Citation in 2017 and recognition in 2016 as Griffith University's Teacher of the Year. Kevin is the 2018 University Teacher of the Year as well as the recipient of the 2018 Award for Teaching Excellence (Social and Behavioural Sciences).

CHAPTER 1

THE CHANGING FACE OF SCHOOL MATHEMATICS

There is a need for new approaches to teaching mathematics. Sadly, mathematics is still rated by many students as one of their least favourite subjects at school, partly because mathematics teaching is too often typified by students working individually with their desks in rows facing the board, and the teacher demonstrating procedures from a textbook or an interactive whiteboard (IWB), with students completing worksheet exercises. With the vast amounts of research that have been generated since mathematics education became recognised as a discipline, there is now a strong research base to inform change in teaching school mathematics. Also, wide changes are occurring rapidly in societies, both nationally and internationally. Change impacts students, and there are new theories of how students learn in contemporary times. Research has also contributed to changing perceptions of mathematics as a discipline.

Mathematics education and society

The mathematics curriculum has not been created in a vacuum. Mathematics in schools, and the way it is taught, are the product of broader factors that extend beyond the classroom. Such factors include employers, lobby groups, government policy, parents and professional organisations.

External authorities, such as education departments and other statutory bodies, may develop a curriculum or syllabus that provides guidelines teachers are expected to use to develop their work programs, and to undertake assessment and reporting of students' learning. The development of curriculum documents is influenced by demands from groups including parents, employers and governments. In the Australian context, the Australian federal government developed an agreement among the states and territories for the development and implementation of a national curriculum, which began in mathematics in 2012. Documents relating to mathematics in the Australian national curriculum can be accessed through the Australian Curriculum, Assessment and Reporting Authority (ACARA) website (www.australiancurriculum.edu.au/f-10-curriculum/mathematics/). Documents relating to the New Zealand National Curriculum can be found at the New Zealand Curriculum Online website (http://nzcurriculum.tki.org.nz/The-New-Zealand-Curriculum/Mathematics-and-statistics). When all variables are considered, the curriculum guidelines that appear in mathematics classrooms have been created through a highly negotiated (and often hotly contested) process. In addition to curricula, many jurisdictions around the world now utilise some form of high-stakes testing. Examples include, but are not limited to, the Trends in International Mathematics and Science Study (TIMSS), the Programme for International Student Assessment (PISA) and, in Australia, the National Assessment Program—Literacy and Numeracy (NAPLAN). The use of these external, high-stakes assessments affect how teachers teach, and students learn, mathematics.

In some cases, curriculum shaping is a reciprocal process, where the benefits of change are two-way. Consider the impacts of technology within society, and of numeracy expectations. In the current context, employers are demanding that students exit schools with high levels of numeracy (and literacy). As a result, there is a much heavier emphasis on numeracy in education. Similarly, schools recognise the value of technology as a learning tool that enables students to exit schools with a strong appreciation of how technology can be used to enhance mathematical work and thinking.

Teaching mathematics in modern society

We use the term 'modern society' throughout this book to refer to the contemporary context of education. 'Modern' is more appropriate than terms such as 'Western' because it does not support a notion that Western views and approaches are more valid than those of Eastern or indigenous cultures. Instead, it suggests that the curriculum reflects a contemporary view of education that embraces new approaches to teaching, such as the use of digital resources (the internet, computers, tablet devices, and so on).

In the times in which our students live, technology, globalisation, the information age and very different patterns of family, leisure and work have brought changes to society, work, schools and life. We use the term 'modern society' to provoke thinking about the age in which we live and the quite different lives of contemporary young people, and to consider how these have changed since 'the old days'. Educational researchers have underscored that modern society is different. Many cultures—such as those that embrace Eastern philosophies, or indigenous cultures seeking to gain access to contemporary ways of thinking and learning—also live in the modern world. Curricula in schools must reflect the changes occurring in the wider society to ensure that schools adequately prepare students for the world beyond compulsory schooling.

Mathematics classrooms

Most young people are now growing up in technology-rich environments. They do not remember a time when you had to physically get up to change the television station (remote controls do that)—indeed, many are now watching television using streaming technologies. Cooking in pre-programmed microwaves happens at the touch of a button. Today's young people are generally technologically savvy.

One of the biggest growth areas in employment is self-employment, which means that many young people will be creating jobs for themselves in positions that will not even exist when they exit school. Our students are growing up immersed in an information-rich society—they no longer have to search through the relatively few books in a school or a local library, but can instead undertake searches on computers or handheld devices that may yield thousands or even millions of hits. The skills they

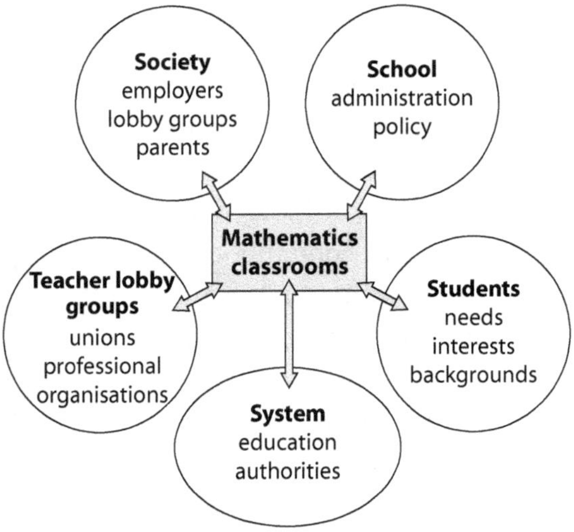

require in order to be able to search for and identify key information are very different from those they needed when only 'page-based' texts were available.

Students growing up in this technology-rich world have become used to multiple sources of information input—they are constantly bombarded with short bursts of infotainment, as well as brief snippets of information from television and other media. They are able to fragment and reconstruct images (such as maps) in ways unimaginable in former times (Lowrie 2003), as well as practise algebraic thinking through the study of curves made possible by dynamic modelling software (Padula 2014). Commonly used terms to describe contemporary students, such as 'cyberkids' or 'digital natives', although contested, recognise that their dispositions towards learning have been formed by the wider social conditions in which they have grown up. Traditional models of teaching and learning need to reflect, and sometimes to challenge, these changed circumstances.

The mathematics education of the students of modern society must be considered in light of this. Students need to develop mathematical ways of seeing and interpreting the world; they need to develop strong problem-solving skills; they need to be numerate; and, most importantly, they must have a disposition towards using mathematics to solve the problems they confront. School mathematics needs to adopt pedagogies that will cater for diversity within a classroom. The old models of seated individual work—found in what might be termed 'traditional mathematics teaching'—are possibly contributing to the problems that emerge as students progress through school. For considerable numbers of students in the years of upper primary and lower secondary school, the teaching that they encounter can lead to many negative feelings and misleading learnings about mathematics (Larkin & Jorgensen 2016).

Eastern philosophies that focus more on the wellbeing of learners are impacting the pedagogies used in Western classrooms. Notions of happiness and mindfulness, as per Buddhist philosophy, are gaining traction in teaching approaches. Yoga is finding a place in the classroom, too.

These approaches are assisting learners to focus and be 'in the moment', rather than be distracted by the numerous stimuli that abound in the contemporary world.

New models of teaching mathematics

The mathematics curriculum encountered by students prior to the 1960s focused on arithmetic and operations. Most of the mathematics education developed after that time in Western countries emerged post-Sputnik, when the race to the moon had become a race for intellectual superiority, with mathematics seen as the linchpin of success. The 'New Mathematics' contributed to a lock-step approach to teaching mathematics, with hierarchies in orders and sequences of teaching (Brown et al. 1998). The 1970s witnessed a boom in the ways in which mathematics curricula were organised; most were not research-based, but rather were influenced by arguments of logic and reason. A hierarchical approach to mathematics teaching ('skill, drill and kill', before application and problem-solving) was implemented in most Western classrooms, and for many teachers and systems, such an approach has become a way of life. Brown and colleagues (1998) argue that much of what was written in terms of mathematics curriculum reform had very little research base, thus raising questions about the validity of the curriculum itself.

In more recent times, there has been a growing awareness that such approaches are not resulting in positive learning outcomes. Indeed, as Clements (1989) argues, all that students learn from ten years of compulsory schooling (which, in most countries, may be extended to twelve to thirteen years) is that they cannot do mathematics! However, not all countries have bought into this approach to teaching. Research, while somewhat dated, emanating from these countries, particularly the Netherlands (see Anghileri 2001, 2006; Beishuizen 1999; Buys 2001; Treffers & Beishuizen 1999; van den Heuvel-Panhuizen 2001), has focused on developing new methods and approaches to teaching mathematics. This research is now being adopted in many countries, as it has been shown to enhance the understandings of numbers, particularly seeing numbers, in a very flexible way (Anghileri 2006; Revina & Leung 2018). The program, referred to as 'Realistic Mathematics' (van den Heuvel-Panhuizen & Drijvers 2014), has sought to develop deep number sense from real-world examples.

The Netherlands did not embrace the New Mathematics movement, instead focusing its efforts on how students think mathematically (Treffers 1991). There is now a substantive body of knowledge drawing on students' thinking that has not been constrained by New Mathematics. From this, curricula have been developed that draw on students' understandings, build on them, and move progressively towards abstract and formal mathematical processes. Dutch mathematics reformers call this 'progressive mathematisation'. Occurring in parallel with this work has been the development of constructivist theory and a general awareness that students actively construct meaning from their experiences. This latter work has had a powerful influence on mathematics education, with it increasingly being recognised that students' individual understandings are based on their lived experiences.

These twin movements have emerged at a time when it was being recognised that many of the old, 'traditional' methods of teaching mathematics were failing too many students. This made the moment ripe for identifying more valid methods of teaching mathematics, and many countries, states and provinces adopted new methods of mathematics teaching and learning. More recently, methods such as direct or explicit instruction have seen a resurgence in traditional behaviourist approaches to teaching, and bring their own set of challenges for the authentic learning of mathematics. The use of approaches such as direct instruction, which is founded on a strong behaviourist approach to teaching, is guided by the strong influence of conservative politics advocating a return to 'the basics', as well as the ongoing questioning of teacher quality. Teacher standards and curriculum/assessment requirements are now placing unprecedented demands on teachers. This makes the work of the mathematics teacher very complex and bound by many compliancy regulations.

Content and pedagogy

Contemporary approaches to teaching mathematics need to encourage two aspects: content and pedagogy.

Content is the intellectual integrity of the subject. It is where students learn, apply and appreciate mathematics, and where deep learning and deep knowledge are paramount to learning experiences. Importantly, students are able to make connections between the mathematics they learn and other curriculum areas, as well as with the world beyond school. It is important for them to develop an appreciation of how mathematics is an informing discipline that has importance and relevance to many spheres of life.

Pedagogy relates to developing supportive environments where student diversity is recognised, and practices are developed that value and build upon the different backgrounds and knowledges that students bring to the mathematics classroom. Good pedagogy is about teachers developing inclusive practices to build and extend their students' knowledge and confidence in using and applying mathematics. Classrooms are places where students understand the expectations teachers have of them, and the work they are to undertake. In developing inclusive practices, intellectual integrity should also be preserved. Teachers need to value students and to believe that *all* students can learn mathematics.

Research on productive pedagogies (Education Queensland 2001; Mills & McGregor 2016) states that good pedagogy is about high intellectual engagement and helping students to see and make connections; it is learner-centred, with each individual's knowledge and culture valued, and students feeling supported in their learning. Many teachers bemoan issues of behaviour management in mathematics classes, but in a large longitudinal study on teaching (Education Queensland 2001), it was found that many of the elements of productive pedagogies were absent from the 2000-plus classrooms observed. Students were not engaged in deep learning about and through mathematics. Often, it was found that students undertook busywork during mathematics lessons (e.g. sticking butterflies on paper, rather than engaging in discussion about area) but did not engage in much, if any, deep mathematical learning. It must be asked whether students in mathematics classes engage in behaviour that is subsequently construed as 'misbehaviour' because they are bored or because the pedagogy is unsound.

Teachers can make a difference

Teachers and teaching can make a significant difference to students' learning outcomes in school mathematics. In a large study of effective teachers, Hattie (2014) reported that it was not the school that made the difference to students, but individual teachers. Teachers have a powerful influence over what and how students learn. Through the provision of an appropriate learning environment in which content and pedagogy match the backgrounds, needs and interests of individual students, *all* students can learn mathematics.

The power of teachers' beliefs

In a study exploring the characteristics of effective teachers of numeracy, Askew and colleagues (1997) concluded that one of the most important influences on learning was the teacher's belief that all students could learn mathematics. Often, values and stereotypes influence how behaviours and actions are interpreted and implemented. Teachers who believe that some students, due to their backgrounds or behaviours, are unable to learn mathematics will ultimately create learning environments that construct the expected outcomes. This has been shown to be the case in many studies. For example, in a seminal study (Rosenthal & Jacobsen 1969), it was shown that when students were assigned scores randomly as they commenced study in a new class at the beginning of term, the teacher, believing that these were the students' academic scores, interacted with different students in particular ways. By the end of the teaching term, the students' results for that class correlated closely with the scores they had been assigned randomly at the term's beginning.

This study (and many subsequent ones) highlights how powerfully teachers' views of their students influence the ways in which they organise learning experiences. In an influential study undertaken in New Zealand, Bishop and Berryman (2006) investigated how to improve the educational achievement of Māori students, finding that if a teacher believes that all students can learn mathematics, then learning environments are likely to reinforce this belief. The important work undertaken by Sarra (2012) and others (e.g. Buxton 2017) in relation to Aboriginal and Torres Strait Islander learners has focused on teachers having high expectations of learners. Similarly, if teachers believe that the best way to learn mathematics is by making authentic connections with real-life examples, or by incorporating digital technologies to support students' mathematical understanding, then their teaching and learning environment will reflect these beliefs. The classroom is likely to be peppered with equipment and displays demonstrating the links between mathematics and the world beyond schools. In good teaching, teachers must be mindful of the influence of their beliefs about teaching and learning mathematics, and of the influence these have on their students' learning outcomes.

What is mathematics?

When a member of the general community is asked the question, 'What is mathematics?', they invariably focus on number and operations. However, the modern mathematics curriculum is far more complex than arithmetic alone. School mathematics has changed over time; during different periods, different mathematics has been taught. A century ago, mathematics in upper-primary classes entailed computation of tasks involving a large amount of numbers, long division, square roots of non-square numbers, and so on. This form of curriculum remained in place until the early 1960s, when the implementation of the New Mathematics represented a considerable shift in mathematics curriculum in most Western countries. New topics were included, and new forms of thinking mathematically (e.g. set theory) became part of the new syllabi.

Since the 1970s, other reforms have influenced the curriculum, including problem-solving in which students were expected to be more creative in their thinking. In the 1980s and, increasingly, since then, technology has played a role in the curriculum. While the software program LOGO® had a strong influence in the early years of technology-aided learning, it has now been replaced by other technological aids, including new software programs, spreadsheets and graphic calculators, as well as, more recently, a multitude of applications (apps) available on tablet devices. Students are expected to be far more creative in their thinking, and to deal with much more knowledge and complexity than in the pre-1960s era.

Mathematics is the study of patterns and relationships

One of the defining characteristics of mathematics is awareness of recurring ideas and relationships between and among mathematical ideas. The knowledge learned in one area of mathematics links to other areas—for example, the number facts do not have to be learned as discrete facts; rather, the student who can see that when you add 3 and 5 there is a total of 8 can also recognise that if you have 8 items and 3 are lost, then there will be 5 remaining. The place value system can be seen as a patterning of 10, so that allowing young students to play on a calculator or a tablet device with the constant function set for the addition of 10 means they can see the pattern of adding tens occurring in the tens column. Seeing these types of patterns and relationships is a key factor in how students come to learn and appreciate mathematics.

Mathematics is a way of thinking, seeing and organising the world

Students who learn mathematics as a dynamic discipline through which much of the world can be interpreted are able to make sense of a wide range of experiences. They are able to organise and analyse events in systematic ways. Sometimes this might be numerically, but it could equally be spatially or through logical reasoning. Viewing the world through a mathematical lens can make everyday tasks, such as remembering phone numbers, easier, because a pattern can be seen and people are thus able to solve problems more efficiently and effectively. Mathematics moves beyond the memorisation of a vast body of facts and procedures in a rote manner to become a more systematic and insightful process. This ability to think, see and organise the world using mathematics is particularly important in this age of 'big data', where we are swamped with information.

Mathematics is a language

In past times, mathematics and language were seen as two disparate disciplines. Today, it is seen that, in order to learn, appreciate and understand mathematics, students need to learn the language of mathematics—complete with its unique words, grammar and symbols. Knowledge of the language of mathematics allows complex ideas to be communicated easily. Language involves communication, and mathematics is a very particular and precise language that is communicated in particular ways. It involves the contraction of lengthy tasks into concise 'sentences'. For example, a shopping expedition during which three items were purchased in one outlet and five in another becomes $3 + 5 = 8$. Abstract symbols, such as 3 and 5, represent specific concepts.

Mathematics is a tool

People use mathematics to solve problems every day. The more competent a person is with mathematics, the more efficiently they can solve problems and, in many cases, the more they can thrive in the world beyond school. Deciding which goods to purchase, checking bank balances, being able to budget, and deciding which phone plan or mortgage to take out or how much concrete to purchase for a patio all require the use of mathematics. Mathematics enables people to make sound decisions and judgements, and to solve problems.

Mathematics is a form of art

For many students, mathematics is a grind that has to be done under duress. In contrast, for those who have been fortunate enough to engage with mathematics, its internal beauty is awesome to behold—it has the consistency of a fine artwork. The logical coherence of Fibonacci's numbers, for example, is almost magical. (The Fibonacci sequence is generated by adding the two previous numbers in the sequence to get the next number: 1, 1, 2, 3, 5, 8, 13 . . .). Likewise, the tessellations incorporated into the artistic work of Escher remind us of the beauty of mathematics. This aspect of mathematics is perhaps the most elusive for many people, since they equate mathematics with the school process of 'doing mathematics' rather than viewing it as something to be appreciated.

Mathematics is power

Mathematics has been behind most inventions—good and bad—in modern history. It has enabled us to walk on the moon, but also to invent nuclear weapons with their potential for mass destruction. Mathematics also gives you the power to take control in your life by ensuring you are given the right change in money transactions, questioning information presented in mathematical terms, and so on. Those students who succeed in mathematics are more likely to live successful lives than those who do not. Mathematics is the social filter that facilitates the access of some students to professions of high status, wealth and power while excluding others. Developing nations actively seek access to mathematics for their youth because they know that such knowledge will benefit them in the future. Many Western countries are recognising the problems inherent in the decreasing number of students undertaking formal studies in mathematics, which is the foundation for many other forms of powerful knowledge. STEM-related disciplines such as computing, science, engineering and technology, for example, all rely heavily on a solid understanding of mathematics.

Where does mathematics come from?

While there are a number of theories regarding the generation of mathematics, two key but opposing views permeate the discipline. While there are other views—dealt with in considerable detail by Ernest (1991)—it is worth considering the two main views and how they impact both the ways in which mathematics curricula are constructed and how the discipline is taught. Just as theories of how humans developed create considerable debate, so too do theories about the origin of mathematics. There are both Platonists and Constructionists in mathematics. Some of their opposing notions are listed in the table below.

Opposing mathematical notions in Platonist and Constructionist theory

Platonist	Constructionist
Mathematics is pre-existing	Mathematics is a construction of knowledge
The role of the mathematician is to 'discover' facts	Mathematicians create new knowledge in order to explain or expand ideas
Mathematics is about truths and objectivity	Mathematics is about creativity and subjectivity
Teaching involves learning of facts	Teaching involves creating learning environments for students to create their own mathematical understandings
Assessment is objective and tests for factual knowledge; e.g. $\frac{3}{10} + \frac{4}{10} = \frac{7}{10}$ is assessable and true—it is either right or wrong	Assessment is subjective; e.g. $\frac{3}{10} + \frac{4}{10}$ should be $\frac{7}{10}$ but if a student responds with $\frac{7}{20}$, question why. It may be that the student sees the question as being about test results from two exams and not the addition of two fractions

The mathematics curriculum

Teachers developing school-based programs or shared units of work can attest to the debates that develop in seeking to form a coherent learning scheme. The mathematics curriculum is a socially negotiated and constructed document; as such, it represents particular viewpoints of various interest groups. The curriculum that is presented by a particular education authority or through a commercial venture is the product of a group of people working towards the development of an agreed document.

The meaning of the term 'curriculum' is highly contested, ranging from the collection of knowledge that will be implemented formally in a classroom through to the planned and unplanned learning of students coming to school. The more widely agreed notion of curriculum is that of knowledge being implemented so that it resonates with the notion of syllabus. This is then supported through teaching and assessment practices.

Curriculum design

When planning mathematics curricula, departments of education and other peak bodies tend to organise them around a number of key features. In many cases, these are seen primarily as the content areas, but mathematics encompasses far more than this. Consequently, these documents often include pedagogical approaches to teaching and assessment; principles of equity and diversity within the mathematics classroom and how best to cater for this aspect of mathematics education; and contemporary shifts in practice that reflect the dynamic and changing nature of not only the content but also ways of teaching and assessing mathematics. Often, the role of various technologies in mathematics is also included.

Curriculum organisation

Contemporary terminology for organising mathematics sees the curriculum broken down into content areas and levels of difficulty. These general principles for organising curricula have filtered through to most contemporary international curriculum documents, and in Australia since the early 1990s, topics within the mathematics curriculum have generally been organised under five main headings, identified as *strands*. Before 2012, the five strands of the curriculum were Number, Space, Measurement, Chance and Data, and Algebra. The curriculum frequently includes reference to 'working mathematically' to capture the processes and attitudes that the mathematics curriculum seeks to promote. In Australia, a new national mathematics curriculum was launched in 2012. It is organised according to six strands, which have been paired: Number and Algebra, Measurement and Geometry, and Statistics and Probability. The overarching idea of 'working mathematically' has been replaced by four proficiencies: Understanding, Fluency, Reasoning and Problem-solving. Similarly, in New Zealand the curriculum is organised according to three strands: Number and Algebra, Geometry and Measurement, and Statistics (including probability). Increasingly, the role of digital technologies in mathematics is acknowledged in curriculum documents. For example, in the Australian Curriculum: Mathematics, the use of digital technologies appears in a number of content descriptors across the primary years.

A general overview of the sorts of topics assigned to these strands is shown in the following table.

NUMBER
- Pre-number work around patterning, one-to-one correspondence
- Number work—cardinal numbers, nominal numbers, ordinal numbers, counting, place value
- Whole-number operations (addition, subtraction, multiplication and division)
- Rational numbers—fractions, common, decimals, percentages, mixed
- Estimation, mental computations
- Number sense
- Ratio and proportion

ALGEBRA
- Pre-algebra patterning
- Equality
- Algebra
- Functions

STATISTICS
- Data representation and interpretation

PROBABILITY
- Chance
- Statistics and data—collect, organise, display

PROFICIENCIES
- Understanding, Fluency, Reasoning and Problem-solving
- Attitudes and values
- Mathematical inquiry
- Choosing and using mathematics (modelling, applications)
- Communication, representation and language
- Problem-solving
- Connections with other strands, curriculum areas, beyond school

MEASUREMENT
- Length
- Area
- Volume and capacity
- Mass
- Time
- Money
- Temperature

GEOMETRY
- Location and arrangement—directions, grid work, coordinates (topological)
- Shape—2D shapes, 3D solids, structures, properties, tessellations
- Transformation, symmetry—changes in position, location
- Geometric reasoning

A model for organising curriculum content areas

Number

This refers to the study of pure number. It involves the early study of number and patterning. In order to develop good number sense, understanding how patterns are developed is an important part of mathematical thinking. The early years of schooling place considerable emphasis on developing a good sense of number. Formal study of number and operations occupies a considerable amount of space in the timetable. Teaching of whole numbers and operations on whole numbers generally precedes work on part numbers. Given the importance of number, it is likely that a large proportion of a teacher's time will involve the teaching of number.

Algebra

In some countries, algebra does not appear in the primary-school years, and is only introduced when students move into secondary school. In part, this could stem from the Piagetian ideology that primary-school students are not able to think abstractly; since algebra is highly abstract, it is seen as being beyond the scope of younger students. However, algebra is about patterning, and being able to identify and describe patterns is integral to the mathematics curriculum and lays the foundation for algebraic thinking (Maj-Tatsis & Tatsis 2018).

Measurement

This strand seeks to measure objects. There is potential for considerable slippage in this strand, since the concept of measurement relates to other strands as well. For example, angle could rest as easily in the measurement strand as in the geometry strand. In some documents, money—a measurement of value—is placed in the number strand because it links strongly with students' experiences with number, and notions of decimal fractions are applied easily to number. Similarly, chance and data can be measured, so in some documents they are located in this strand. These differences show the arbitrariness of curriculum construction.

Geometry

This strand, the study of how space is filled and described, typically includes shape, geometric reasoning, location and arrangement. This element of the curriculum tends to be slightly more fragmented than the other strands, since the sub-strands are not so connected, and hence are less open to a coherent framework for development. A considerable body of research documents the importance of geometry in later school achievement in mathematics and beyond.

Statistics

This is the strand that works from data to make inferences. It requires students to identify appropriate ways to collect, analyse, interpret and make inferences from data. An increasing sophistication is expected of learners as they progress through the strand, so that they can analyse data and critically appraise statistical information provided through a range of media. A branch of mathematics that is commonly associated with this strand is one referred to as 'Critical Mathematics'. This branch of mathematics seeks to identify the politics behind the use of statistical data.

Probability

Probability is often based on data collected and used to predict outcomes. In order to make effective predictions, data are usually collected or inferred mathematically, and presented in a format that makes sense of the data so that informed decisions can be made about the predictability of an event occurring. There are two main approaches to probability—experimental and theoretical—which students apply to situations and assess the likelihood of events occurring. Both approaches are used in the teaching of probability. Since data are collected and the chance of an event is predicted, some curriculum documents have statistics and probability as topics in the measurement strand.

Working mathematically (proficiencies)

Contemporary thinking sees mathematics as more than a collection of knowledge and skills. It is instead viewed as a way of thinking and working—a new dimension that is being included in most modern curriculum documents. Debate on the inclusion of mathematics as a separate curriculum strand is something that teachers need to address. The argument for mathematics to be integrated is founded on the belief that it is very difficult to undertake much mathematics without working mathematically; thus, having it as a separate strand would result in considerable doubling up of content. For example, an operation such as multiplication is often raised in a problem-solving context; while the two items are covered simultaneously, the teacher may focus on only one outcome. Not seeing the operation holistically results in considerably more work.

Diversity in curriculum organisation

While there is considerable diversity in the organisation of mathematics curricula in various states within individual countries, and across different countries, there is enough synergy among them to talk in general terms. Typically, curriculum documents are broken down into a number of organising concepts. The content areas discussed above are not meant to be comprehensive; rather, they provide some guidelines about what could typically be expected in a strand. Rather than being perceived as a static organiser, such an outline should be seen as a set of organisers that interrelate with each other. For example, multiplication links strongly with area and volume; negative integers link with temperature; and money is a measure, but also links with decimal understanding (the Number strand). Because documents are working and evolve over time, different manifestations of them can reasonably be expected, since writers and developers work under very different constraints. As curriculum documents are developed, demands are placed on writers to conform to other expectations—such as those of teachers who will be using the documents, policymakers, systems administrators, parents, employers and other lobby groups. The curriculum should be seen as a representation of particular values and interests within the community from which it emerges.

As indicated earlier, this process occurred in Australia from 2012 via the development of a new organising framework for the mathematics curriculum, reducing the five strands to three paired strands (Number and Algebra, Measurement and Geometry, and Statistics and Probability). The rationale for this organisation was that it would enable greater connectivity between topics within strands. The Australian mathematics curriculum has integrated proficiency strands that, in a sense, envelop but extend a strand of working mathematically. There are four proficiency strands: Understanding, Fluency, Problem-solving and Reasoning. The *Understanding* strand emphasises the conceptual aspects of the content, and *Fluency* emphasises associated skills. *Problem-solving* is about strategies, communicating mathematically and interpretation of problems, while *Reasoning* incorporates processes of justifying, explaining, analysing, and so on. Curriculum organisation is thus changing and shifting, and reflects the changing society in which we live and the social construction of mathematics curriculum documents.

Technology for thinking mathematically

The changes in the wider society towards a much more technology-oriented lifestyle have resulted in significant changes in the ways in which mathematics is developed. There is a substantive body of knowledge that demonstrates how computers and calculators encourage much deeper levels of mathematical thinking than are possible with pencil-and-paper work (e.g. Gowland 1998). For example, in their work with calculators, Groves (1995) has shown that students can gain a greater sense of number in the early years through guided play with calculators. Whereas most curriculum documents demarcate the first year for number study of 1 to 10 (or maybe 1 to 20), the following year for numbers to 100, and the subsequent year for numbers to 1000, Groves has shown that students in their first year of school can gain a strong grasp of numbers up to 1000. In part, this is due to a reduction of other demands—such as writing numerals—and a greater emphasis on thinking mathematically. Other advocates of digital tools have shown that their effects have been significant, both as a computational aid and as a learning tool (Bennison & Goos 2010). However, opponents of calculators (and, more recently, tablets and computers) have argued that they reduce students' capacity to complete even simple tasks. Longstanding research demonstrating the capacity of digital tools has been resisted by educators and policymakers who, unfortunately, see the digital tools as a replacement for mathematical thinking rather than as a tool to assist and extend learning.

The role of graphic calculators and graphics software in secondary-school contexts has also been debated on much the same grounds. Studies have shown that graphic calculators have enormous benefits in improving deep understanding of functions and algebra (Doerr & Zangor 2000). This is due to the calculators' capacity to show the graphs, as well as how the gradients and intercepts change with changes to the function. This dynamic process allows students to see the effects of changes quickly; old methods focused on physically plotting the graphs, which meant that few graphs could be undertaken, thereby restricting the dynamic visualisation process.

The computer and its applications have had similar effects on mathematical thinking, fostering deeper understandings of mathematical ideas. For example, the graphing function allows students to construct bar graphs or pie graphs painlessly. Constructing pie graphs by hand is very tedious. Spreadsheet software allows students to explore which graphs are best for which purposes in ways not possible (or at least restricted) when constructing graphs through pencil-and-paper methods. Similarly, using spreadsheet functionality requires students to develop simple instructions for calculations—such as 'add cell B1 to cell B2'. This process requires some algebraic thinking, and provides a strong rationale for using algebra.

See Calder (2010) for further discussion on the usefulness of spreadsheets in supporting mathematical thinking.

More recently, digital manipulatives have become increasingly available for use in mathematics classrooms (Bray and Tangney 2016). Two useful sites for teachers to access manipulatives are Illuminations (https://illuminations.nctm.org/) and the National Library of Virtual Manipulatives (http://nlvm.usu.edu/en/nav/vlibrary.html). Many of these manipulatives are now also available for use on tablet devices. The advent of more sophisticated digital tools makes it possible to pose deeper questions about mathematics—questions about patterns and place value, as well as higher-order questions such as 'Why?', 'What if?', 'What happens when?' and 'Why is this happening?' The focus shifts from the process of construction to understanding the reasons, purposes and rationales for undertaking particular work. For instance, research by Larkin, Kortenkamp, Ladel and Etzold (2018) investigated the potential of young children, supported by an application (https://apps.apple.com/au/app/place-value-chart/id568750442), to understand place value at a deeper mathematical level than that currently proposed in many curricula.

Young people in modern society have grown up in a technology-rich environment. Their familiarity with technology makes it an amenable tool for considerable innovations in mathematics education. Software such as spreadsheets can be open-ended, offering considerable potential for use in many ways to support various aspects of mathematics. Other packages are closed and restricted to particular topic areas. Examples include specific software packages for learning multiplication facts or other operations (e.g. Maths Blaster®). Such packages can simply be upmarket worksheets, and hence encourage rote learning rather than mathematical thinking. Packages such as LOGO®, Geometers Sketchpad®, GeoGebra or Calibri may be oriented towards particular topic areas—in many cases, geometry—but still have an open-endedness to them, allowing students to explore other aspects of mathematics, particularly working mathematically. When selecting software, teachers should be cognisant of what they contain and mindful of their shortcomings. Scootle (www.scootle.edu.au/ec/p/home), an online database of digital resources, provides a reliable source of activities for teachers that are mapped against curriculum topics. Overall, the research suggests that digital technologies, including calculators, are powerful tools for supporting, enhancing and extending students' mathematical understandings when used appropriately in the classroom.

Integrating curriculum

The mathematics curriculum gains its benefit and purposefulness as a discipline that informs other disciplines. For example, within the social studies curriculum, where students study environments and demographics, there is frequent referral to graphs and charts; without the contributing knowledge gained in mathematics, students would be restricted in their capacity to interpret this information. One of the major concerns of mathematics education is the fragmentation of knowledge, which leads to teachers and students not being provided with learning experiences that allow them to make connections between concepts. For students, this can result in their perceiving mathematics as a large discipline comprising a lot of disconnected facts. For teachers, it can result in the planning of many activities so that considerably more time is spent achieving outcomes. Thus, it is important to see the connections between the various strands of the mathematics curriculum, the connections to other curriculum areas and the connections to the world beyond school.

Relationships with other areas of mathematics

Each of the various strands of a mathematics curriculum should not be seen in isolation from the other strands or sub-strands; their links to other strands or sub-strands should be considered. For example, it is difficult to teach measurement concepts and skills unless students understand other number concepts (such as 1000), understand multiplication processes and have some sense of number. Similarly, the relationships between the sub-strands of volume and capacity can be explored (many students are not aware of the relationship between mL and cm, for example). One of the disadvantages of breaking knowledge into compartments, or fragments, is that such connections are at risk of not being noted.

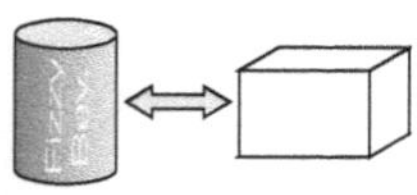

The relationship between liquid and solid measures

Integrating mathematics with other curriculum areas

Many teachers practise using an integrated curriculum model, and subject integration is a signature practice of the middle years of schooling (or

MYS) pedagogy. This means that teachers link a number of curriculum areas together. There are three reasons for working in this way:

1. *It works in a pragmatic sense.* With the increasing amount of content to be covered in primary schools, and the increasing demands on teachers' time, teachers have fewer resources available to cover essential learnings. Integration across curriculum areas allows them to ensure that they make the best use of their time. For example, if a teacher is covering report-writing in English, it is possible to link this to science and expect students to write a report for an experiment that has been conducted. Such an experiment may have involved considerable mathematics in measuring results, so links can be made to the measurement strand. Such cross-curriculum links allow teachers to assess and access students' understanding in three areas, rather than one.
2. *Mathematics cannot exist on its own.* One of the criticisms of mathematics teaching can be that it is divorced from other areas of learning. Making links to other curriculum areas allows students to see how the skills they are learning in mathematics are relevant to other areas of study. For example, students may be monitoring the water quality in a local creek. This is a part of the environmental science unit. The data that have been collected each day or week can be recorded as tables or graphs and then interpreted to determine the quality of the water. Using their mathematics skills in collecting and recording data, students are able to see their graphing skills as having a real purpose. In another example, students plotting points on a globe require a working knowledge of spherical geometry; thus, the application of mathematics is central to being able to identify locations through coordinates. This is a critical aspect of good teaching of mathematics—students being able to see why they need to learn mathematical skills and knowledge.
3. *Mathematics connects to the world beyond school.* Studies on the transfer of mathematics from the school to the workplace have shown that very little is, in fact, transferred between the two contexts. Workers often fail to use the knowledge that they learned in school mathematics to help them in their work and home environments, and this can restrict the effectiveness and quality of their lives. For example, calculating bank

balances or working out the best buys in the supermarket requires the knowledge and skills learned in school; these are experiences in which students can make important links to the world beyond the mathematics classroom. Being able to use and apply mathematics in daily activities empowers people by enabling them to make informed decisions.

Critical Mathematics

There is an approach in mathematics that specifically considers the political implications of mathematics education. Mathematics, especially when viewed from an objectivist or Platonic view, is often seen as something apolitical. That is, mathematics is an objective set of facts. However, from a Critical Mathematics standpoint, mathematics can be very political. When teaching fractions, for example, it is seen to be mathematically appropriate that sharing is an equal division of quantity. However, in some cultures sharing is based on the status of the people, and older people may get a larger share than younger people. The teaching of per cent is often done within the context of the marking up (or down) of prices. The inflation of prices to reflect profit and loss masks the capitalist ideology that profit is a normal part of society. Critical Mathematics questions the notion of normalising profit, and asks questions of who wins and who loses in these contexts. A similar example is students undertaking a large investigation focused on litter or waste. Rather than leaving the investigation as a set of tables or charts to show the findings, a further action-oriented step is undertaken to address the issues identified—such as writing to a member of parliament or taking other direct action in relation to the issue. Many examples of Critical Mathematics can be found in web-based searches.

Review questions

1.1 What are your experiences of school mathematics? How would you categorise your school mathematics experiences: Platonist or Constructionist? Give examples of experiences to support your position.

1.2 How is the mathematics curriculum organised? Write a summary of the key emphases for each strand and/or sub-strand within the mathematics curriculum in your educational context.

1.3 What is the place of written algorithms, such as long division, in your school curriculum documents? What do the current syllabus documents say about the teaching of written algorithms? What is your view on this?

1.4 What is the place of technology in the school mathematics curriculum? How has technology impacted the way you learn mathematics? What do new technologies mean for school mathematics?

Further reading

Historical curriculum documents are a good way to learn more about curriculum organisation, content and emphases. You should also look at the following:

- Curriculum documents of your locality. These may be local, state/territory or national documents. Frequently, state/territory authorities will publish curriculum or syllabus documents that interpret national documents, to which teachers are expected to adhere as they plan their teaching. All teachers need to be familiar with the materials relevant to their contexts.
- Other support materials. These can be from commercial publishers and/or can be departmental resource materials developed by local authorities to support teachers in schools.
- Materials from other regions (nationally and internationally) so that a broader sense of curriculum can be developed.
- Australian Education Council (1991). *A national statement on mathematics for Australian Schools.* Melbourne: Curriculum Corporation.

In addition to curriculum and support documents, these readings will also be useful in supporting your understanding of mathematics:

Bishop, R., Ladwig, J. & Berryman, M. (2014). The centrality of relationships for pedagogy: The whanaungatanga thesis. *American Educational Research Journal, 51*(1), 184–214. doi:10.3102/0002831213510019.

Skovsmose O. & Borba, M. (2004). Research methodology and critical mathematics education. In P. Valero and R. Zevenbergen (eds), *Researching the socio-political dimensions of mathematics education.* Mathematics Education Library, vol 35. Boston, MA: Springer.

References

Anghileri, J. (2001). Intuitive approaches, mental strategies and standard algorithms. In J. Anghileri (ed.), *Principles and practices in arithmetic teaching: Innovative approaches for the primary classroom* (pp. 79–94). Buckingham: Open University Press.

Anghileri, J. (2006). A study of the impact of reform on students' written calculation methods after five years' implementation of the national numeracy strategy in England. *Oxford Review of Education, 32*(3), 363–80. doi:10.1080/03054980600776506.

Askew, M., Brown, M., Rhodes, V., Johnson, D. & Wiliam, D. (1997). *Effective teachers of numeracy: Report of a study carried out for the Teacher Training Authority 1995–96.* London: School of Education, King's College.

Beishuizen, M. (1999). The empty number line as a new model. In I. Thompson (ed.), *Issues in teaching numeracy in primary schools* (pp. 157–68). Buckingham: Open University Press.

Bennison, A. & Goos, M. (2010). Learning to teach mathematics with technology: A survey of professional development needs, experiences and impacts. *Mathematics Education Research Journal, 22*(1), 31–56. doi:10.1007/BF03217558.

Bishop, R. & Berryman, M. (2006). *Culture speaks: Cultural relationships and classroom learning.* Wellington: Huia.

Buxton, L. (2017). Ditching deficit thinking: Changing to a culture of high expectations. *Issues in Educational Research, 27*(2), 198–214.

Bray, A. & Tangney, B. (2016). Enhancing student engagement through the affordances of mobile technology: A 21st century learning perspective on Realistic Mathematics Education. *Mathematics Education Research Journal, 28*(1), 173–97.

Brown, M., Askew, M., Baker, D., Denvir, H. & Millett, A. (1998). Is the national numeracy strategy research-based? *British Journal of Educational Studies, 46*(4), 362–85.

Buys, K. (2001). Progressive mathematization: Sketch of a learning strand. In J. Anghileri (ed.), *Principles and practices in arithmetic teaching: Innovative approaches for the primary classroom* (pp. 107–18). Buckingham: Open University Press.

Calder, N.S. (2010). Affordances of spreadsheets in mathematical investigation: Potentialities for learning. *Spreadsheets in Education, 3*(3), 45–62.

Clements, M.A. (1989). *Mathematics for the minority: Some historical perspectives of school mathematics in Victoria.* Geelong: Deakin University Press.

Doerr, H.M. & Zangor, R. (2000). Creating meaning for and with the graphing calculator. *Educational Studies in Mathematics, 41*(2), 143–63. https://doi.org/10.1023/A:1003905929557.

Education Queensland (2001). *The Queensland longitudinal reform study.* Brisbane: GoPrint.

Ernest, P. (1991). *The philosophy of mathematics education: Studies in mathematics education.* London: Falmer Press.

Gowland, D. (1998). Calculators: Help or hindrance? *Mathematics in School, 27*(1), 26–8.

Groves, S. (1995). The impact of calculator use on young children's development of number concepts. In R.P. Hunting, G.E. Fitzsimons, P.C. Clarkson & A.J. Bishop (eds), *Regional collaboration in mathematics education* (pp. 301–10). Clayton: Monash University.

Hattie, J. (2014). *Visible learning for teachers: Maximizing impact on learning.* London; New York, NY: Routledge. doi:10.4324/9780203181522.

Larkin, K., Kortenkamp, U., Ladel, S. & Etzold, H. (2019). Using the ACAT framework to evaluate the design of two geometry apps: An exploratory study. *Digital Experiences in Mathematics Education, 5*(1), 59–92. doi:10.1007/s40751-018-0045-4.

Larkin, K. & Jorgensen, R. (2016). 'I hate maths: Why do we need to do maths?' Using iPad video diaries to investigate attitudes and emotions towards mathematics in year 3 and year 6 students. *International Journal of Science and Mathematics Education, 14*(5), 925–44.

Lowrie, T. (2015). Digital games, mathematics and visuospatial reasoning. In Lowrie, T. & Jorgensen (Zevenbergen), R. (eds), *Digital games and mathematics learning: Potential, promises and pitfalls* (pp. 71–92). Springer.

Maj-Tatsis, B. & Tatsis, K. (2018). The use of variables in a patterning activity: Counting dots. *CEPS Journal, 8*(2), 55–70. doi:10.26529/cepsj.309.

Mills, M. & McGregor, G. (2016). Learning not borrowing from the Queensland education system: Lessons on curricular, pedagogical and assessment reform. *The Curriculum Journal, 27*(1), 113–33.

National Council of Teachers of Mathematics [NCTM] (2018). Illuminations. Retrieved from https://illuminations.nctm.org/.

National Library of Virtual Manipulatives [NLVM] (2018). Retrieved from http://nlvm.usu.edu/en/nav/vlibrary.html.

Padula, J. (2014). Graphs and networks for years 7 to 10: Reasons for and ways of using digital technologies to teach algebra and the standard normal curve. *Australian Mathematics Teacher, 70*(3), 16.

Revina, S. & Leung, F.K.S. (2018). Educational borrowing and mathematics curriculum: Realistic mathematics education in the Dutch and Indonesian primary curriculum. *International Journal on Emerging Mathematics Education, 2*(1), 1–16.

Rosenthal, R. & Jacobsen, L. (1969). *Pygmalion in the classroom.* New York, NY: Holt, Rinehart & Winston.

Sarra, C. (2012). *Good morning, Mr Sarra.* Brisbane: University of Queensland Press.

Treffers, A. (1991). Realistic mathematics education in the Netherlands 1980–90. In L. Streefland (ed.), *Realistic mathematics education in primary school* (pp. 11–20). Utrecht: Freudenthal Institute.

Treffers, A. & Beishuizen, M. (1999). Realistic mathematics education in the Netherlands. In I. Thompson (ed.), *Issues in teaching numeracy in primary schools* (pp. 27–38). Buckingham: Open University Press.

van den Heuvel-Panhuizen, M. (2001). Realistic mathematics education in the Netherlands. In J. Anghileri (ed.), *Principles and practices in arithmetic teaching: Innovative approaches for the primary classroom* (pp. 49–64). Buckingham: Open University Press.

van den Heuvel-Panhuizen, M. & Drijvers, P. (2014). Realistic mathematics education. In Lerman, S. (ed.) *Encyclopedia of mathematics education* (pp. 521–5). Dordrecht: Springer.

CHAPTER 2

THEORIES OF LEARNING MATHEMATICS

Why study theories of learning mathematics?

To be able to plan how to teach mathematics effectively, there needs to be some understanding of how students learn mathematics. A major United Kingdom review of effective teachers of numeracy (Askew et al. 1997) has shown that theory is one of the key factors in developing high-quality practices and outcomes in numeracy learning. The authors argue that 'teachers' beliefs and understandings of the mathematics and pedagogical purposes behind particular classroom practices seem to be more important than the forms of practices themselves' (p. 3). That is, it is not so much the ways in which teachers use particular practices in the classroom (e.g. whole-group work, mental maths, direct teaching or other methods) that are critical, but rather the beliefs they hold towards their teaching approaches. The role of theory in underpinning practice is an essential element of quality teaching in mathematics. The ways in which teachers interpret students' mathematical responses and behaviours are shaped by their theoretical assumptions of learning (Jorgensen & Larkin, 2017). The same responses can be interpreted quite differently by teachers depending on how they see the learners and learning. One of the clearest examples of the impact of theory is the attempt to understand why some students are more at risk of failing mathematics than others (see Chapter 4). If a teacher subscribes to a cognitivist view of learning, they may ascribe failing to some innate (and deficient) thinking process. In contrast, if a teacher subscribes to a more social theory of learning, they may recognise that learners come to school mathematics with very different backgrounds and languages, and that these influence what is valued in mathematics; hence, there is a disjuncture between school and home knowledge systems, and this is the potential cause of failing. Coming to a better understanding of the roles of theory in shaping thinking is central to better understandings of practice. In many cases, theory may not be stated explicitly or known, and will remain at a level of intuitive understandings. Articulating these understandings, in the form of theory, is the work of good teaching.

While a number of theories are observable in mathematics education, it is possible that various theories may be applied at different times, or in tandem with one another. For example, in the teaching of a skill or process where the ultimate aim is to achieve a high degree of automaticity (fluency) (e.g. the multiplication facts), behaviourist approaches might be applied during practice activities. In contrast, prior to practising number facts, where there is a need for deep learning of number (e.g. developing a sense of number), a more prudent approach to teaching would be to adopt a constructivist approach, where the need is for deep, interconnected understandings. (This is the approach outlined in Chapter 9.)

By having an idea of how students learn, teachers are better able to plan and anticipate in particular ways, and to create learning environments to facilitate better learning. Three significant classes of theory have had a strong influence on our understandings of how students come to learn and understand mathematics:

1. cognitive theories that focus on students' thinking
2. sociocultural theories that seek to understand cognition within a social context
3. social (or socially critical) theories.

This chapter provides an overview of key ideas associated with these theories. Theories associated with learning in modern society are also discussed.

Behaviourist theories

While behaviourist theories had, since the 1980s, been marginalised in favour of constructivist theories, there has been a resurgence in behaviourist approaches in recent times. In a recent study of primary-school teachers, researchers (Kashefi et al. 2017) reported that as many as 75 per cent of primary-school teachers subscribed to behaviourist models of learning. In direct teaching models, the approach is for the teacher to pass knowledge on to learners—the 'empty vessel' model. The assumption in this theory is that learning can be achieved through passing knowledge from one person to another—a memorisation process.

A number of key features of a behaviourist approach are relevant to mathematics education. These can be readily observed in what is often labelled 'traditional' mathematics teaching. The traditional mathematics teaching approach was for the teacher to undertake direct instruction to the whole class and then provide students with examples on which to practise. The examples were often hierarchical in design—that is, simple examples leading to more complex ones. Students would practise these worked examples. The conclusion of the lesson typically involved the teacher reading out the correct responses, with the students checking their work. Frequently, the number of correct answers attained by students was announced in a very public manner. The following elements are examples of the principles of a behaviourist approach to learning:

- Learning is hierarchical, whereby students need to achieve the lower levels before moving on to higher levels. This is observed in mathematics education through a very linear curriculum.
- Learning is achieved through a process of reinforcement—either positive or negative. Typically, learning occurs when a particular behaviour is exhibited and rewarded, so that behaviours are shaped. This is observed in mathematics when students receive overt reinforcement such as ticks, praise or scores as forms of reward for their work.
- Learning occurs through practise. This is observed in mathematics in worksheets and texts where students practise particular exercises. A good example is learning the multiplication facts through memorisation rather than through understanding.

Some reforms in education that are being adopted in countries such as Australia are heavily imbued with behaviourist theories. An example is

'Direct instruction' (Cape York Academy 2017), a program developed in the United States in the 1960s in which teachers are provided with scripted lessons and students work through levelled worksheets. Students move through the levelled program based on their success with the worksheets. These programs are gaining considerable political traction in disadvantaged communities, including indigenous and low socio-economic status communities. The call for a return to 'the basics', which is being echoed by many policymakers, is a reflection of the behaviourist approach to learning.

The behaviourist model of learning has been heavily criticised for its emphasis on rote learning as opposed to understanding. Students might be able to repeat their multiplication facts but not understand what 9 × 9 is, so that when they make a mistake, they do not recognise the error. Similarly, the models of operations in which a student goes through the rote process of adding the 1s, and trading any tens for a 1 in the next column, may produce correct answers, but, again, when an error is made, students may fail to either see or understand it.

Cognitive theories

The influence of Piaget

The work of Jean Piaget (1972) has had a significant impact on mathematics education. Piaget's work on two aspects of how ideas were formed spanned a considerable part of the twentieth century. His early and late periods were dominated by the active construction of meaning, where he proposed that, through the twin processes of accommodation and assimilation, schemas were constructed. These ideas are foundational to the significant impact of constructivism in mathematics education. Piaget's theory proposed that as students encountered new knowledge, they would incorporate it into an existing schema until there was a problem with the assimilation because the schema would not stay coherent. When this happened, the student would then adjust the schema to something different to make better sense of the concept. For example, when learning about fractions, such as halves, students build a schema that something is broken into pieces—usually two. They can keep making sense of halves as being two pieces even when the pieces may not be even—such as having two unequal pieces to share—with the logic being that one person deserves or likes the item more than the other, so that the uneven pieces are not a problem. Eventually, a new schema must be created to incorporate the new understanding that half means one of two equal pieces.

Piaget's middle period was dominated by stage theories, through which he tried to develop an idea of how young students went through particular stages in their thinking patterns. Piaget's writings on children's cognitive and affective development were extremely influential in education, particularly from the 1950s onwards. His stage theory of cognitive development, for example, strongly influenced early childhood and primary education in the 1960s and 1970s. Stage theory also had a very strong influence in mathematics. Piaget's early work influenced the notion that young students should have concrete experiences so that they can see and build knowledge from concrete examples. Meaning would be created as the student moved from concrete to abstract reasoning. Only as the student matured could more abstract reasoning be introduced. This is evident in the progression in the teaching of number, where the early years focus on concrete resources (e.g. six blocks), which are removed as the student develops a more abstract sense of number that has been extracted from the concrete experiences (the numeral 6), before more abstract reasoning in the adolescent years when abstract reasoning (algebra) is introduced.

In more recent years, the influence of this theory has declined because of the view that it tends to highlight what children cannot do, rather than what they can do. As well, stage theory, in the forms in which it has been translated (from French), interpreted and applied, is criticised on the basis that it highlights notions of readiness as if there is a biological basis to learning. In doing so, it can hold back particular aspects of instruction at the expense of teaching particular content. It is recognised that young children are capable of abstract thinking in certain conditions. As such, algebraic thinking in the early years is now part of most modern curricula.

One important focus of Piaget's work that is highly relevant to our context is the development of logico–mathematical knowledge. In particular, Piaget made a major contribution to the understanding of the development of number concepts in young children, as well as to the development of concepts relating to logic, time, geometry, movement and speed. Since the early 1980s, Piaget's theories have constituted one of the main bases for the development of constructivist theories in education.

Constructivism

'Constructivism' is a term that has been used in education and educational psychology with increasing frequency since the late 1970s. Today, any serious discussion of learning theory related to mathematics, science or literacy, for example, would include a detailed discussion of constructivism. As outlined by Cobb (2000):

> A range of psychological theories about learning and understanding falls under the heading of constructivism. The common element that ties together this family of theories is the assumption that people actively build or construct their knowledge of the world and of each other. (p. 277)

Cobb describes how constructivists reject notions of stimulus–response theory (behaviourist learning theory) and how remembering is more than direct retrieval, since thought processes are in operation. The individual mind of the learner is central to constructivism.

The impact of Piaget's work in contemporary mathematics is obvious in the ways in which constructivism, in its numerous versions, has been taken up by teachers and curriculum writers. There are a number of different forms of constructivism, but underpinning all versions are three premises:

1. Rather than being received passively, knowledge is actively constructed by students.
2. Mathematical knowledge is created by students as they reflect on their physical and mental actions. By observing relationships, identifying patterns and making abstractions and generalisations, students come to integrate new knowledge into their existing mathematical schemas.
3. Learning mathematics is a social process whereby, through dialogue and interaction, students come to construct more refined mathematical knowledge. Through engaging in the physical and social aspects of mathematics, students come to construct more robust understandings of mathematical concepts and processes, through processes of negotiation, explanation and justification.

Constructivism recognises that mathematics must make sense to students if they are to retain and learn mathematics. For students to develop appropriate knowledge, they must be provided with rich learning experiences so that their constructed meanings and understandings will be in keeping with the discipline of mathematics.

The importance of dialogue and argumentation

Within the constructivist paradigm, the role of language and dialogue is central to fostering learning environments. Providing appropriately organised experiences in which students can talk with their peers allows students to explore ideas in language and concepts similar to their own. This enables higher-achieving students to practise their control of language, and lower-achieving students to hear ideas being modelled in a language that is more likely to be in a genre that they can access. For example, when talking about the properties of three-dimensional objects, the formal language of edges, faces and vertices may be introduced. High-achieving students may find this language useful if they have been confused by the use of the term 'side'—does it refer to the face or an edge? The appropriate language aids their learning. Other students, however, may be grappling with notions of three-dimensional shapes. Having students explain their ideas to their peers often supports both sets of learners.

The provision of opportunities for dialogue (as opposed to rote and drill learning) is integral to many quality practices. This approach is seen in many classrooms where teachers pose 'good questions' that promote discussion and argumentation. Good questions and/or tasks provide a rich environment for learners to discuss, clarify, question, challenge and extend mathematical ideas. The dialogic context also allows students to build and extend their mathematical language. Good questions should be planned carefully so that they will foster active dialogue among students.

TEACHING TIP

At the conclusion of a lesson, small groups should report back to the whole class. The whole group will thereby learn protocols for supportive questioning that promote deep conversations on mathematical ideas.

Constructivism in the mathematics classroom

Within a constructivist classroom, the teacher acknowledges that students will have constructed a range of understandings from any given interaction on the basis that they have entered the context from a range of different perspectives and experiences. A constructivist perspective recognises that it is not correct to assume that the teaching of a concept relates to the development of the ideas proposed by the teacher, and that there will be a multiplicity of understandings constructed by the range of students in the classroom. A constructivist teacher realises that having taught something does not mean that students have learned exactly what was envisaged. It is important for the teacher to use a range of tools and techniques to assess what the students have constructed.

By identifying what the students have constructed, the teacher is then able to identify constructions akin to the objective of the lesson, as well as to identify misconceptions. It is the misconceptions that allow the teacher to access what the students have come to construct, and thus to develop teaching strategies that will move the students into more appropriate constructions.

When teaching many concepts and processes in mathematics, the constructivist teacher is aware that there are likely to be numerous (and different) constructions being made across the classroom. In an early-years classroom, for example, the teacher may be teaching about sorting and classifying. Typically, the teacher may want the students to sort objects by colour or shape, so that the groups will be very clear; often, though, a student creates groups of different colours or shapes. In such a situation, rather than assume the student is wrong, the constructivist teacher will seek to understand the student's thinking behind their groupings. The student may have created groups that are all different. While this is not what the teacher had intended, the student's understanding of grouping is solid—just different from the teacher's. However, if the student is unable to justify their grouping in a way that is coherent and demonstrates a sense of grouping/classifying, then the teacher needs to work on that understanding.

Sociocultural theories: the influence of Vygotsky

Lev Vygotsky is regarded as the founder of sociocultural theory, or what can be described as the sociohistorical or sociocultural approach in psychology (e.g. Cole 1996; Moll 1990). Vygotsky's work, which is embodied in the literature on sociocultural theories of learning mathematics, has gained increasing importance in theorising how students learn mathematics. Vygotsky saw that students internalised complex ideas (Daniels 1990), but he extended the general constructivist approach by arguing that the internalisation of knowledge could be achieved more effectively when students were guided by good, analytic questions posed by the teacher or supported by appropriate technologies.

Zone of proximal development

The expert teacher is central to Vygotskian theory. The teacher's role is to identify the student's current mode of representation and then—through the use of good discourse, questioning, or learning situations—provoke students to move forward in their thinking. A student's representation or thinking was recognised as their zone of proximal development (ZPD), and the teacher's actions to support learning were described as scaffolding. When working in the ZPD, particular attention is paid to the language being used, since students' language influences how they will interpret and build understandings (Bell & Woo 1998). Within a Vygotskian approach, it is seen to be important that teachers use and build considerable language and communication opportunities within the classroom environment in order to build mathematical understandings.

The ZPD is a very useful construct in the classroom. In most if not all classrooms, levels of student understanding are diverse, and the teacher needs to identify where students are located along the learning continuum, and then provide good questions to move the students along the continuum.

TEACHING TIP

Magic numbers

Provide students with a sheet that has a range of number challenges—what two numbers add up to my number, what number comes before/after my number, and so on. But the magic number may vary across the class—perhaps some students work with numbers up to 10, while others work with 100s or fractions. Such activities are good for differentiation, since the sheets look the same for all students. They can also support all students with numbers that extend their current levels of work.

Scaffolding

Good teaching involves teachers knowing students' current thinking about mathematical concepts, and then understanding how to move students towards more complex, complete and robust constructions through the use of organised learning activities and environments. Good questions are important in facilitating learning. Typically, good questions are those that foster deeper levels of learning, as opposed to simple recall. Increasingly, the use of computer-based learning has gained a foothold in mathematics. As a result, there is a need to consider how digital tools can be used to appropriately scaffold the learning of mathematics (Belland 2017).

Self-concept theories

We have opted to discuss a cluster of theories that focus on the internalisation of perceptions and beliefs of oneself. How students see themselves as learners of mathematics can significantly impact how they learn and their motivation to learn. These factors are often described as non-cognitive. In a large study of non-cognitive factors, it was found that the best predictors of performance in mathematics were confidence, self-efficacy, anxiety and self-concept (Stankov & Lee 2017). Negative self-concepts in mathematics have been found to be formed as early as Year 3 (7–8 years old) (Larkin & Jorgensen 2016).

Locus of control and attribution theory

There are two loci of control: internal, where attributions are directed at the self, and external, where attributions are made to something external. It is known that when girls succeed, they are more likely to attribute it to an external factor—they passed the mathematics test because the test was easy. In contrast, failure is internalised, so that failing the test might be attributed to their level of intelligence or lack of preparation. The opposite is the case for boys: failure is attributed to an external factor, whereas success is internalised.

Affective domain of learning mathematics

Constructs of confidence, affect, anxiety, self-belief, emotions, attitudes and belonging are all internalisations of how students come to see themselves in relation to mathematics. The trends in this field of research suggest that when a student has a positive image or disposition towards mathematics, they tend to have more positive experiences in mathematics than peers who have negative dispositions or self-concepts (Hannula 2014).

Promoting social and emotional wellbeing

One way to ensure that students have positive self-concepts has been for teachers to ensure that mathematical learning environments promote positive wellbeing. Some pedagogy related to wellbeing focuses on building psychological constructs associated with wellbeing. Certain approaches have focused on the development of a strong mathematical identity or sense of self (Briede 2016), while other pedagogies focus on the inclusion of marginalised learners and their ways of knowing (Vale et al. 2016) and/or values in mathematics education (Seah et al. 2016).

Sociological theories

Sociological theories, and particularly critical sociology, have gained increasing importance in mathematics. These theories shift the focus of mathematics education. Aside from students with learning disabilities, these theories relate to indigenous students of almost all nations, students from working-class (or low socio-economic status) families, students who live in remote or rural areas, and students whose first language is not English (in English-speaking countries). When gender is considered in concert with these variables, differences are exacerbated (Walkerdine 1988, 1989). It has long been recognised that girls have been particularly disadvantaged in the study of mathematics as a result of gendered practices in teaching and assessment (Fennema & Meyer 1989; Leder 1992; Lowrie & Jorgensen 2018). However, it is now recognised that this applies not to girls per se, but rather to girls from particular social and cultural backgrounds. When considering the ways in which the practices of mathematics work to exclude girls, it becomes important to recognise that some girls (generally, middle-class girls) are more likely to be successful than their peers from working-class backgrounds (both girls and boys); thus, gender is not the sole variable, but must be considered in concert with other variables. Rather than assume that success is due to some innate 'mathematical ability', socially critical theories focus their attention on the practices of school mathematics.

Socially critical theories explore the practices of mathematics education to see how they are implicated in the reproduction of inequities, and in so doing challenge such practices to change. Assessment, mathematical language (Zevenbergen 2000; 2001) and classroom talk, textbooks (Dowling 1998) and ability grouping (Boaler 1997) are some of the areas that have been examined critically in terms of the ways in which social, cultural, linguistic and gender differences are reproduced through mathematics education. These studies have illustrated the very subtle ways in which school mathematics contributes to, and legitimates, the failure of particular groups of students. This book devotes a full chapter to the study of equity from this perspective (see Chapter 4).

Mathematics is one of the most important subjects in the school curriculum, and it serves a particular role (among others) as a social filter. The work of Lamb (1997) shows that success in school mathematics is the best predictor of success in life. More recently, it has been reported that spatial reasoning is the best predictor of whether students will end up in a STEM-related career (Wai et al. 2009). Spatial skills are closely related to mathematics skills. When these findings are correlated with Lamb's work showing that mathematics is the best predictor of post-school success, the importance of giving all learners the best opportunities to learn mathematics, and to succeed in it, becomes apparent.

People who come to believe that they are 'not good' at mathematics tend to accept their position in life. Thus, it is important for all students to succeed in school mathematics—regardless of background, gender or language. By knowing how practices are implicated in the construction of differences, teachers can change their practice in order to produce more equitable classrooms and outcomes.

Adopting a socially critical theory in the classroom acknowledges that students' backgrounds impact learning. When students come to school speaking a language that is different from the language of instruction—usually Standard Australian English (SAE), in the Australian context—teachers can create strategies to build bridges between the two languages. In remote communities in Australia, 'both ways' education often sees students' first three years of schooling being done in their home language as well as in SAE, so that students' home languages are recognised and used in the classroom, which helps with students' transition to English. Teachers in these schools work closely with their teacher assistants, who are Indigenous people from the community and speak the local language(s). The teaching assistants are instrumental in supporting the use of home languages in the classroom. In New Zealand, the use of Māori language is very common, including in schools that speak only Māori. A vocabulary of mathematics has been created in Māori. When learners come to school having already been exposed to a range of spatial play—such as block play (e.g. Lego), a rich spatial language, and experiences with maps and jigsaws—they are more likely to have a head start on peers who have not had these experiences. Mental transformation was seen to be the hallmark of spatial intelligence. In some cases—indeed, in most cases—these gaps persist through the years of schooling.

Modern society

Current educational theory has moved to a stronger recognition of how society impacts learning. Postmodern theory has drawn considerable attention to how current times are very different from previous times (e.g. the Industrial Age) largely due to the use of technology. As early as 30 years ago, educationalists were commenting on how television was influencing attention rates and suggesting that young students had much shorter attention spans than in the past. Similarly, the advent of television meant that reading books was being replaced by viewing television, and there was seen to be a consequent decline in reading skills and motivations. More recently, the use of digital technologies such as desktop computers, laptop computers and tablet devices has been seen to have considerable influence over students' thinking and behaviours—so much so that terms such as 'cyberkids', 'digital natives', 'technoliteracy', and 'screen time' have become part of the educational discourse (Hirsh-Pasek et al. 2015; Larkin 2011). 'Bring your own device' has also become an accepted practice in many schools (Parsons & Adhikar 2016). The world for which schools are preparing students is fundamentally different from that of even a few years ago. The term 'modern society' has been used to represent the very different social, economic, political and educational factors of the society in which young people now live (Gee 2002). Many of the new terms in educational theory are prefaced with the word 'new' to represent this thinking. Students who have grown up in a technological age are less intimidated by technology, and hence its use in mathematics education is an important change.

Technology and modern society

In terms of learning, 'modern society' embraces the use of technology (both digital and non-digital) so that the tedium often associated with working calculations ('doing sums') can be replaced with technologies to support the development of mathematical thinking. This change is fundamental in preparing students for life in modern society.

Old curricula have been challenged from a technological perspective by the work of Stacey and Groves (1996). In an old curriculum, students in the first year or two of schooling would only work with numbers to 20. Through the use of calculators, young students' number sense can now be developed to numbers with four or more digits (Groves 1995). More recently, apps such as TouchCounts (www.touchcounts.ca) and Place Value Chart (www.cermat.org/projects/place-value-chart-0) use the affordances of touchscreen technology to encourage early subitising, counting skills and place-value concepts. Thus, rather than digital technologies replacing skills, they are seen as tools to enhance mathematical thinking.

Within this framework, consideration must also be given to what is seen as mathematics. Often, this is framed within basic skills, with the wider society bemoaning young graduates who are unable to calculate using written algorithms or mental computation. Within modern society, the emphasis becomes somewhat different due to the saturation of technology. Forty or so years ago, shop assistants could enter only the price of the goods into the cash register—calculations were done on paper, and change given using a counting-on method. Today's society is much richer in technology—most if not all stores have registers that scan items and calculate the amount of refund (change) required; thus, no data need be manually entered, with the assistant generally being required only to press a button for the particular item of purchase (e.g. McDonald's). Other stores do not accept cash payments and instead use only 'tap and go' credit card purchases. Thus, old basics have been superseded by new basics—sales assistants need to be able to estimate, to problem-solve and to validate in order to evaluate whether items have been scanned or not, and thus to verify the validity of the total amount or the change given (in case an incorrect amount is entered for the amount tendered).

In other words, they need some of the old computational basics, but technology has brought with it a range of new skills that they require. These skills are different from, but not exclusive of, the attributes that would be sought in mathematics.

Within the framework of modern society, the teaching of mathematics therefore involves a much greater emphasis on the use and integration of digital technologies (such as computers, graphic calculators and apps). Since most Western students have been exposed to such technologies, they are more likely to think and respond to these forms of teaching.

Furthermore, there is a renewed emphasis on the mathematics curriculum containing new forms of mathematics that reflect the needs of the world beyond school. Modern society—the era of big data—offers a supersaturation of information, so students need to be more literate in analysing, interpreting and being able to critique the texts to which they are exposed. This often means being able to use and apply their mathematical thinking and analysis to texts containing data—such as graphs, measures of central tendency, and so on. They not only need skills to construct or calculate such measures; they also, within modern society, need to be able to better interpret such information. This demands a refocusing of the mathematics curriculum into such areas.

Theory into practice

The value of a good theory is its capacity to enable teachers to develop good practice that supports and enhances student learning. Teachers need to have a strong theoretical basis to their work. By understanding how students learn, teachers are able to organise the learning in ways that will enhance the capacity for learning. Rather than advocate one theory as being superior to another, it is more appropriate to consider what is being learned and take the most appropriate theory of learning to apply to that context.

Review questions

2.1 What are the main characteristics of the constructivist theories underpinning mathematics teaching and learning?

2.2 Discuss how you would organise learning when using scaffolding. Provide an example.

2.3 Describe how you would teach Cartesian coordinates, and outline the theories of learning that you would draw upon.

2.4 In what ways does a socially critical approach to learning mathematics differ from a cognitive or sociocultural approach? Outline the main tenets of such approaches.

2.5 List some of the challenges faced by teachers in the teaching of mathematics in modern society.

Further reading

Cape York Aboriginal Australian Academy (2017). *Annual report, 2017*. Retrieved from https://cyaaa.eq.edu.au/Supportandresources/Formsanddocuments/Annual%20 reports/annual-report-2017.pdf.

Chambris, C. (2018). The influence of theoretical mathematical foundations on teaching and learning: A case study of whole numbers in elementary school. *Educational Studies in Mathematics, 97*(2), 185–207. https://eric.ed.gov/?id=EJ1167556.

Mathews, J.M. (2013). *Place-based design: An instructional design theory for supporting community-based inquiry and design projects* Available from ERIC. https://eric.ed.gov/?id=ED554283.

Mousley, J. (2001). Theories of learning: What's next? From construction to activity theory. *Vinculum, 39*(3), 8–13.

Muijs, D., Kyriakides, L., van der Werf, G. Creemers, B., Timperley, H. & Earl, L. (2014). State of the art teacher effectiveness and professional learning. *School Effectiveness and School Improvement, 25*(2), 231–56. https://eprints.soton.ac.uk/362878/2/SOA%2520TEACHER%2520EFFECTIVENESS%2520SESI.pdf.

Muijs, D. & Reynolds, D. (2017). *Effective teaching: Evidence and practice*. London: SAGE Publications.

Reys, B. & Long, V.M. (1995). Teachers as architects of mathematical tasks. *Teaching Children Mathematics, 1*(5), 296–9.

Rowan, T.E. & Robles, J. (1998). Using questions to help children build mathematical power. *Teaching Children Mathematics, 4*(9), 504–9.

Schön, D.A. (2017). *The reflective practitioner: How professionals think in action*. Routledge.

Stephens, M., Montgomery, P. & Waters, M. (1997). Creating a climate for effective project work in mathematics. *Australian Primary Mathematics Classroom, 2*(2), 6–10.

Stronge, J.H. (2018). *Qualities of effective teachers*. Alexandria, VA: ASCD.

References

Askew, M., Brown, M., Rhodes, V., Johnson, D. & Wiliam, D. (1997). *Effective teachers of numeracy: Report of a study carried out for the Teacher Training Authority 1995–96.* London: School of Education, King's College.

Bell, G. & Woo, J.H. (1998). Probing the links between language and mathematical conceptualisation. *Mathematics Education Research Journal, 10*(1), 51–74.

Belland, B.R. (2017). *Instructional scaffolding in STEM education: Strategies and efficacy evidence.* Dordrecht: Springer.

Boaler, J. (1997). *Experiencing school mathematics: Teaching styles, sex and setting.* Buckingham: Open University Press.

Briede, L. (2016). The relationship between mathematics teachers' teaching approaches and 9th grade students' mathematical self. *Journal of Teacher Education for Sustainability, 18*(1), 34–47.

Cobb, P. (2000). Constructivism. In A.E. Kazdin (ed.), *Encyclopedia of psychology* (vol. 2, pp. 277–9). Washington, DC and New York: American Psychological Association and Oxford University Press.

Cole, M. (1996). *Cultural psychology: A once and future discipline.* Cambridge, MA: Harvard University Press.

Daniels, H. (1990). Number competence and communication difficulty: A Vygotskian analysis. *Educational Studies, 16*(1), 49–59.

Dowling, P. (1998). *The sociology of mathematics education: Mathematical myths/pedagogical texts* (vol. 7). London: Falmer Press.

Fennema, E. & Meyer, M.R. (1989). Gender, equity and mathematics. In W. Secada (ed.), *Equity in education.* New York, NY: Falmer Press.

Gee, J.P. (2002). New times and new literacies: Themes for a changing world. In B. Cope and M. Kalantzis (eds), *Learning for the future: Proceedings of the learning conference* (pp. 3–20). Spetses: Common Ground.

Groves, S. (1995). The impact of calculator use on young children's development of number concepts. In R.P. Hunting, G.E. Fitzsimons, P.C. Clarkson and A.J. Bishop (eds), *Regional collaboration in mathematics education* (pp. 301–10). Melbourne: Monash University.

Hannula, M.S. (2014). Affect in mathematics education. In S. Lerman (ed.), *Encyclopedia of mathematics education* (pp. 23–7). Dordrecht: Springer.

Hirsh-Pasek, K., Zosh, J., Michnick Golinkoff, R., Gray, J., Robb, M. & Kaufman, J. (2015). Putting education in 'educational' apps: Lessons from the science of learning. *Psychological Science in the Public Interest, 16*(1), 3–34. doi:10.1177/1529100615569721.

Jorgensen, R. & Larkin, K. (2017). Analysing the relationships between students and mathematics: A tale of two paradigms. *Mathematics Education Research Journal, 29*(1), 113–30. doi:10.1007/s13394-016-0183-1.

Kashefi, H., Ismail, Z., Mirzaei, F., Tak, C., Obeng, S. & Ching, T. (2017). Teaching and learning theories applied in mathematics classroom among primary school teachers. *7th World Engineering Education Forum (WEEF).* doi: 10.1109/WEEF.2017.8467070.

Lamb, S. (1997). *Longitudinal study of youth labour markets.* Melbourne: Australian Council of Educational Research.

Larkin, K. (2011). You use! I use! We use! Questioning the orthodoxy of one-to-one computing in primary schools. *Journal of Research on Technology in Education, 44*(2), 101–20.

Larkin, K. & Jorgensen, R. (2016). 'I hate maths: Why do we need to do maths?' Using iPad video diaries to investigate attitudes and emotions towards mathematics in year 3 and year 6 students. *International Journal of Science and Mathematics Education, 14*(5), 925–44. doi:10.1007/s10763-015-9621-x.

Leder, G. (1992). Mathematics and gender: Changing perspectives. In G. Grouws (ed.), *Handbook of research on mathematics teaching and learning: A project of the National Council of Mathematics Teachers*. New York, NY: Maxwell Macmillan.

Lowrie, T. & Jorgensen, R. (2018). Equity and spatial reasoning: Reducing the mathematical achievement gap in gender and social disadvantage. *Mathematics Education 30*(3), 65–75. doi:10.1007/s13394-017-0213-7.

Moll, L.C. (1990). *Vygotsky and education: Instructional implications and applications of sociohistorical psychology*. Cambridge: Cambridge University Press.

Parsons, D. & Adhikar, J. (2016). Bring your own device to secondary school: The perceptions of teachers, students and parents. *Electronic Journal of e-Learning, 14*(1), 66–80.

Piaget, J. (1972). *To learn is to invent*. New York: Grossman.

Seah, W.T., Andersson, A., Bishop, A. & Clarkson, P. (2016). What would the mathematics curriculum look like if values were the focus? *For the Learning of Mathematics, 36*(1), 14–20.

Stacey, K. & Groves, S. (1996). Redefining early number concepts through calculator use. In J. Mulligan and M. Mitchelmore (eds), *Children's number learning* (pp. 205–25). Adelaide: Australian Association of Mathematics Teachers.

Stankov, L. & Lee, J. (2017). Self-beliefs: Strong correlates of mathematics achievement and intelligence. *Intelligence, 61*, 11–16. doi:10.1016/j.intell.2016.12.001.

Vale, C., Atweh, B., Averill, R. & Skourdoumbis, A. (2016). Equity, social justice and ethics in mathematics education. In K. Makar, S. Dole, J. Visnovska, M. Goos, A. Bennison and K. Fry (eds), *Research in mathematics education in Australasia 2012–2015* (pp. 97–118). Singapore: Springer.

Wai, J., Lubinski, D. & Benbow, C.B. (2009). Spatial ability for STEM domains: Aligning over 50 years of cumulative psychological knowledge solidifies its importance. *Journal of Educational Psychology, 101*(4), 817–35.

Walkerdine, V. (1988). *The mastery of reason: Cognitive development and the production of rationality*. London: Routledge.

Walkerdine, V. (1989). *Counting girls out*. London: Virago.

Zevenbergen, R. (2000). 'Cracking the code' of mathematics: School success as a function of linguistic, social and cultural background. In J. Boaler (ed.), *Multiple perspectives on mathematics teaching and learning* (pp. 201–23). New York, NY: JAI/Ablex.

Zevenbergen, R. (2001). Mathematics, social class and linguistic capital: An analysis of a mathematics classroom. In B. Atweh and H. Forgasz (eds), *Socio-cultural aspects of mathematics education: An international perspective* (pp. 201–15). Mahwah, NJ: Lawrence Erlbaum.

CHAPTER 3

LANGUAGE AND MATHEMATICS

For a long time, mathematics was seen as a subject area that remained independent from language. That view has been challenged considerably over the years, to the point that it is now recognised that language plays a central role in mathematics. Students need to learn the language of mathematics in order to participate in and learn the discipline. Mathematics has a very particular language, and being able to read, interpret and respond in that language are all central to being an effective mathematics learner.

The language of mathematics

Typically, coming to learn and understand mathematics involves a pairing of written words, oral sounds or abstract symbols with particular meanings. For example, the word 'ruler' is used to refer to a measuring instrument that is usually used to measure lengths. Similarly, the '+' sign is used to refer to the addition concept/process, where, in most cases, two sets are combined to form a superset. In learning mathematics, students need to learn the particular relationships between signs and meanings. Many new words are introduced, some of which will be familiar and some of which are unique to mathematics. To be competent and effective in mathematics, students need to learn this language. While some words may create difficulties due to their linguistic complexity (e.g. tessellation) or the concept to which they refer (e.g. parallel), others create problems because they are familiar from everyday discourse (e.g. irrational, ruler, odd) but take on specific and/or different meanings in mathematics.

When teaching mathematical ideas, teachers need to be sure that students understand the very particular language of mathematics. Often, what are seen as misconceptions emerge due to the use of lay terms and meanings. When working with fractions, the numerator is often interpreted as the 'top number' and the denominator the 'bottom number'. This 'misconception', while working well with fractions, cannot be transferred to other contexts such as subtraction, where we have heard teachers say, 'The numerator is taken away from the denominator.' This is incorrect, and shows the misappropriation of language.

The development of a specialised mathematical language

As noted above, part of the challenge in understanding mathematics is learning the specific language of mathematics. This may involve learning complex mathematical terms; however, students often have more difficulty when terms familiar to them in everyday language are used in more specific ways in mathematics. For example, consider the student who does not like odd numbers 'because they are strange'. Here, the student transfers the everyday meaning of 'odd' to the mathematics context, resulting in a novel interpretation of a mathematical idea. Other words that can lead to confusion include ruler, mass, root and vulgar. Even within mathematics, the same word can be used in different ways. To communicate well mathematically, students need to understand the language of mathematics, thus part of learning mathematics is the progressive refinement of language. For example, when working with shapes, the common words 'sides' and 'corners' are used in the early years. But this becomes problematic when three-dimensional objects are introduced. For two-dimensional shapes, 'sides' is used early to draw attention to the lines that make up the shape, but when talking about three-dimensional shapes the term 'side' can become confusing—does it refer to the line or to the shaded area? Similarly, the term 'corner' becomes problematic.

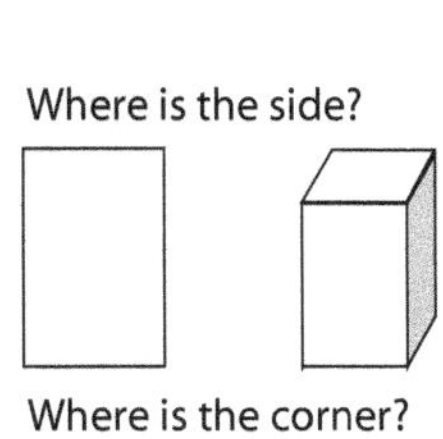

As a result, the very specific language of shapes and objects needs to be introduced. Learning the language of mathematics can be very empowering for students as they progress through school, and teachers need to pay particular attention to teaching the language of mathematics. In being aware of potentially problematic areas, teachers will be able to better organise effective learning environments for all students. This is particularly important when considering students whose first language is not the language of instruction. Teachers need to remember that this includes English-speaking students whose language is different from the middle-class English normally used in the classroom.

Ambiguity in terminology

Mathematical terms that are ambiguous in meaning (i.e. have multiple meanings) can create difficulties for students. Imagine the confusion of young students who hear of the exciting lesson about to begin on rulers. Thinking that they are going to hear about kings and queens, princes and princesses, emperors and empresses in the first part of the lesson, the students will be awash with confusion as the teacher brings out a 30 cm piece of wood and talks about it. The confusion, and loss of learning, has occurred due to a misinterpretation of the term 'ruler'. Three types of words fall into this category of ambiguity: homonyms, homophones and polysemes.

Homonyms: shifting between the everyday and school mathematics

Homonyms are words that look and sound the same, but have two or more meanings—such as mass, vulgar, irrational and odd. Students can become confused when they hear and interpret one meaning but the teacher is intending the term to be interpreted in a different way. An example can occur when students are asked to calculate 'the *volume* of a box'. Some students will interpret this as a three-dimensional task, while others will become confused as they wonder how they will calculate the 'noise' of the box. Teachers need to be aware of the homonyms used in a lesson to avoid confusion.

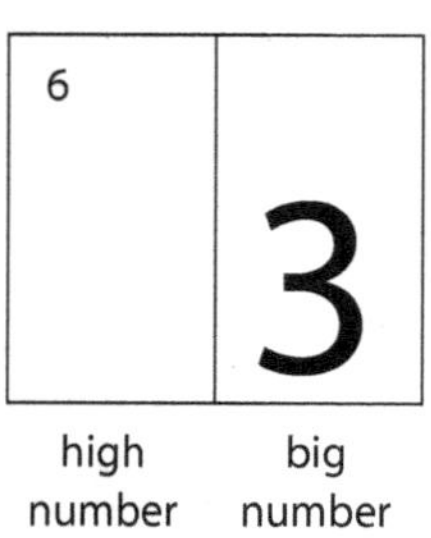

It can be particularly confusing when spatial terminology is tied in with number study. In the following examples, students were asked to write a 'high' number and a 'big' number. 'High' and 'big' were meant to infer three-digit numbers, but the students interpreted them in a spatial sense rather than in terms of magnitude.

There is considerable potential for other areas of mathematics language to create confusion. Consider the question, 'What is the length of time between 9.05 a.m. and 9.55 a.m.?' How do students make sense of this type of question when they have a construct of length as a unit of measure that can be measured with a ruler?

Homonyms common in mathematics

angle	face	odd	right
average	figure	parallel	root
base	improper	point	sign
below	leaves	power	similar
cardinal	left	product	square
change	make	proper	table
common	mean	rational	times
degree	model	real	unit
difference	natural	record	volume

The list of homonyms given in the table does not include every homonym found in mathematics; it merely highlights some of those most commonly used. When planning lessons, teachers should always consider the language being used in the lesson, paying specific attention to these types of words. When they appear in the teaching episode, teachers should overtly draw attention to the specific mathematical meaning being used in the immediate context.

Homophones

Whereas homonyms are words that have more than one meaning, homophones have different spellings but the same sound. In teaching the addition of fractions, it is common to hear phrases such as 'two halves make a whole'. For example, in a lesson where two halves are being placed together to make a full item, if a student is

hearing 'two halves make a *hole*' but no hole is being created by combining the two objects, the student could be left wondering what the teacher is talking about (unless a doughnut is being used as an example, which could develop further misconceptions!). This will result in considerable talking past the student, who will have difficulty making sense of what is happening despite expending much effort to understand.

Common homophones

for—four	pie—pi
hole—whole	route—root
one—won	to—too—two

When terms such as those in the table appear in a teaching episode, potential confusion can be avoided by writing both words on the board. Since the words are different in appearance (i.e. spelling), students can then see the differences.

Polysemes

The third type of confusing word is those that have two or more related meanings. The term 'product' is an example of a polysemous word. Similarly, students can become confused by the term 'base', which means foundation. Having developed a construct of base as the bottom of an object (such as a pyramid), number study then brings in a different idea of base, as in base 10. Students will not necessarily make the connection between the two meanings if they do not have a sufficient grasp of language to see base as a foundation.

Clearly, where lexical ambiguity exists, teachers need to develop an acute awareness of the language they will use in a lesson. When planning a teaching episode, language should be an important consideration—both in terms of the new language to be learned and any language that may become problematic. As discussed in Chapter 5, language is a key consideration in the overall planning of a lesson.

Prepositions: the little words that mean a lot

The language of mathematics is very specific and concise. The types of reading skills that students develop as they progress through school may mean that many of the 'little' words are skipped over. This can become a difficulty in mathematics, since there is little redundancy in mathematical language. Students need to read mathematical texts very carefully, especially in regard to prepositions. Prepositions often have a very specific function in mathematics. Typically, a preposition is a word that indicates the relationship of a noun (or pronoun) to another word in the statement or sentence. For example, the ball is *on* the table, or the cat is *under* the tree. Often, there is a spatial element to the preposition. Some common prepositions used in mathematics are shown in the table below.

about	below	inside	over
above	beneath	into	past
after	between	near	through
among	down	of	toward
around	for	off	underneath
before	from	on	upon
behind	in	out	with

Consider the small words 'off' and 'of'; they can affect responses significantly. When working with percentages, '25 per cent *of* a price' and '25 per cent *off* a price' result in two very different answers. Paying attention to this level of detail is important. As a teaching point, it is vitally important that attention is drawn to these small but significant terms.

Students develop strategies through which they search for key information in other subjects and use this to solve tasks, but they often pay little attention to prepositions. In mathematics, this strategy can be flawed. It is particularly problematic for students whose command of English is not strong, and where prepositions are not easily learned. In teaching students to read mathematically, the teacher needs to draw attention to the words being used, and to encourage students to read mathematical texts differently, and more carefully, than non-mathematical texts.

Prepositions can be challenging in mathematics, particularly for students for whom English (the language of instruction) is not their first or home language. Learning how prepositions impact mathematics can create misunderstandings. When two objects are placed side by side, many prepositions can be used to describe the relationship—near, next to, beside, to the right. So how do learners make sense of the statements?

For students for whom English is a second (or other) language, attention needs to be directed to support the teaching of these words. In the early years, big books can be created to show relationships. Using photographs of the students to illustrate the preposition is often a very positive strategy (Jorgensen 2015); for example, 'Billy is under the play gym', 'Margarite is on top of the slide' and 'Tran is near the gate'. Using examples that are familiar to the learners also helps to show the relevance of mathematics to their everyday worlds.

There are many practical ways for teachers to build the rich vocabulary of mathematics. The first step is to recognise the impact of words in mathematics, so that attention can be drawn to the vocabulary. Drawing on the rich literature of ESL learning, the same strategies can be applied to mathematics:

- Create big books for early-years learners.
- Create word walls. These can have just words, words with meanings/definitions or even pictorial representations of the words.
- Create student dictionaries. Students may take definitions from a dictionary or create their own meanings.
- Create displays, such as posters, and draw attention to these while teaching so that students refer to the displays when they need assistance.

When creating resources such as these, there is a need for direct teaching that is related to the resources, so that students use them.

Word problems

Within the literature on language and mathematics, a significant body of research has been undertaken using word problems. In attempting to link mathematics to the world beyond schools, many arithmetic tasks have been inserted into worded problems. For example, $3 + 4 =$ becomes 'If I went shopping and bought three apples and four oranges, how many pieces of fruit would I have?' Part of the reasoning behind this is to make mathematics more realistic. However, it can create another layer of complexity for students through the language of the problem.

One of the problems with mathematics teaching and learning is that students come to see school mathematics in particular ways. They develop a sense of the culture of the classroom, in which answers are expected. They also learn strategies for reading word problems. Most of these strategies work well for the standard questions being posed. The work of Freebody (1992) and Barton (2017) illustrates the importance of the roles of the literate student, which are very evident in the mathematics classroom. The ability to decipher, interpret and respond to word problems effectively is an important skill for students to acquire. For example, when posed a nonsensical problem such as 'There are 60 ice-cream flavours. If the shopkeeper has $20 in his pocket, how much would an ice-cream cost?', students may attempt to answer the question by playing with the numbers and not addressing the task. This strategy of taking key words and numbers, then trying to provide an acceptable answer, is common in mathematics classrooms, and highlights the inappropriate literacy strategies employed by students. A greater emphasis on reading for meaning (as opposed to 'doing') needs to be an integral part of the teaching of mathematics. Using the literacy model, a competent reader of a mathematical text is one who is able to 'break the code' to make meaning of the text, so as to use the text in an appropriate manner and critically analyse it (Freebody 1992). These elements apply to the types of literacy demanded in mathematics classrooms.

Heller and Greeno (1978) classification of word problems

Within the word problems posed in the mathematics classroom, a particular form of question is often used. (Other word problems are discussed later in the chapter.) Heller and Greeno (1978) classify and order word problems based on the actions undertaken and the level of difficulty, and propose a taxonomy of word problems. They identify three main types of problem—change, combine and compare:

1. *Change questions* involve a dynamic process whereby an event alters the value of the quantity—for example, Michael has five oranges;

Kelly gives him two more oranges. How many oranges does Michael have now?

2. *Combine questions* relate to static situations where there are two amounts. These are considered either as separate entities or in relation to each other—for example, Steven has four oranges; Michelle has two oranges. How many oranges do they have altogether?
3. *Compare questions* involve the comparison of two amounts and the difference between them—for example, Stephanie has five oranges; Alice has two more oranges than Stephanie. How many oranges does Alice have?

The arithmetic is very simple, but there is complexity due to the semantic structure of the problems. The levels of difficulty depend on the operation to be undertaken (addition or subtraction), the type of word problem (change, combine or compare) and the location of the unknown set.

Problems involving addition are usually solved more frequently than similar problems involving subtraction. Similarly, problems where the unknown set is the solution are solved more frequently than those where the set is at the start of the problem. These trends have been confirmed across numerous studies (Halford 2014; Heller & Greeno 1978; Lean et al. 1990; Shi et al. 2015; Verschaffel et al. 2010); Zevenbergen et al. 2001). Some studies have found that simply rewording the problems makes them more accessible to students (Davis-Dorey et. al. 1991; Halford 2014; Verschaffel et al. 2010).

With these sorts of problems, the demands on students are not highly complex insofar as the arithmetic is concerned; the complexity comes about through the structuring of the questions. The complex organisation of the sentences and the order of the operations to be undertaken are potentially confused. In most cases, the examples of word problems students encounter in classrooms present as clear problems where students do not have to rely on deciphering the linguistic demands of the task. Rather, they rely on cues that they have built up from earlier experiences—for example, with the commonly posed problem form, 'I had six apples; Jenny gave me three more. How many do I have altogether?' Students have had numerous experiences with this type of question, so they can generally rely on a method whereby they seek out key words—such as 'more' and 'altogether'—and assume that the task requires addition. However, what has been lost in this process is a textual or literacy component, where students need to read and work out what the question is actually demanding. When posing problems of this kind to students, teachers need to be aware that, while the arithmetic may not be difficult, the *structuring* of the problem may cause difficulties. Teachers need to recognise the ways in which these types of word problems are structured, as well as any inherent complexity in the language used.

Realistic word problems

Another form of word problem occurs in attempts to embed mathematics into realistic (or real-world) contexts. There is some debate as to whether or not such questions are real-world ones. Problems of this format look something like, 'A farmer has six hectares of prime land. It is divided into four equal-sized paddocks. If he can graze six cows per hectare, how many cows can he put in each paddock?' In these problems, the teacher tries to embed the mathematics into a context that appears to be a realistic one—a task of division into a farming context. In part, this is to show the rationale for doing this type of arithmetic. (See Baroody et al. 2017.)

In order to complete these tasks, students need to be able to decipher the text for what is being asked, and to then respond in an appropriate manner. In the following section, a tool is provided for working through the errors that students make in responding to these tasks. However, sociological theories of school performance indicate potential difficulties for students about which teachers need to be aware when posing this type of question (Lerman & Zevenbergen 2004). Consider the following question: 'There are 315 students at the sports carnival. If a bus holds 60 people, how many buses are needed to get all the students back to school?' Within a mathematics classroom, the students are expected to divide 315 by 60, find that there is a remainder and then round up to the next whole number. The answer 6 would be seen as the desired response. However, some students may offer the answer '5 rem 15', which indicates that they did not realise there was a need to round up. Others may offer responses of '5, and put some kids so there are 3 per seat' or '5, and some kids go home with the teachers in their cars'. The last two responses suggest that the students have not read the context of the problem correctly, interpreting it as a real-world problem (as opposed to a school problem) and responding within that discourse.

When using real-world problems such as those described above, teachers need to be aware that there is layering of complexity. Not only do students need to have the literacy skills to decode the question in order to respond; they also need to be able to critically appraise the context of the question in order to decide that the question is really a mathematics task masked in an everyday context, and that the correct response is one that is embedded in a mathematics discourse as opposed to an everyday discourse. The correct identification of the appropriate discourse in which to reply may require teachers to undertake explicit instruction in the language game that is being lived out in such teaching episodes.

Language error analysis

Many of the tasks set in mathematics classrooms have been embedded in some word-based problem where students need to read the problem and work out a response. When a student makes an error, there are a number of levels at which the error could be made that do not necessarily represent a mathematical error. Consider the problem posed in the previous section: 'There are 315 students at the sports carnival. If a bus holds 60 people, how many buses are needed to get all the students back to school?' Here, the work of Newman (1983) provides insights. In her work on error analysis and its adaptations (Ellerton & Clements 1992), she shows that there are five levels to be undertaken before a student is able to offer the desired response of six buses. Newman and her advocates propose that students:

1. must be able to read the question (can they read the wording?)
2. should comprehend what the task is asking (can they identify the task as asking how many buses are needed to transport students?)
3. need to be able to translate this into the mathematical demands (can they identify that the task involves division and rounding up?)
4. must undertake the necessary mathematical operations, or whatever process is demanded by the task, in order to calculate (can they divide 315 by 60 and then round this to the next highest whole number?)
5. should represent the answer as a meaningful construct (in this case, six buses).

Frequently, incorrect responses have been assumed to be a result of a student not being able to do the mathematics—since the problem is a mathematical one. In contrast, this model indicates that there are both literacy and mathematical demands within the tasks. These five levels, or steps, indicate different areas in which students can demonstrate misunderstanding.

Teachers need to be aware that the production of an incorrect response may be due to an error occurring at any of these steps. In considering the bus example, Newman's error analysis gives educators a method for deciphering where students may be making errors when they undertake tasks of this kind. It moves away considerably from a view that the errors are purely in the mathematics, indicating there are other sites for errors—namely, in reading and comprehending the task. It therefore indicates both literacy and numeracy demands.

This is a very useful diagnostic tool for helping teachers to decipher how and where students may be making errors. It is useful in providing a sequence through which teachers can track progress in learning and comprehending. By posing relevant questions, teachers are able to identify the level at which the students are experiencing difficulties. The first three levels can be seen as relating to literacy demands, whereas the last two tend to be more mathematical. What this model demonstrates is that placing arithmetical tasks into contextual problems places a greater demand on students—particularly in terms of readability—and that this, in turn, creates another level of difficulty for them.

Mathematical literacies

The term 'literacy' is one that causes considerable debate about its depth and application. The term 'multiliteracies' has gained some recognition in attempting to capture this debate, and the particular and more contemporary forms of literacy needed for the emerging societies of the twenty-first century (Barton 2017; Unsworth 2002). For example, the literacies needed in the digital environment, the literacies needed for communication within various social, work, educational or other contexts, and the literacies needed for reading and comprehending modern forms of communication such as media and television (Luke 2005) can all be encompassed within the more traditional forms of literacy. These new forms of literacy are evident in the teaching and learning of mathematics, where students are expected to be digitally literate in order to undertake successful engagement in activities related to various software packages, including spreadsheets that aid computational thinking (Sanford & Naidu 2016; West 2000), specialised geometry packages (e.g. Cabri or GeoGebra or Cinderella) or apps on mobile devices (Larkin et al. 2018). When it comes to reading maps, students have developed new forms of digital map reading through the electronic games that many play (Lowrie 2003) and the use of various mapping software (GPS, Google Earth, etc.).

Within digital gaming contexts, while students only have access to the immediate screen at any point in time, they are often able to visualise screens than are not immediately seen, and predict movements in these unseen spaces. Another form of literacy involves appropriate ways of acting and responding in the classroom. As the mathematics classroom is a particular social and cultural context, students need to be able to identify appropriate ways of reading and interpreting social situations in order to be effective participants in the classroom.

In teaching students to apply mathematics to other contexts and situations, teachers often embed mathematics tasks into real-world problems—such as the bus example cited earlier. Textbooks and teacher resources abound with these examples. Two levels of literacy are needed in order to respond to these tasks. The first is the functional literacy noted earlier, where students need to be able to read and decode a text. The second level is where students need to develop a social literacy in order to decipher the unspoken 'rules of the game' in order to respond to such tasks. Students often misinterpret questions as 'trick' questions, whereas it is more likely that they did not interpret the unspoken rules of the game correctly.

Making language explicit

In a large study of classroom-based research, Sullivan and colleagues (Sullivan et al. 2002) focused on teachers making aspects of mathematics language explicit to students. Many of the issues identified in this chapter are not taught explicitly to students, but rather are expected to be learned through a process akin to osmosis. This results in some students 'cracking the code' of classroom literacies (Zevenbergen 2000, 2005) and others making little sense of the game being played. Unfortunately, a failure to crack the code can have serious consequences. For ethical teaching, teachers need to be aware of the ways in which language is implicated in teaching mathematics, and make the game explicit to their students. This can be undertaken in relatively simple ways—for example, by writing new words on the board or by explaining that the game being played is mathematics (rather than a pseudo-real context). By using literacy strategies and strategies used in ESL teaching, teachers can support their students in learning the language of mathematics—and its multiple forms of literacy. These are simple strategies, yet they have profound effects for learners. There are two main considerations: the language of mathematics, and the language through which mathematics is taught and learned (instructional language).

Having an initial awareness of the language implications, and then teaching the explicit language of mathematics, is an important first step. This includes the words and phrases specific to mathematics. The second step is the pedagogic relay through which mathematics is taught. This relates to the language systems that the teacher uses to teach mathematics. For example, when using instructions (direct teaching) or group work, the teacher will talk in different ways to students. Posing questions is another aspect of the instructional language. When posing problems in mathematics that have a real-world context, students need to identify that the question may be not in the real-world but in the mathematics context. Many rounding questions are posed in real-world contexts—for example, if a can of paint is 4 litres in capacity and can cover 20 square metres, how many cans would I need to buy if I have to cover 82 square metres? This is a rounding question, so students would be expected to reply with 5 cans. However, in the real world, a painter may buy 4 big cans and a small can, or water the paint down. So identifying that this is a mathematics problem, not an everyday problem, is critical for responding correctly.

Review questions

3.1 What are some of the key language issues that can create difficulties for students? Provide examples for each.

3.2 In terms of test items, what are some key areas that can create difficulties for students?

3.3 How might you go about teaching mathematics language explicitly in the mathematics classroom?

3.4 Reflect on your teaching experiences and identify an example in which students had difficulties responding to your questions due to recognition and realisation rules.

3.5 Using Newman's (1983) error analysis, work with a student to identify where they may be making mistakes when responding to word problems.

3.6 There are three different types of word problem. List each type and provide an example, outlining why it belongs to the category.

Further reading

Ellerton, N.F. & Clements, M.A. (1991). *The mathematics of language: A review of language factors in school mathematics*. Geelong: Deakin University Press.

Goos, M., Vale, C. & Stillman, G. (2017). *Teaching secondary school mathematics: Research and practice for the 21st century*. Crows Nest: Allen & Unwin.

Orton, A. (1996). *Learning mathematics: Issues, theory and classroom practice* (2nd ed.). London: Cassell.

Pimm, D. (1991). Communicating mathematically. In K. Durkin and B. Shire (eds), *Language in mathematical education: Research and practice* (pp. 17–24). Philadelphia, PA: Open University Press.

Redish, E.F. & Kuo, E. (2015). Language of physics, language of math: Disciplinary culture and dynamic epistemology. *Science & Education*, *24*(5–6), 561–90.

van den Heuvel-Panhuizen, M. & Drijvers, P. (2014). Realistic mathematics education. In Lerman, S. (ed.), *Encyclopedia of mathematics education* (pp. 521–5). Dordrecht: Springer.

References

Baroody, A. et al. (2017). Division problems and the concept of rate: A study of the performance of second-grade children on four kinds of division problems. In *A Focus on Multiplication and Division: Bringing Research to the Classroom* 74(3), (pp. ix–xii). Portsmouth, NH: National Council of Teachers of Mathematics.

Barton, D. (2017). *Literacy: An introduction to the ecology of written language*. Carlton, Victoria: Blackwell Publishing.

Davis-Dorey, J., Ross, S.M. & Morrison, G.R. (1991). The role of rewording and context personalization in the solving of mathematical word problems. *Journal of Educational Psychology*, *83*(1), 61–8.

Ellerton, N.F. & Clements, M.A. (1992). Implications of Newman research for the issue of 'What is basic in school mathematics?' In K. Owens (ed.), *Space: The first and final frontier* (pp. 276–84). Sydney: Mathematics Education Research Group of Australasia.

Freebody, P. (1992). A socio-cultural approach: Resourcing the four roles of the literacy learner. In A.J. Watson and A.M. Badenthorp (eds), *Preventing reading failure* (pp. 48–60). Sydney: Ashton Scholastic.

Halford, G.S. (2014). *Children's understanding: The development of mental models.* Washington, D.C.: Psychology Press.

Heller, J.I. & Greeno, J.G. (1978). Semantic processing of arithmetic word problems. *Paper presented at the annual meeting of the Midwestern Psychological Association.*

Jorgensen, R. (2015). Language, culture and access to mathematics: A case of one remote Aboriginal community. *Journal of Intercultural Education. 26*(4) 313–25.

Larkin, K., Kortenkamp, U., Ladel, S. & Etzold, H. (2018). Using the ACAT Framework to evaluate the design of two geometry apps: An exploratory study. *Digital Experiences in Mathematics Education.* https://link.springer.com.article/10.1007/s40751-018-0045-4.

Lean, G.A., Clements, M.A. & Del Campo, G. (1990). Linguistic and pedagogical factors affecting children's understanding of arithmetic word problems: A comparative study. *Educational Studies in Mathematics, 21*, 165–91.

Lerman, S. & Zevenbergen, R. (2004). The socio-political context of the mathematics classroom: Using Bernstein's theoretical framework to understand classroom communications. In P. Valero and R. Zevenbergen (eds), *Researching the social dimensions of mathematics education: Theoretical, methodological and practical issues.* Dordrecht: Kluwer.

Lowrie, T. (2003). Posing problems in ICT-based contexts. In L. Bragg, C. Campbell, G. Herbert and J. Mousley (eds), *Mathematics education research: Innovation, networking and opportunity. Proceedings of the 26th Annual Conference of the Mathematics Education Research Group of Australasia* (vol. 1, pp. 499–506). Geelong: MERGA.

Luke, C. (2005). Cyber-schooling and technological change: Multiliteracies for new times. In *Multiliteracies: Lit learning* (pp. 77–98). Routledge.

Newman, M.A. (1983). *Strategies for diagnosis and remediation.* Sydney: Harcourt.

Sanford, J.F. & Naidu, J.T. (2016). Computational thinking concepts for grade school. *Contemporary Issues in Education Research, 9*(1), 23–32.

Shi, S., Wang, Y., Lin, C.Y., Liu, X. & Rui, Y. (2015). Automatically solving number word problems by semantic parsing and reasoning. In *Proceedings of the 2015 Conference on Empirical Methods in Natural Language Processing* (pp. 1132–42).

Sullivan, P., Zevenbergen, R. & Mousley, J. (2002). Contexts in mathematics teaching: Snakes or ladders? In M.O.J. Thomas (ed.), *Mathematics education in the South Pacific: Proceedings of the 25th Annual Conference of the Mathematics Education Research Group of Australasia* (vol. 2, pp. 649–56). Auckland: MERGA.

Unsworth, L. (2002). Changing dimensions of school literacy. *The Australian Journal of Language and Literacy, 25*(1), 11–26.

Verschaffel, L., Van Dooren, W., Greer, B. & Mukhopadhyay, S. (2010). Reconceptualising word problems as exercises in mathematical modelling. *Journal für Mathematik-Didaktik, 31*(1), 9–29.

West, P. (2000). Introduction to spreadsheets for Grade 3. *COM 3 Journal, 23*(3/4), 27–9.

Zevenbergen, R. (2000). 'Cracking the code' of mathematics: School success as a function of linguistic, social and cultural background. In J. Boaler (ed.), *Multiple perspectives on mathematics teaching and learning* (pp. 201–23). New York, NY: JAI/ Ablex.

Zevenbergen, R. (2005). The construction of a mathematical habitus: Implications of ability grouping in the middle years. *Journal of Curriculum Studies*, *37*(5), 607–19.

Zevenbergen, R., Hyde, M. & Power, D. (2001). Language, arithmetic word problems and deaf students: Linguistic strategies used by deaf students to solve tasks. *Mathematics Education Research Journal*, *13*(1), 204–18.

CHAPTER 4

DIVERSITY AND EQUITY

Students enter mathematics classrooms with a wide range of background knowledge, experiences and dispositions. These differences—which arise both before they come to school and while they are in school—create different orientations and learning experiences. For some students, their experiences will help them in their learning of mathematics, while for others considerable input will be needed to support their learning. How teachers work with this diversity is informed by their own ideologies of how differences come to exist. Some teachers will see such differences as something biological or innate in students, whereas others will see them as something constructed and reified through school practices. These two extreme positions represent the nature/nurture poles of one of the most perennial debates in education.

Before continuing this chapter, we would like you to write down some answers to the questions below. Retain them for reflection after reading about the practices of the mathematics curriculum found in most classrooms:

- Why do disproportionate numbers of indigenous students (Aboriginal, Torres Strait Islander, Māori, First Nation, Inuit) fail mathematics in comparison with their non-indigenous peers?
- Why are students from working-class backgrounds more likely to fail mathematics than their middle- or upper-class peers?

- Why are students who live in regional, remote or rural settings more at risk of failing or performing poorly in school mathematics than their urban peers?
- Why are some students whose first language is not English more at risk of performing poorly in school mathematics than their peers whose first language is English?
- Why is it that when these variables are combined, the risk of failing is even higher?

Equity and equality

When considering how one deals with diverse classrooms, two different positions exist. Some teachers advocate an approach whereby each child is treated the same, as they regard this as fair. Others advocate an approach that recognises differences are evident at the start of schooling, and that students need to be treated differently in an attempt to redress these differences. When consideration is given to the disparate outcomes for an equity target group—for example, Indigenous students—different ideologies come into play. It can be that a teacher may hold one view at a particular time but move to another view under different circumstances.

Equality view

- *Assumption 1:* It is important that all students be given a fair chance at succeeding, so all students should be given the same opportunities. When students are provided with the same opportunities, those who are successful are seen to be the students who have chosen the options for success. The equality view is one of choice whereby all students are provided with the same opportunities. That is, while everyone has the opportunity for success, only some students choose the options to bring about that success.
- *Assumption 2:* Discourses of equality subscribe to principles of fairness and choice.

Equity view

- *Assumption 1:* Disadvantaged students are more likely to perform poorly at school due to different home circumstances and the practices of school mathematics not aligning with the knowledge, skills and dispositions the students bring to school.

- *Assumption 2:* If disadvantaged students are to succeed in mathematics, different opportunities need to be made available to them in order to redress the differences in school and non-school experiences.
- *Assumption 3:* Discourses of equity subscribe to principles of justice and difference.

In considering the two very different views outlined above, it becomes clear that if students enter school with very different learning experiences—some of which are recognised in school, others not—their chances of success are different. In the case of white, middle-class students whose experiences mean that they likely enter the mathematics classroom with a knowledge of colours and shapes, and ability to recognise and write numerals, the chances of success in the first year of schooling are greatly enhanced compared with some of their peers who enter the first year of mathematics without these experiences. This raises the issue of what teachers then do with these two very different groups of students. In the case of the equality teacher, there is every likelihood that this teacher will follow set curriculum guidelines for Foundation/Kindergarten and Year 1 and teach this content, making slight adjustments to support the learning of disadvantaged students. The equity teacher is more likely to adopt a teaching practice whereby the background knowledge of the students becomes the centre of the curriculum, and learning experiences are based around those experiences so that learning is built from there. The school curriculum will be adjusted to suit the needs of the students, but with every expectation that the students will learn—that is, the curriculum will not be watered down.

Cognitive models of difference

Cognitive models of learning, such as constructivism, recognise the ways in which different experiences orientate students towards constructing different meanings from classroom work. For example, the student who enters the classroom with an understanding of a square having more than three sides will construct very different meanings from their interactions with the teaching process than the peer who enters the classroom with an understanding of a square having four sides; this will be different again from the student who enters the classroom with an understanding of a square having four equal sides. For a considerable time in the early years, all three meanings will work well. Only when mismatches arise in students' experiences with rectangles and squares will such knowledge become problematic for them. Constructivism allows teachers to see how students construct their own meanings, so that when students enter the classroom their pre-existing knowledges and experiences are framed within cognitive terms.

A purely cognitive approach can engender a victim-blaming model whereby the students who fail or perform poorly in mathematics are seen as problematic. Terms such as low ability, low motivation, cognitively inferior, delayed growth and so on are often used to explain why some students do not progress as expected. By placing the blame on the student (or their families or peer groups), teachers reflect a belief that they can do little to change learning outcomes.

Social models of difference

Rather than see students' constructions of mathematical ideas and processes as purely cognitive, social and cultural models seek to understand diverse backgrounds, and the experiences brought about through those backgrounds, and the ways in which these will impact students' construction of meanings. In viewing the mathematics curriculum as a construction of social knowledge, social models of difference see the curriculum as representing some knowledges and excluding others. Mathematics education represents some cultural values and denies others. Under these models, when teaching practices are seen as problematic and contributing to the failure of many students, teachers are able to change their practices in order to enhance learning outcomes for all students. In order to change their practice, teachers need to recognise differences in students, their backgrounds and their orientations towards schooling and mathematics. Rather than such differences being viewed as deficits, they should be seen simply as differences. Social, cultural, gendered, linguistic and geographic differences are critical factors when it comes to success in school.

In the cognitive model, the focus is on developing cognition. Social models extend this position, at the same time recognising the social factors that affect success in mathematics. Social factors include those brought to the learning environment by students and those of the learning environment itself. Learning mathematics is a social activity, and as such it has particular unspoken rules. The rules of the mathematics classroom will be more accessible for some students than for others. Teachers need to be cognisant of these rules so that all students will be able to participate in classroom practice.

Apart from the cohort of students with intellectual disabilities, the students most at risk of failing (or underperforming in) mathematics are those who come from working-class (or low socio-economic status) backgrounds, students from indigenous cultures, students whose first language is not English, students living in remote or rural/isolated areas, students with disabilities, and particular gender groups (this is tied closely to the previous variables). There is a widely held belief that boys are better at mathematics than girls; there is some evidence to support this claim, yet it is very esoteric. (Boys may be more likely to outperform girls on some spatial tasks—those involving mental rotation—but this is insufficient to support a biological basis for gender differences.)

Students coming from these backgrounds are disproportionately represented in the lower percentiles of most test measures. This is not to say that all students from such backgrounds will perform poorly—only that they are disproportionately represented in the lower ends of measures. This chapter discusses ways in which mathematics education is implicated in the production of such outcomes. By better understanding the hidden biases and assumptions built into the teaching of mathematics, teachers can alter their practices to cater for the diversity of students in the classroom.

Teachers' beliefs

Teachers' beliefs about learners and learning have a powerful impact on learning outcomes. The power of teachers' beliefs is profoundly evident in the study *Pygmalion in the Classroom* (Rosenthal & Jacobsen 1969), in which students were randomly assigned rankings and placed into a new classroom. The teacher believed that the marks were the students' academic rankings. Even though the rankings were totally random, at the end of the teaching year students confirmed the ranking originally given. This study showed how the teacher's beliefs about the students created different expectations and learning environments, which in turn produced the expected outcomes. The implications of this work are that when a teacher believes a student to be able or not able, then learning experiences will be provided that will confirm this expectation. A large-scale study in the United Kingdom (Askew et al. 1997) showed that the most effective teachers of numeracy were those who believed that *all* students could learn mathematics. This view has been reinforced by many other subsequent studies.

The myth of ability: challenging the practice of ability grouping

Perhaps one of the most pervasive practices in mathematics—and one not as evident in any other curriculum area—is founded on a belief in ability. This belief is manifested in the use of streaming (or setting) students in groups according to 'ability'. Teachers of mathematics are more likely than those in any other curriculum area to support the notion of ability grouping. At the end of a teaching year, unsurprisingly, results tend to reinforce the ability groupings made, leading teachers to believe that such groupings were appropriate. What is alarming in education and policy is that there is very little (if any) evidence to support the grouping of students by ability; indeed, most research suggests that the practice is detrimental for learners, yet it persists (Francis et al. 2017). In a number of studies of ability grouping in mathematics in the United Kingdom, the United States and Australia (Boaler 1997; Boaler et al. 2000; Zevenbergen 2001), some consistent findings emerged:

- Students in high streams were most advantaged by having good teachers, being exposed to extensive curriculum and entering examinations with a good knowledge, which allowed them to succeed in the exams. In the Australian study (Zevenbergen 2001), students reported that they would enter the exam feeling as if they already had a pass grade and were working towards higher grades. In the United Kingdom study (Boaler 1997), students in the high stream also reported that the pacing of lessons was very quick, because teachers expected that they were the 'bright students' and could cope with the work. The ethos of these classrooms was work-oriented.
- Students in the low streams were exposed to low levels of curriculum, were in classes with considerable management issues, covered minimal content, entered exams with low levels of knowledge, frequently had teachers who were not specialist mathematics teachers, and could only gain low marks.

These learning environments produced very different orientations towards mathematics, as well as different achievement levels. The high-stream students were more likely to report a positive attitude towards mathematics and an intention to continue with study in the area. Students in the low streams reported a frustration with mathematics, an inability to move out of the low streams due to the subject organisation (rather than their desire) and a dislike for the subject. These different learning outcomes can be viewed as a consequence of learning environments based on a belief in ability. The ultimate outcomes reinforced teachers' beliefs about the students' abilities in mathematics—in fact, questions need to be posed about the teaching practices and their role in the construction of 'ability'.

Similar studies have shown that teachers hold powerful beliefs about students' backgrounds as a factor in determining success. In a study of teachers' beliefs about why students from socially disadvantaged backgrounds tended to underperform in school mathematics, Jorgensen (2018a) found that the teachers generally blamed personal attributes of the student or family for lack of success. In this study, teachers cited their experiences with minimal resources—namely, calculators. In one case, the teacher described the students as having no respect for school property or themselves. When a structured borrowing system was put in place, teachers believed that students would damage or steal the class set of calculators. In another school, teachers saw students as needing to be respected and valued as people; here, the students could borrow the calculators on an honour system. In this school, all calculators were returned undamaged at the end of the year, despite no policing of equipment. These two experiences starkly highlight the ways in which teachers' beliefs about students can impact practice significantly.

Mathematical mindsets

The term 'mindsets' has emerged as one of the new buzzwords in education generally, and in mathematics in particular. It originally came from the work of Dweck (2017), in which she identified that students would either accept failure or bounce back. From this, she identified two mindsets: fixed and growth. When students have a *fixed* mindset, there is a belief that ability is something innate and that they cannot do mathematics. In contrast, those with a *growth* mindset see their capacity to learn mathematics as a process of growth. The latter mindset is more likely to see that with effort and hard work, the learner can achieve their goals in mathematics. Boaler (2015) has established a strong following in mathematics, one centred around growth mindsets. One of her arguments is that errors are good because they make the brain grow. While this mantra may appeal to educators and policymakers, there is a high degree of scepticism about whether the brain science behind it is scientifically valid. What is important with the mindset approach is that the myth of a fixed ability in mathematics is challenged, and that good teaching practices support students to learn and grow their mathematical understandings. While there is significant hype around mindsets, there is a need for caution. Good teaching in mathematics, in which students can access the content in engaging ways that allow them to construct a positive sense of themselves as learners, should be the inspiration for teaching. Locking learners out of mathematics through teaching practices that allow students to see themselves as failures should be avoided. While this is now framed in discourses around 'growth' and 'fixed' mindsets, the practices are longstanding and are often framed within many other discourses. The neuroscience that is used to legitimate the claims of growth mindsets is not needed, and is not advanced enough, to promote teaching approaches. Good pedagogy in mathematics is the key to success.

Creating classroom environments that promote among students positive perceptions of themselves as learners of mathematics is a key pedagogical tool for good teaching. Irrespective of whether this is framed within discourses of self-concept, self-esteem, subjectivity, identity, mindsets or habits of mind (habitus), the goals in mathematics teaching are to bring about success in mathematics and in how students come to see themselves as learners of mathematics. Larkin and Jorgensen (2016) reported that learners as early as Year 3 are constructing a 'self of themselves' as learners of mathematics—more frequently in negative ways. Sadly, these conceptions of themselves correlate with social, gender, geographic and cultural backgrounds.

Home–school differences

Students from socially disadvantaged backgrounds have been found to be exposed to very different literacy practices than their middle-class peers. In a study of preschool interactions, Walkerdine and Lucey (1989) found that mothers interacted very differently with their children depending on the mother's social background. Similar generic studies conducted by Heath (1989) and Delpit (1995) have shown similar patterns of interaction. Where patterns of interaction at home are language-rich—as in, for example, 'Please get your yellow jumper. It is on the top shelf of the middle section of your wardrobe, just next to the pile of T-shirts', the language is rich not only in the genre of school interactions but also in the language of mathematics. The studies cited have shown that interactions in the homes of socially disadvantaged students are more likely to consist of declarative statements depleted of adjectives—for example, 'Go and get your jumper.' Those in the first group of students are exposed to a rich spatial language— 'top', 'middle', 'next to'—as well as to the language of colour, which is one of the key tools for early-years sorting and classifying activities. Many teachers bemoan the deficits of disadvantaged students when they come to school unable to write their numerals and not knowing colours or basic shapes. It is not that these students have lower levels of cognitive development; it is that their family environment does not use school English.

Students whose home environments expose them to the rich language that prepares them for school are better prepared for many aspects of school mathematics. It is not that they are more intelligent, but that the practices in their homes resonate with the practices of school mathematics, so that the transition into and through schooling is much easier for some students than for others. In other words, some students have been exposed to home practices that will prepare them more effectively for school than their less advantaged peers. It is important to recognise that the families are not framed within deficit models of thinking, but rather models of difference. Many Indigenous learners, for example, may come to school unfamiliar with English but fluent in a number of home languages. So, while they may not have the rich repertoire of English spatial, colour or number language, they are strong in many languages. This difference-based model of thinking about the skills, knowledges and dispositions students bring to school is important for transitioning learners into mathematics. Sarra (2013) espouses the importance of having high expectations of learners. In his case, he is advocating for Indigenous learners, but his message applies to all equity target groups—having expectations that *all* students can learn mathematics creates very different potential for teaching and learning than holding deficit views of learners and learning.

There is now a very strong push to 'reduce the gap' in the early years so that learners will come to school with greater alignment with the practices and discourses of schooling. More children from socially, culturally and linguistically diverse families are encouraged to attend day care and early childhood settings to ameliorate some of the differences prior to school (Perry et al. 2016). The early years—prior to school and in the first few years of schooling—are foundational to mathematics learning.

Other studies have shown similar differences in specific aspects of language. Cross-cultural studies have indicated that the language used by different cohorts of students varies significantly even when they speak English. In her work with young students, Walkerdine (1992) found that middle-class parents were more likely to use the terms 'more' and 'less' in their interactions with their children—'Would you like a little more or a little less meat with your dinner?'—whereas working-class parents were more likely to only use the signifier 'more'—'Would you like some more meat with your dinner?' or 'Don't you want any more?' Similar cross-cultural work has found the same differences in language use among many indigenous peoples. This subtle difference has considerable manifestations for early work in mathematics. In these strands, when undertaking mathematics work teachers will be asking questions such as 'Which number is 2 more than 3?', 'What is more—3 or 5?', 'What number is 5 less than 8?' and 'Which column has more than this one?' Such questioning is integral to most early teaching of mathematics. However, when young students from disadvantaged backgrounds are not exposed to these forms of language, and when this is not seen as problematic, many of these students may miss significant chunks of the content due to their unfamiliarity with the terms 'less' and 'more'. Rather than seeing these students as being deficient in their understandings, teachers should recognise that the language of the students' backgrounds inhibits their potential to make sense of the lessons.

More subtle differences emerge when other aspects of mathematics teaching are considered. Reforms, such as the use of open-ended tasks, change the culture of mathematics, but this is never made explicit to students. In studies of curriculum reform using open-ended tasks, various aspects of pedagogy were found to be problematic for young students. Lubienski (2000) found that her low socio-economic status students preferred working on closed tasks, whereas middle socio-economic status students reported a preference for open-ended questions. Many students have constructed a sense of mathematics as being a discipline in which there is one correct answer, but when this is changed, as in the case of open-ended tasks, teachers need to be cognisant of the need to make the criteria explicit to students. Similarly, when using group work or setting students to tasks such as 'discuss the problem in small groups', teachers need to explain the rules and expectations (Sullivan et al. 2003). While this may be seen as simply good classroom practice, it is particularly important for students from socially disadvantaged backgrounds, for whom the purpose of the tasks may not be as obvious as it is to their peers. Without an explanation of the expectations of the task, students may be at risk of not undertaking the task in a way envisaged by the teacher, due to their inability to recognise the implicit rules of group work rather than the mathematical demands.

Many of the tasks set for students have been embedded in word problems with the intention of making the task 'real'. However, the use of contexts can create difficulties. Most obviously, the immediate readability of the text can be problematic for students whose grasp of English is not strong. However, another layer of difficulty is added, as there are now two competing discourses operating: students need to identify whether the task is a school mathematics task or a real task (Cooper & Dunne 1999). Working with students in the upper primary grades, Zevenbergen and Lerman (2001) reported that in order to be able to respond correctly, students need to identify the task as mathematical, select the elements that will enable a correct response to be generated, and then respond in a format that is seen to be appropriate in that context.

In a large-scale study of similar questions posed on a testing scheme in the United Kingdom, Cooper and Dunne (1999) found that working-class students were more likely than their middle-class peers to respond in ways that suggested they interpreted the questions as real problems rather than school mathematics, and thereby to offer incorrect responses. Middle-class students were more likely to see the problem as a school mathematics task and to answer in that mode. Cooper and Dunne suggest that middle-class students are more able to code-switch between the contexts of home and school, whereas working-class students are more likely to work within everyday contexts, and thereby to misinterpret the questions and ultimately offer incorrect responses. When follow-up interviews were undertaken, the students were able to calculate the answers correctly once they realised that their interpretation of the question was incorrect, thus suggesting that the problem is not so much mathematics but rather the registers within which the questions are posed, and that the contexts are not made explicit to students. In her work in culturally diverse mathematics classrooms, Jorgensen (2018b) showed how teachers can create environments that bridge the gap for learners whose home background is very different from the cultural norms of the mathematics classroom. Creating practices that recognise these differences is central to establishing equitable practices.

Knowledge and worldviews

One of the reasons for differences in school performance is the worldviews with which students enter school—that is, their particular ways of seeing and viewing the world. In most Western societies, considerable emphasis is placed on quantifying things. Young children are indoctrinated with these worldviews from the time they enter the world. How old is your friend? How many brothers and sisters does she have? What date is your birthday? Everyday events are talked about in numbers, so the idea of numbers, counting and measuring is an integral component of the world in which most students live.

For those who live in a quantifying society, many of the ideas used in mathematics may be seen as unproblematic. Take percentages. They are often taught in the contexts of profits, mark-ups or discounts as if these are part of the natural order. Making a profit—as represented in the idea of mark-ups—is legitimised through the teaching of percentages (Gill 1988). It is taken for granted. But is this so for everyone? In egalitarian and many agrarian cultures, and for those of the Muslim faith, there is no notion of profit or mark-up, and the idea of profiting from your peers is the antithesis of community.

The cultural knowledges with which people are familiar can be illustrated by the 'fish in water' analogy. When in the water, the fish assumes that the world is as it should be; however, when the fish is removed from the water, the world is very strange. Worldviews work in a similar way. Even within Western cultures, things are done differently in different countries—how to catch a bus, how to queue, how to pay for items. The differences become even greater when the cultures are even more diverse. Consider the assumptions that are built into Western economics of profit and the need to make money. Islamic banking is a non-interest banking that is founded on the principles of Islam and shaped by Islamic economics, and so it strongly contradicts the assumptions built into Western banking and economics—which are an integral part of the school mathematics curriculum.

The base 10 counting system is commonly used in Western cultures—to the extent that it is seen by most members of the public as the 'natural' way to count. Yet, while school mathematics shows other counting systems based on different number bases, other 'natural' counting systems exist in other cultures. For example, in his extensive study in Papua New Guinea, Lean (1996) recorded hundreds of different counting systems, many of them based on body parts. It is argued that more complex counting systems arise from the need to count large numbers of objects. In many nomadic cultures, there is no need for large numbers because the society hunts and gathers with little trading. When cultures begin to trade, a need for a more complex number system arises. In Western society, the demise of the Roman numeral system resulted from its complexity when dealing with large numbers; the Hindu-Arabic system used in most countries allows for large numbers and undertaking work with those numbers.

Measuring time is a construct unique to quantifying societies, as represented by phrases such as 'time is money' and 'saving time'. It is a construct that has arisen from a need to communicate about time. In order for Western society to work well, points in time and the passage of time need to be measured. Such constructs are often taken for granted, as if they were natural phenomena, whereas in fact they are social and historical artefacts (Poole 1998). Even within Western societies, there are differences—for example, consider the seasons. In some countries in the Southern Hemisphere, spring and autumn begin on the first days of September and March; in other countries, they begin on the equinox (when day and night are of equal length), around the 21st to 23rd of those months. This indicates that the timing of the seasons is merely a convention.

In some Indigenous Australian cultures, seasonal attributes are considered from a qualitative perspective. In her work with Australian Aborigines, Harris (1990) notes that the equivalent of a calendar was discussed in terms of qualitative features—the weather, the flora and fauna—so that in any one year a season would be as long as the qualities were apparent. Rather than seeing summer as being three months in the period from December to February, she suggests that summer would last as long as particular qualities remained apparent.

The subtle power of how people come to see the world can be seen when teaching this area. In this section of the curriculum, common examples relate to students undertaking calculations centred on areas of various shapes. Sometimes decisions need to be made about choices, which are inevitably made on attributes of size. For example: 'A farmer is given the choice of two blocks of land: one is a rectangle of 7 km by 2 km, the other a square with sides of 4 km. Which one would he choose?' The expectation would be that the students would calculate the area and choose the bigger block. Students with a knowledge of farming might also decide that the square would be a better proposition because it would require less fencing material. When similar problems were posed to students in Papua New Guinea, Bishop (1988) found that students' reactions were not based on quantification; rather, they posed more questions about the quality of the land—how fertile it was, the slope (too much would create erosion), and so on. These examples suggest that students' worldviews may position them very differently in terms of how they will interpret the problems that are posed, and how they are generally orientated towards working mathematically.

There is a common perception that many Indigenous cultures have a limited counting system. Some desert cultures have a counting system of 'one, two, three big mob'. There are many very practical reasons why the counting system has remained at this level. These Indigenous people can see greater amounts, but do not have the language to talk about them. The counting systems that these learners bring to school mean that the teacher(s) need to develop quality processes to teach larger numbers, place value and impart number as well as operations concepts. The practices used in urban settings will need to be different to cater for the number worlds of the learners. Conversely, many Indigenous learners have an acute sense of space and location, one that far exceeds that of their urban peers. Recognising the differences in what is brought to school and how this influences teaching and learning is the first step in developing curriculum and pedagogy to meet the learners' needs (Jorgensen 2018a).

How people come to see their worlds significantly impacts how they think and perform mathematically. It is not that there are differences in innate ability; rather, there are differences in cultural knowledges. For some students, it is not simply a case of learning concepts and processes that extend their familiar knowledges, but something much more complex. This applies not only to students from very diverse cultures, as in the cases cited here, but also to students whose gender, social class and geographic location (i.e. rural vs. urban) are different.

How is difference constructed through the practices of school mathematics?

There are many ways in which teaching practices are implicated in the construction of differentiated outcomes of learners. It is important for teachers to be aware of the subtle (and overt) ways in which the practices they adopt can reify the social, cultural, gender and language diversity of society. In a comprehensive account of textbooks, Apple and Christian-Smith (2017) have shown how race, class and gender are represented in textbooks, and how these practices include and exclude different groups of students. In an analysis of gender, for example, Sleeter and Grant (2017) showed how differences were represented and the potential for exclusion or inclusion of particular groups of learners.

Implications for school mathematics

When students from diverse social and cultural groups come to the mathematics classroom, they bring with them a wealth of knowledge and experiences. Some of these experiences will be acknowledged, and others will be denied. The counting systems of Western cultures are seen as legitimate mathematics. Where Indigenous counting systems are taught (which is infrequently), they are seen as novelties rather than as legitimate ways of knowing.

Indigenous students working with a number board

When a teacher becomes aware of differences in knowledges and how these relate to school mathematics, there is a significant opportunity to adjust practice in ways that can allow students from diverse backgrounds to come to learn and understand mathematics. For example, when teaching ratio, one of the most common teaching examples is for students to calculate the best buys of commonly purchased items. In attempting to be inclusive, a teacher may assume that students from wealthy backgrounds would purchase expensive items and students from economically disadvantaged backgrounds cheaper items, and so use economically differentiated items for the worked examples. However, when considering the backgrounds of disadvantaged students, it is not uncommon to find that they engage in fishing, an inexpensive activity and one that can support the family. It is also one rich in ratio knowledge—the hook needed for small fish such as whiting is very different from the hook needed for large fish such as mackerel, for example—which is seldom used in mathematics examples. This knowledge would be familiar to many disadvantaged students and would lead to success in understanding ratio, since it has relevance and purpose. It also legitimates the knowledges of the students, so that they feel they are a valued part of the mathematics culture.

In a large study aiming to identify successful practices that teachers use in the classroom, Jorgensen (2018b) outlined a series of enacted norms that underpin successful numeracy/mathematical practice:

1. All students can learn mathematics—to high levels.
2. Embedding mathematics—in thoughts as well as contexts—is critical for understanding.
3. Mathematics is as much about language as it is about mathematical concepts.
4. Transparency in learning and teaching mathematics enables students to access the 'secret knowledge' of school mathematics.
5. Mathematics lessons should engage learners at their levels of understanding, then extend learning into new levels.

For learners who are more at risk than other students, perhaps the most important factor in attaining success is for teachers to have high expectations of them. Many educators working with Aboriginal students, for example, advocate strongly that teachers must have high expectations of their students. This will ensure that the teacher provides a rich learning experience that meets and exceeds expectations, and is based in a high-quality curriculum with strong scaffolding to enable students to learn. When teachers work with deficit thinking, they are more likely to offer a low-level, poorly stimulated learning environment that holds back students' progress.

Review questions

Consider the questions that were posed at the start of this chapter. What are your reactions to them as a consequence of reading this chapter and the further readings?

4.1 In what ways does the mathematics curriculum exclude some students from participating fully in classroom activities?

4.2 Imagine you are sent to a remote area where there are high numbers of socially and/or culturally diverse students. What strategies would you put in place that might support students' learning in mathematics?

4.3 In what ways could you adjust the curriculum so that you could include (in a legitimate and appropriate way) different cultural knowledges that have a mathematical orientation?

4.4 Reflect on what you learned about language and mathematics, and apply it to what you have learned from this chapter. How are mathematics, language and diversity implicated in the construction of social and cultural differences?

Further reading

Becker, J.P., Silver, E.A., Kantowski, M.G., Travers, K.J. & Wilson, J.W. (1990). Some observations of mathematics teaching in Japanese elementary and junior high schools. *Arithmetic Teacher, 38*(2), 12–21.

Hessler, G.L. (2001). Who is really learning disabled? In B. Sornson (ed.), *Preventing early learning failure* (pp. 21–36). Alexandria, VA: Association for Supervision and Curriculum Development.

Koshy, V. (2001). Identifying mathematically promising students. In V. Koshy, *Teaching mathematics to able students* (pp. 19–28). London: David Fulton.

Milton, M. (2001). Who helps children who have learning difficulties in numeracy? *Australian Primary Mathematics Classroom, 6*(3), 6–10.

Montis, K.K. (2000). Language development and concept flexibility in dyscalculia: A case study. *Journal of Research in Mathematics Education, 31*, 541–56.

Robert, M. (2002). Problem solving and at-risk students: Making 'mathematics for all' a classroom reality. *Teaching Children Mathematics*, *8*(5), 290–5.

Sun, W. & Zhang, J. (2001). Teaching addition and subtraction facts: A Chinese perspective. *Teaching Children Mathematics*, *8*(1), 28–31.

Weaver, L.R. & Gaines, C. (1999). What to do when they don't speak English: Teaching mathematics to English-language learners in the early childhood classroom. In J.V. Copley (ed.), *Mathematics in the early years* (pp. 198–204). Reston, VA: National Council of Teachers of Mathematics.

Zazlavsky, C. (2001). Developing number sense: What can other cultures tell us? *Teaching Children Mathematics*, 7(6), 312–19.

References

Apple, M. & Christian-Smith, L. (eds). (2017). *The politics of the textbook*. New York, NY: Routledge.

Askew, M., Brown, M., Rhodes, V., Johnson, D. & Wiliam, D. (1997). *Effective teachers of numeracy: report of a study carried out for the Teacher Training Authority 1995–96.* London: School of Education, King's College.

Bishop, A.J. (1988). *Mathematical enculturation: A cultural perspective on mathematics education.* Dordrecht: Kluwer Academic.

Boaler, J. (1997). *Experiencing school mathematics: Teaching styles, sex and setting.* Buckingham: Open University Press.

Boaler, J. (2015). *Mathematical mindsets: Unleashing students' potential through creative math, inspirational messages and innovative teaching.* San Francisco, CA: JosseyBoss.

Boaler, J., Wiliam, D. & Zevenbergen, R. (2000). The construction of identity in secondary mathematics education. In M. Santos (ed.), *Proceedings of Second International Mathematics Education and Society Conference* (pp. 192–202). Lisbon: Universidade de Lisboa.

Cooper, B. & Dunne, M. (1999). *Assessing children's mathematical knowledge: Social class, sex and problem solving.* London: Open University Press.

Delpit, L. (1995). *Other people's children: Cultural conflict in the classroom.* New York, NY: New Press.

Dweck, C.S. (2017). *Mindset: Changing the way you think to fulfill your potential.* New York, NY: Random House Publishing.

Francis, B., Archer, L., Hodgen, J., Pepper, D., Taylor, B. & Travis, M.C. (2017). Exploring the relative lack of impact on research on 'ability grouping' in England: A discourse analytic account. *Cambridge Journal of Education, 47*(1), 1–17.

Gill, D. (1988). Politics of percent. In D. Pimm (ed.), *Mathematics, teachers and children: A reader* (pp. 122–5). London: Hodder & Stoughton.

Harris, P. (1990). *Mathematics in a cultural context: Aboriginal perspectives on space, time and money.* Geelong: Deakin University Press.

Heath, S.B. (1989). *Ways with words: Language, life and work in communities and classrooms* (2nd ed.). Cambridge: University of Cambridge.

Jorgensen, R. (2018a). Building the mathematical capital of marginalised learners of mathematics, *ZDM: The International Journal on Mathematics Education, 50*(6), 987–8.

Jorgensen, R. (2018b). *Celebrating success: Numeracy in remote Indigenous contexts. Final report of the Remote Numeracy Project.* Canberra: University of Canberra.

Larkin, K. & Jorgensen, R. (2016). 'I hate maths: Why do we need to do maths?' Using iPad video diaries to investigate attitudes and emotions towards mathematics in year 3 and year 6 students. *International Journal of Science and Mathematics Education, 14*(5), 925–44.

Lean, G.A. (1996). *Counting systems of Papua New Guinea and Oceania.* Unpublished doctoral thesis. Lae, Papua New Guinea.

Lubienski, S.T. (2000). Problem solving as a means towards mathematics for all: An exploratory look through a class lens. *Journal for Research in Mathematics Education, 31*(4), 454–82.

Perry, R., Hampshire, A., Gervasoni, A. & O'Neill, W. (2016). 'Let's count': Improving community approaches to early years mathematics learning, teaching and dispositions through noticing, exploring and talking mathematics. *Paper presented at the 39th Annual Mathematics Education Research Conference.* Adelaide: MERGA.

Poole, R. (1998). Calendars, calibration and culture. *Mathematics in School, 27*(5), 14–16.

Rosenthal, R. & Jacobsen, L. (1969). *Pygmalion in the classroom.* New York, NY: Rinehart & Winston.

Sarra, C. (2013). *Good morning, Mr Sarra.* Brisbane: University of Queensland Press.

Sleeter, C.E. & Grant, C.A. (2017). Race, class, gender, and disability in current textbooks. In M. Apple and C.E. Sleeter (eds), *The politics of the textbook* (e-book). New York, NY: Routledge.

Sullivan, P., Mousley, J., Zevenbergen, R. & Turner Harrison, R. (2003). Being explicit about aspects of mathematics pedagogy. In N. Pateman, B. Dougherty and J. Zillox (eds), *Proceedings of the 28th Annual Conference for the International Conference of the Psychology of Mathematics Education 27* (vol. 4, pp. 267–74). Honolulu, HI: PME.

Walkerdine, V. (1992). Reasoning in a post-modern age. Unpublished paper presented to the *5th International Conference on Thinking*, Townsville.

Walkerdine, V. & Lucey, H. (1989). *Democracy in the kitchen: Regulating mothers and socialising daughters.* London: Virago.

Zevenbergen, R. (2001). Is streaming an equitable practice? Students' experiences of streaming in the middle years of schooling. *24th Annual MERGA Conference*, Sydney, July 2001.

Zevenbergen, R. & Lerman, S. (2001). Communicative competence in school mathematics: On being able to do school mathematics. In M. Mitchelmore (ed.), *Numeracy and beyond: Proceedings of the 24th Annual Conference of the Mathematics Education Research Group of Australasia* (vol. 2, pp. 571–8). Sydney: MERGA.

CHAPTER 5

PLANNING FOR TEACHING

Planning is an integral component of the teaching and learning process.

In order to develop a quality learning environment, the teacher needs to plan for short-, medium- and long-term goals. Without good planning, teaching risks becoming ad hoc, and thus threatening the gains that students can make. Three key questions underpin effective teaching in mathematics:

1. What am I going to teach?
2. How will I teach it?
3. How will I know whether I have been successful in teaching it?

When a teacher is able to provide answers to these questions, the task of teaching becomes somewhat simpler.

In contemporary education contexts, where there is increased accountability, teachers are held responsible for providing learning for their students. This focus represents a substantive shift from the recent past, when the emphasis was more on teaching and on providing activities that could engage students. Often, students could be seen to be doing 'busywork', but there was little intellectual quality to these activities—or, if there was, it was not made explicit to students. More recently, the emphasis has shifted to student learning and high expectations being placed on teachers to provide learning environments that will produce the desired learning outcomes. Current reforms recognise the importance of focused, high-quality learning experiences. In some places, this has resulted in governments providing guidelines to teachers on how to prepare quality lessons and quality discussions within those lessons. More than ever, teaching in contemporary times demands that teachers plan for their students' learning.

Planning

In planning short- through to long-term teaching episodes, certain salient, overt factors that are needed for successful teaching and assessment of mathematics are inherent in the planning process. Two factors impact planning:

1. *The individual's subjective orientation towards teaching.* This includes beliefs about how students best learn mathematics. If the teacher subscribes predominantly to a particular view of learning, this is likely to be reflected in their planning (including teaching and assessment).
2. *The objective or structural aspects of teaching mathematics.* This relates to aspects of mathematics education that may be perceived as being beyond the control of the teacher, including set curriculum documents or schemes and testing regimes.

Levels of planning

Three key levels of planning guide teachers' work in mathematics:

Teacher display in a classroom showing class units of work

1. *Long term*—where the teacher may develop plans for the full year's work. These can come from, and may be guided by, school policy documents, government education documents and materials from commercial publishers. Teachers may use all of these more fully developed products to greater or lesser degrees, depending on how well they link in with the school and the individual classrooms.

2. *Medium term*—where the teacher may plan for an extended period of time, such as a school term or a period of time within a term. Planning at this level involves units of work that may be thematic or concept-based.
3. *Short term*—where the teacher plans for a single lesson or a short unit of time, such as a week. This can be undertaken by pre-service teachers as they embark on the principles of planning and the minutiae of organising for effective learning and teaching.

Planning can be undertaken at the level of the individual teacher, across teams and/or across the school. Much of this planning will be influenced by school policies.

Why plan?

For experienced teachers, early planning becomes an internalised process, and is often barely visible. For novice teachers, the early planning of lessons is not an end product, but rather the introduction to the planning process—the development of planning principles. Planning is an essential part of good teaching, and has the following purposes:

- It ensures that teachers have deliberate intentions of what they want to achieve and in a particular timeframe. Teachers need to have clear goals about their teaching. Being focused on what is intended over a given period of time allows teaching to remain focused on particular outcomes.
- It avoids repetition. The mathematics curriculum is a very crowded body of knowledge, so teachers need to ensure that they do not repeat material unnecessarily. At the same time, they need to allow adequate time for revision of work when required.
- It ensures that all content is covered. Progression through the primary school demands that students cover the nominated content. Planning allows teachers to see the big picture of their year's work, allowing them to ensure that appropriate amounts of time are spent on topics so that all key content is addressed.
- It ensures that content is covered in an ordered and appropriate time-frame. Seeing the big picture of a year's work, or a smaller amount of work, allows teachers to determine the best order in which to deliver

content, and the most effective way to achieve the breadth of what is stated in curriculum documents.

Specifically, planning:

- allows teachers to cater for the diversity within a classroom
- ensures that the lesson/unit will be motivating for students
- creates a sense of confidence in what is going to be achieved
- assists teachers when they reflect on how their teaching can be improved.

Planning: what do I want to teach?

Identify the intended learning for students

Central to quality learning is knowing why particular experiences are beneficial for students' mathematics learning. Consider a task in which brightly coloured paperclips are placed on various items, and students are required to count and record the number of clips used. This activity, which has the loose purpose of bringing about the idea of informal units of length, might keep students busy—but the real point of the activity must be questioned. Is it just a counting and recording activity, since the students are going to be introduced to various forms of tables? Is it about measurement? If so, measurement of what? Length? Area? Using informal or non-standard measurements? Is it a situation in which students are expected to problematise the use of informal measurements in preparation for the use of formal units of measure? With so many questions raised, the activity has considerable potential for diversion, resulting in lack of focus. Making connections is a critical aspect of teaching, but if there is no sense of the purpose of the lesson, it will be difficult to make coherent connections for students.

Depending on the teaching episode, there is a range of potential foci for a lesson. Not all teaching episodes will focus on mathematical knowledge alone; they may consider other aspects of mathematical activity, including:

- *conceptual*—where the knowledge of mathematics is considered (e.g. fractions, subtraction)
- *process*—where the 'how to' is the focus (e.g. problem-solving, inferring, using technology); this can be related to thinking and working mathematically

- *affective*—where students' feelings are considered (e.g. perseverance, enjoyment, motivation)
- *social*—where social dynamics are considered (e.g. group work, collaboration)
- *language*—where specific aspects of mathematical language or modes of communication are considered.

Specific learning outcomes

Once the main outcome for a lesson or unit is identified, syllabus documents provide extra guidance by suggesting indicators of learning. Within the new approaches to curriculum development, whereby learning outcomes form the basis of the reform, support documents have broken down these outcomes into smaller, manageable units that are usable (and assessable) by teachers. These can be identified and used for planning. Various systems have adopted different formats for their syllabus documents, so teachers need to become familiar with those adopted in their own locality.

The context

Teachers need to consider the specific context within which they will teach. This will vary from classroom to classroom, school to school and year to year. It is important to consider the interests, needs and background knowledge of the students in any given classroom. This is particularly the case for diagnostic work and extension work.

Background knowledge of students

When planning any learning, consideration of what students already know is central to decision-making. Considerable time can be wasted if students continually repeat content they already know. Similarly, planning around their interests is important. Good planning involves recognising what students bring to the learning situation.

The learning continuum

Identify the concepts that come before and after a particular activity. By knowing what the prerequisite knowledge and follow-up knowledge are, the teacher is able to cater for the diversity in the classroom, as well as be better prepared for the unexpected. Quite often, students may not have remembered earlier work, or they may be learning beyond what has been intended. By knowing what would extend the focus of the teaching episode, or knowing where there may be potential errors or gaps in learning, the teacher is better able to adjust the lesson to suit the students' current levels of understanding. For all types of lessons and teaching episodes, teachers often need to plan for at least three different levels of achievement to ensure that all students can engage through revision activities, consolidation activities and extension activities (see diagram).

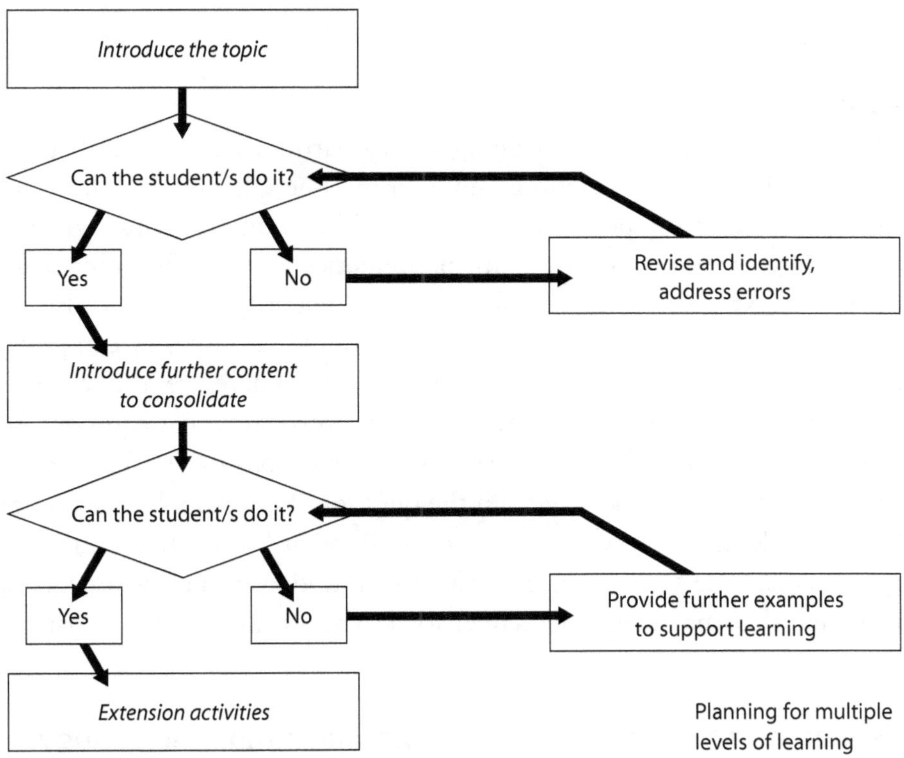

Planning for multiple levels of learning

Common misconceptions

A knowledge of the common errors and misconceptions that occur in learning particular concepts is useful when preparing learning experiences. There is a very rich literature on the common errors and misconceptions in mathematics (Hansen et al. 2017), and also many resources on how to address error patterns. For example, a common error that students make when learning subtraction (with decomposition) is to take the smaller number away from the larger number, regardless of its position. The novice teacher may interpret this type of action as showing that 'the students do not really understand the practice of subtraction'. Experienced teachers anticipate such errors, and commonly plan to avoid their development. Often described as *error patterns*, these mistakes can be due to teaching practices. In this example, the error can be due to the students internalising a process of 'taking the little number from the big number', which has worked well in their experiences up to this point, so that when they encounter this sum, they implement their existing construct. The misinterpretation of the process can occur due to the way in which the teaching has been undertaken up to this point.

$$\begin{array}{r} 42 \\ -\,28 \\ \hline 26 \\ \hline \end{array}$$

Language

Mathematics is recognised as being a very particular language that has many implications for teaching and learning (see Chapter 3). There are many words that are specific to mathematics (e.g. numerator, pi), whereas others have specific meanings in mathematics that are different from other contexts (such as mass, ruler and base). Identifying the words that will be used in a lesson or unit is very important, because they can prove to be another source of confusion for students. There are also ways in which words are organised in sentences and questions that can confuse learners. The lexical density of mathematics questions means that a lot of supportive text is taken away from the question. For example, the task of 'find *x*' is lexically very dense, since the implicit but unstated question is, 'Calculate the value of *x*.' Anticipating and identifying the aspects of language that may cause challenges for learners is an important part of the planning process.

Resources and information and communication technology

Identifying the resources and information and communication technology (ICT) needed for a lesson or unit is important. Resources should be chosen carefully to reinforce the learning intent of the lesson. When equipment is missing—such as protractors, thermometers or an IWB—or when digital sites are not accessible due to systemic or local restrictions, teaching potential is diminished because the lesson is altered or valuable teaching time is lost hunting for the missing resources. As more advanced forms of technology (such as computers, calculators and bring your own devices) permeate the teaching of mathematics, their use and application need to be included in planning for learning. Learning through, and with, digital technology is an important and growing area of teaching in modern society, and is increasingly being recognised in curriculum documents.

Inclusion of all students

Ensuring that *all* students are included in the learning is a high priority in planning. Planning must consider the students who need help, those who are likely to finish early and those whose English (or the language of instruction) may not be strong. It should also incorporate students' cultural or social experiences into the learning. At the same time, the focus must remain on mathematics learning. It is insufficient to allow early finishers to colour in or undertake activities that are not related to mathematics. Authentic experiences related to the goals of the lesson or unit should be developed so as to include and extend all students.

Transdisciplinary planning

As teachers' work becomes more intensive and there are greater demands to make curricula more relevant to realistic settings, the demand to integrate curricula becomes more important. When planning, it is useful to investigate other curriculum areas and to make links between them. Linking of data collection and analysis (statistics in the mathematics curriculum) is achieved easily with the social studies curriculum. Monitoring of the environment can be undertaken through science, while recording of body function rates can occur in physical education lessons. Links can be made with English when writing reports of mathematical activities, patterns can be explored through music, and time can be studied through social and environmental studies. Teachers in modern society need to make as many cross-curricular links as possible in order to meet some of the curriculum demands placed on them. As attempts are made to make the curriculum more relevant to the world beyond schools, links across curriculum areas are essential if students are to see that learning mathematics has purpose and relevance to the world beyond the mathematics classroom.

Teaching: how will I teach?

Having identified important pre-planning issues, it is time to consider the best means of achieving the identified goals. There should be a balance between the different approaches to teaching to cater for the diversity in the classroom, as well as the different demands of the curriculum. Some aspects of mathematics lend themselves better to certain approaches than to others. For example, many measurement activities are suited to hands-on, materials-based learning, whereas some areas of number may be suited better to whole-class, directed learning. Likewise, the use of digital technologies needs to be planned carefully—many are useful in supporting the reinforcement and practice of strategies and skills, but fewer are useful in supporting early conceptual development.

Types of lesson

Three main types of lesson exist: problem-based learning, teacher-directed learning (or direct teaching) and review–teach–practise. Each lesson form has particular strengths and weaknesses. The most common is the review–teach–practise format. However, the Trends in International Mathematics and Science Study (TIMSS) reported that this was a common method in many countries that scored poorly on the tests. Countries such as Japan, which scored highly on the test, were more likely to use problems or investigations. Such data support a common belief that the review-and-revise format may not be the best method for teaching mathematics—particularly new concepts.

Problem-based learning

This type of lesson is most beneficial when encountering new content, developing problem-solving skills, applying understandings and consolidating learning. It most aligns with the ways in which people learn, use and apply knowledge in the world beyond schools. Providing a motivation for the task (such as using newspaper clippings, video extracts or a problem relevant to students' interests) engages students in purposeful learning. Having engaged in the task, students may work individually, in pairs or in small groups to solve the problem. In this type of lesson, it is important that significant time is allowed for a synthesis of ideas and processes at the completion of the investigation. The teacher should provide a rich context in which students can evaluate each other's work in terms of the ideas, processes and content used by their peers. One strategy that works with this type of lesson is the use of problem posing (Singer et al. 2015), where students generate problems that are relevant to their contexts—which are increasingly recognised as important in problem-based learning.

Direct teaching

Unlike problem-based learning, where the students are the centre of learning, direct instruction sees the teacher as central. It is a useful lesson type when there is a need to teach new concepts or a particular language. In this lesson type, the teacher assumes a key role in the initial part of the lesson, and then has students work on the lesson's focus in the later part. The lesson's final phase usually involves a teacher-led summary of the ideas central to the lesson.

Review–teach–practise

Arguably the most common lesson type, this format has been found to be highly ineffective in catering for student learning. The lessons tend to focus on skill development followed by practise of that skill. The approach is evident in many textbook series, where the examples are used for practising skills. Typically, lessons are disjointed, so that there is little connection or application of knowledge and skills to be learned. This hinders students from developing a deep understanding of the purposefulness and application of mathematical ideas. This lesson type can be useful for consolidation of ideas, but its use should be kept to a minimum, with other lesson types (particularly the problem-based format) taking a much more dominant place in planning.

Elements of a lesson

The teaching of a lesson is often broken into three main phases—for both lessons and units—each serving a very different purpose within the overall structure of teaching episodes.

Orientation phase (introduction)

The orientation phase serves three key roles: to introduce the topic, to motivate the students, and to assess whether or not students have the prerequisite knowledge to move into the lesson. With problem-based learning, this phase is used to introduce the topic, to motivate students and to launch the problem. Within the orientation phase, teachers should make known to the learners the lesson's (learning) intent. This could be simply writing a statement of what is intended on the board, or discussing the lesson's mathematical focus.

Enhancing phase (work or consolidation)

This phase of the teaching episode enhances and consolidates student learning. Within a problem-based format, this phase is used for students' exploration of ideas. Planned activities should cater for the diversity in the classroom and ensure that the different needs and understandings of students are addressed. Regardless of the lesson type, this phase is used to develop the ideas or skills that are the focus of the teaching episode. Teaching strategies can vary even with the same outcome(s) being achieved. It is critical that the activities that are planned for this phase link with the intended learnings. If the teacher is not able to make this link, then it is questionable whether or not students will be able to. For example, if the activities are linked to students' understanding of the concept of area, then a range of activities that allow students to 'cover a closed shape' would be a possibility—such as covering a table with pieces of coloured paper, pasting pasta onto a piece of cardboard or placing pattern blocks over a benchtop. Including non-regular shapes, such as stars or kidney bean shapes, can prevent students from developing the misconception that only regular shapes have area. By knowing clearly what is intended, the teacher can ensure that the range of activities addresses those outcomes.

Synthesising phase (conclusion)

The synthesising phase serves to sum up and focus on the key learning of the lesson. It should not be a 'show and tell', but rather should engage students in rich, productive discussion centred on the intended learning(s). Current research and policy in the United Kingdom focuses very heavily on this aspect of teaching, and it is now a core part of mathematics teaching in the United Kingdom that teachers must make explicit links between the purpose of the activities and the proposed learning outcomes. In order to do this effectively, the teacher must be cognisant of the links. It is insufficient to simply ask students to recall what they have done. It is critical that students are able to articulate the learning in what they have done. Constructivist writings have also drawn considerable attention to this aspect of classroom dialogue (Yackel 1991). Debate among students is healthy, since it allows for clarification of misconceptions and development of new (and improved) ideas. It is important that the teacher is able to facilitate discussion in such forums.

Within a unit of work, the synthesising phase may be more developed and be ongoing, both through daily lessons and at the completion of the unit. The purpose of the teaching episode is transparent to both the students and the teacher. The pasting of pasta onto a sheet is no longer an activity that keeps students busy; its explicit purpose is the development of an understanding of the area attribute.

Evaluating teaching: how can I determine success?

There are two major forms of evaluation that teachers must undertake to determine the success of their teaching. The first is the ongoing assessment that occurs throughout a teaching episode—that is, assessment on the run, which is a formative assessment that is used to inform the teaching processes throughout the episode. It involves posing questions, listening and responding to students, and observing students' actions, writing and behaviour throughout the lesson or unit. Experienced teachers undertake this type of assessment as a natural part of their teaching.

'Assessment on action' is a more formal aspect of evaluation. It is summative and occurs at the end of a teaching episode. Self-reflection at the end of a teaching episode is crucial for future planning, to ensure that all teaching episodes are in tune with students' interests and to ensure students' continued motivation to learn. A good habit to adopt is to write a short self-reflection on each lesson as soon as possible after teaching. This could be notes on the lesson plan itself or a reflective journal/diary. The reflective comments should note things that worked well during the lesson, things that didn't work so well and what you could do next time if you were to teach that lesson again. The focus should be primarily on analysing your own actions and practices, rather than finding excuses for why things did not work so well for students. If your planning was inclusive of all learners, was well structured and scaffolded all learners, ensured students were challenged but also engaged in the activities, and took a student-centred approach, then the lesson should have been a great success.

Providing feedback

Students need to have a sense of how well they are travelling in a mathematics lesson. Hattie and Clarke (2018) have argued strongly for the use of feedback. Feedback should be an integral part of planning the lesson and at the time of teaching. The feedback should also be specific. Simple strategies, such as using 'thumbs up' or 'thumbs down', can provide the teacher with a sense of how well students are progressing in their understandings. An example is posing a question such as 'Does this make sense to you?' and then having students reply with their thumbs. Building trust with learners is important if they are to give genuine feedback, so time needs to be invested in building that trust. Using resources such as small whiteboards or cards can allow learners to write their responses to questions and then show the teacher (in the whole-group context). The teacher can then view the responses and see if the students are understanding the work (or not) and then progress (or review) the content.

The feedback provided to learners should also be specific. Generic comments, such as 'That is good work', do not give learners a sense of what they have done well. Be specific, such as by saying, 'That is good, the way you have set out your columns; I can see what you are thinking in the way you have explained what you have done. You have tackled that problem with a lot of logic and drawn in things you have learned to help you solve it.'

Planning the learning environment

Grouping

Three main forms of grouping are apparent in teaching mathematics: whole-class teaching, small-group teaching and individual teaching.

Whole-class teaching is useful when introducing new work, and can save considerable teaching time. The teacher needs to identify which students may not have understood what was being said, and then work with those students in a small-group environment. Small groups of three to six are useful in catering for various learning situations. Groupings can be used to cater for achievement levels, learning styles, interest groups, gender groups, and so on, and both homogenous

and heterogeneous groups can be used. In general, it is often more practical for whole-class teaching to be undertaken when introducing new concepts so that the teacher's workload is consistent, while for consolidating or very differentiated work, small groups are more appropriate. When structured properly, small groups can facilitate greater dialogue among students and enhance students' understandings of concepts, as well as develop social skills. Similarly, individual work can be successful for consolidation of skills where an individualised learning program is clearly needed. The type of grouping employed in classrooms also has implications on the types of ICT that might be used. For example, an IWB might be most useful in whole-class teaching, large-screen computers for small-group teaching and tablets for individual teaching. Likewise, the types of applications or software used will depend on the type of grouping being deployed.

Questioning

Many teachers use questioning rather than direct instruction as the basis for their teaching. Being able to do this requires considerable skill and enables students to be actively engaged in the teaching episode, while simultaneously providing the teacher with valuable information about what students are thinking. When planning teaching episodes, consider the form of the key questions to be used. The questions should move on from closed, low-level forms to include higher-order, open-ended questions. Often, low-level, shallow or closed questions have the stem of 'what', 'where' or 'who', whereas deep questions begin with 'why' or 'how'. The first type only requires a quick response, whereas the latter is aimed at fostering reflection and depth. This is not always the case, as can be seen from the following table.

Surface or shallow questions	Deep questions
What answer did you get? Why did you use . . . ? Why did you use [a particular strategy—e.g. doubling]? Was it easy or difficult? Why do you think the answer is . . . ? Did you write the answer correctly? What did you do?	Why did you do it that way? How did making that [model, table, etc.] help you solve the problem? Why did you use [strategy], and how do you think it has helped you? What other mathematics did you need to help you solve this problem? Why did you use that strategy, and how has it helped you solve the problem?

In considering the use of questions, some researchers in science (Lemke 1990) and mathematics classrooms (Zevenbergen & Lerman 2001) have found that particular questions—such as single-response, yes/no questions—are used for controlling the content and flow of the lesson but have limited value in terms of intellectual engagement or substantive mathematical learning. Questions that encourage students to justify their thinking or working, propose different methods or solutions, or analyse their thinking require a much deeper level of engagement, and hence more time for the formulation of responses; therefore, teachers need to practise patience when posing these questions to allow time for students to form responses. A common tool used for planning for teaching is the (revised) Bloom's taxonomy (Radmehr & Drake 2018). Bloom's taxonomy has six levels, which increase in complexity. Teachers should try to incorporate the different levels in their planning relevant to the content and intent of the lesson.

Good questioning is central to good teaching (Nicol 1998; Sullivan & Clarke 1991; Sullivan & Lilburn 2017) in that it can promote deep learning (Francis 2016). Unfortunately, it has been found that most questions posed by teachers in the mathematics classroom are lower-order questions (Wimer et al. 2001). Greater emphasis and planning thus needs to be devoted to posing good questions. In the general literature on education, good questions are those that foster higher levels of cognition. Typically, such questions are based on Bloom's taxonomy and are of the form that encourage reflection, justification and argumentation (Patterson 1999). In this typology, a good question would be in the form 'Why is that the case?', where students are expected to explain and justify their reasoning. Other types of good question are those advocated by Sullivan and Lilburn

(2017), which relate specifically to mathematics. These authors argue that good mathematics questions provoke deep learning about particular content and processes. Good mathematics questions are open questions.

To change a closed question to an open one, a simple strategy is to look at the answer to a closed question—for example, 'The area is 300 cm^2'—and then work backwards. For example, given that the answer is 300 cm^2, students might have to work out the dimensions. The activity's focus (in this case, area) and process (calculating using multiplication) are the key aspects of the task; these remain preserved, and the task is inverted. There will be a number of correct and different answers to this question.

A common strategy used by many teachers is to ask three good questions for an effective lesson. Developing good questions is not a simple task; time and effort need to be spent on their development. There should be a balance between open and closed questions, since the questions perform very different purposes in the classroom. Posing questions that demand more reflective or analytical thinking may involve less interaction across the class. To ensure participation from a wide range of students, a useful approach can be the *think–pair–share* strategy, whereby students formulate a response, share with a peer in pairs, then move to bigger groups—which could include groupings of four, eight and/or the whole class. This strategy is less threatening, and allows students the time and space to work on their responses. Sharing with peers can also help students to finetune, adjust or rethink their responses and thinking strategies.

TEACHING IDEA

Open-ended questions

- Over a series of six days, Joe caught a mean of eight fish. How many fish could he have caught each day?
- Three numbers add up to 31. What might they be?
- A rectangle has an area of 36 cm^2. What might its dimensions be?
- A solid has six faces. What object might it be?

Materials

Teaching many aspects of mathematics involves the use of materials. It is often assumed that students will make the connections between materials and concepts, yet research has shown that this is not always the case—indeed, when used inappropriately, equipment can create greater confusion if students are unable to see the purpose or connections between concrete and abstract/symbolic work (Howard & Perry 1997; Perry & Howard 1994). The equipment used should be chosen carefully so that it matches the intended learning (Zacharia & Michael 2016). When introducing equipment to students, protocols need to be developed for how it will be used, distributed and packed away. As digital or virtual manipulatives become more common (e.g. apps, geometry packages, etc.), teachers should also consider how these can be used thoughtfully in supporting students to make conceptual connections.

Textbooks

Many primary schools opt for some mathematics scheme whereby students will have their own workbooks. The topics to be covered in a year level may be guided by the writers of these programs rather than another authority, such as state offices or school-based documents. While there is considerable heated debate about the value of textbooks in primary schools, there is certainly a place for high-quality textbooks that support teachers and students in quality learning experiences (Shield 1998).

Textbooks are used badly in lessons when they become the lesson plan. That is, the teacher demonstrates a procedure on the board (taken from the textbook), and the students copy and then complete practice exercises from the textbook. The place for textbooks is to support a teacher's planning, rather than to replace it. When mathematics is dominated by textbook-based lessons, there is little creativity or support for learners at work. This form of teaching is often referred to as 'shallow teaching' (Vincent & Stacey 2008), because it does not allow for the growth of deep knowledge, favouring procedural understandings rather than rich mathematical understandings linked to other aspects of mathematics. Unfortunately, many teachers adopt this approach, which effectively limits the development of richness in mathematical understandings. Likewise, the overuse of student worksheets (both paper-based and, more recently, digital) can limit the development of mathematical understanding.

Planning for substantive learning

In this chapter, the discussion has centred on various aspects of planning. When the teacher is clear about the planned outcome(s) of the lesson or unit, how it will be taught and how to assess it, a coherence between the three phases of planning (orientation, enhancing and synthesising) emerges. Knowing what is the best learning environment to achieve the highest possible learning becomes central to planning in mathematics. It is not sufficient to undertake busywork—the question of substantive learning, in concert with both the ethical and the legal responsibilities of the teacher, is called into question.

Planning to engage students with deep knowledge and deep learning, as opposed to the rote learning of skills and drills, is now seen as a key reform in teaching mathematics. Too many students leave school without an understanding of mathematics—they 'do' mathematics, but do not understand mathematics. Developing practices that foster substantive learning is important to good teaching. This demands that teachers consider seriously what will be learned and how best to organise learning. Encouraging practices through which students are able to debate, contest and challenge can foster lifelong skills needed for life in the wider community. Employing practices whereby students can engage with mathematical thinking and problem-solving of realistic and important problems is vital for the world beyond school.

Review questions

5.1 Why is planning important in mathematics?

5.2 What are some of the key considerations to be made when planning?

5.3 Devise some open-ended questions.

5.4 Briefly outline the four considerations teachers make when planning assessment.

5.5 Compare and contrast the three main forms of grouping in the teaching of mathematics.

5.6 How will you know if your teaching is effective? Draw up a list of questions that will enable you to become a more reflective practitioner.

Further reading

Hattie, J.A., Fisher, D.B., Frey, N., Gojak, L.M. & Moore, S.D. (2016). *Visible learning for mathematics: Grades K–12*. Singapore: SAGE Publications.

Patterson, M. (1999). Questioning techniques. *Teacherlink*, *8*(1), 14–16.

Sullivan, P. & Lilburn, P. (2017). *Open-ended maths activities: Using 'good' questions to enhance learning*. Melbourne: Oxford University Press.

References

Francis, E.M. (2016). *Now that's a good question!: How to promote cognitive rigor through questioning*. Alexandria, VA: ASCD.

Hansen, A., Drews, D., Dudgeon, J., Lawton, F. & Surtees, L. (2017). *Children's errors in mathematics*. London: SAGE Publications.

Hattie, J. & Clarke, S. (2018). *Visible learning feedback*. London: Routledge.

Howard, P. & Perry, B. (1997). Manipulatives in primary education: Implications for teaching and learning. *Australian Primary Mathematics Classroom, 2*(2), 25–30.

Lemke, J. (1990). *Talking science: Language, learning and values*. Norwood, NJ: Ablex Publishing Company.

Nicol, C. (1998). Learning to teach mathematics: Questioning, listening and responding. *Educational Studies in Mathematics*, *37*(1), 45–66.

Patterson, M. (1999). Questioning techniques. *Teacherlink*, *8*(1), 14–16.

Perry, B. & Howard, P. (1994). Manipulatives: Constraints on construction? In J. Geake (ed.), *Challenges in mathematics education: Constraints on construction* (vol. 2, pp. 487–96). Lismore: Mathematics Education Research Group of Australasia.

Radmehr, F. & Drake, M. (2018). An assessment-based model for exploring the solving of mathematical problems: Utilizing revised Bloom's taxonomy and facets of meta cognition. *Studies in Educational Evaluation, 59* (Dec), 41–51.

Shield, M. (1998). Mathematics textbooks: Messages to students and teachers. In C. Kanes, M. Goos and E. Warren (eds), *Teaching mathematics in New Times: Proceedings of the 21st Annual Conference of the Mathematics Education Research Group of Australasia* (vol. 2, pp. 516–23). Gold Coast: MERGA.

Singer, F.M., Ellerton, N.F. & Cai, J. (2015). *Mathematical problem posing*. Dordrecht: Springer.

Sullivan, P. & Clarke, D. (1991). *Communication in the classroom: The importance of good questioning*. Geelong: Deakin University Press.

Thomson, S., Wernert, N., O'Grady, E. & Rodrigues, S. (2017). *TIMSS 2015: Reporting Australia's results*. Melbourne: Australian Council for Educational Research (ACER).

Vincent, J. & Stacey, K. (2008). Do mathematics textbooks cultivate shallow teaching? Applying the TIMSS Video Study criteria to Australian eighth-grade mathematics textbooks. *Mathematics Education Research Journal, 20*(1), 81–106.

Wimer, J.W., Ridenour, C.S., Thomas, K. & Place, A.W. (2001). Higher order teacher questioning of boys and girls in elementary mathematics classrooms. *Journal of Educational Research*, *95*(2), 84–92.

Yackel, E. (1991). The role of peer questioning during class discussion in second grade mathematics. In F. Furinghetti (ed.), *Proceedings of the 15th PME conference* (vol. 3, pp. 364–71). Assisi, Italy: International Group for the Psychology of Mathematics Education.

Zacharia, Z.C. & Michael, M. (2016). Using physical and virtual manipulatives to improve primary school students' understanding of concepts of electric circuits. In M. Riopell and Z. Smyrnaiou (eds), *New developments in science and technology education* (pp. 125–40). Cham: Springer.

Zevenbergen, R. & Lerman, S. (2001). Communicative competence in school mathematics: On being able to do school mathematics. In J. Bobis, B. Perry and M. Mitchelmore (eds), *Numeracy and Beyond. Proceedings of the 24th Annual Conference of the Mathematics Education Research Group of Australasia* (pp. 571–8). Sydney: MERGA.

CHAPTER 6

ASSESSMENT

The new and innovative ways of teaching mathematics discussed in the previous chapters attempt to challenge the dominance of teacher-directed models of teaching. Similar changes are posed in this chapter, where the dominance of pencil-and-paper testing as a means of assessment is challenged. As new theories and research have influenced teaching approaches, assessment has similarly been opened up. Assessment does not exist as an entity in and of itself; it is inextricably bound to the teaching and learning process. So, while this chapter is written as a stand-alone resource, assessment should be seen as integral to teaching rather than as an activity divorced from teaching and learning.

When considering assessment, three aspects need to be considered:

1. *Why assess*—What are the purposes of assessment?
2. *How to assess*—What assessment tools can be used?
3. *Monitoring and reporting assessment outcomes.*

Why assess?

In the past, assessment was undertaken predominantly to report to parents about their child's progress in school. Contemporary assessment practices in mathematics are multifunctional, and assessment is now used to:

- monitor student progress and document the outcomes of student learning
- support student learning and progress through the curriculum
- inform teaching practice and improve programs
- report to governments and funding agencies about student outcomes at state/territory and national levels.

Monitoring student progress

Teachers need to monitor student learning and progress continually through a variety of means. Rather than focusing only on knowledge and skills, a wider range of attributes and attitudes is now monitored. Monitoring is a continual process whereby teachers systematically collect a range of evidential pieces in order to assess student progress. Students are often involved in the monitoring process as well.

A contemporary move in education, where accountability is paramount, has seen assessment move towards more transparent processes. This involves criteria being made explicit so that students are able to see what they will be assessed against. The assessment is then marked against a rubric, where the marks to be allocated are indicated clearly. This process enables teachers to identify whether or not students are moving towards the intended learning outcomes as articulated by the teacher.

Supporting student learning

Assessment is used as an informative process—or formative assessment—whereby information gained through assessment is fed back into the teaching process in order to support and enhance student learning. Assessment becomes a tool for diagnosing what is known, or unknown, and this information is then used to inform subsequent practice to move students' learning in the direction desired by the teacher. Programs such as the First Steps in Mathematics Program (Department of Education and Training Western Australia 2004) and the Early Numeracy Research Project (Department of Education and Early Childhood Development Victoria 2006) focus on the use of developmental or diagnostic frameworks, with students assessed and mapped against the frameworks. These frameworks, which have the goal of achieving the desired learning outcomes, help teachers to understand why students can do certain things and not others, and support teachers in realising why some students may experience difficulties in learning mathematics while others do not. They also show the challenges that students may face in moving their thinking forward in particular areas or with particular concepts, in terms of potential misunderstandings or misconceptions. These programs are also useful in helping teachers to interpret students' responses to the frameworks. This type of assessment is interventionist (Wright et al. 2000), in that it is used as the basis for subsequent action. It is also a form of scaffolding, whereby teachers identify what their students know or can do and then provide appropriate learning environments through which the students are enabled to move towards more complex understandings.

Making informed decisions about instruction and programs

Assessment allows teachers to identify which aspects of their teaching and programs are enhancing (or hindering) student learning. Using assessment to evaluate programs or learning episodes, teachers identify the strengths and weaknesses in their methods and move towards changing and improving their teaching.

Reporting to external agencies

In the past, teachers have used assessment to report to parents about their child's progress in school. Employers have also used school assessments for staff selection. Thus, schools and teachers have used assessment to report to the wider community about the successes of students in their classes. Mathematics has been a high priority within the overall curriculum, because it is a key life skill and an important discipline in most areas of work and life.

In an increasingly economically driven world, more recent trends in education have seen teaching and schools bound by external agencies in terms of reporting. Contemporary reforms by state/territory, provincial and national agencies have seen a greater accountability for government funding, so governments want to see clear outcomes of their spending. This applies in most Western economies, where governments are increasingly demanding documentation of student outcomes. This has resulted in considerably more testing by authorities at the systems level to document student performance.

National testing

Most countries now include some form of national testing of student performance. In Australia, the testing has been focused on literacy and numeracy. Under the auspices of the Australian Curriculum, Assessment and Reporting Authority (ACARA), national curriculum, testing and reporting are a reality for Australian teachers. In terms of assessment, national tests are conducted each May and are referred to as the National Assessment Program—Literacy and Numeracy (NAPLAN). While there is much debate about whether the tests are assessing 'numeracy' or 'mathematics', the reality for teachers is that they are required to supervise testing in Years 3, 5, 7 and 9 under strict testing conditions. Test results are published on the My School website and schools (and anyone in the community) can see how well any school has performed against other schools. This makes the outcomes of these tests open to public scrutiny. This public show of information has been challenged by teacher unions and other groups, but the government has been steadfast in making these results public. Increasingly, there is pressure on schools to enhance the performance of learners, and this testing process has been the basis for government decision-making in this area. Funding is also often tied to testing, causing difficulty for some schools. In addition, parents are increasingly making decisions about the quality of schools based on these scores.

While there is criticism of the value of the tests and what they are testing, they are a reality for schools. Teachers can use the results of these tests to inform their teaching and to gauge whether they are improving or enhancing learning. Similarly, where there are sustained differences over time, questions need to be raised about the performance of students—in particular, if there is sustained failing. Most importantly, the tests have shown that there are consistent patterns of failing among particular groups—remote Indigenous students are the most at risk in Australia. The complexities of why this occurs must be considered by teachers, communities, schools and agencies if there is to be some change to practices in these areas. Issues of language, culture, attendance and relevance all contribute to the lack of success in these and other communities.

Models of assessment

Typically, models of teaching, learning and assessment have been seen as linear, with the teacher planning a lesson or unit, undertaking the teaching and then assessing the overall teaching episode in terms of what the students may have learned. Such assessment may employ a range of techniques, but it generally follows the plan–teach–assess process.

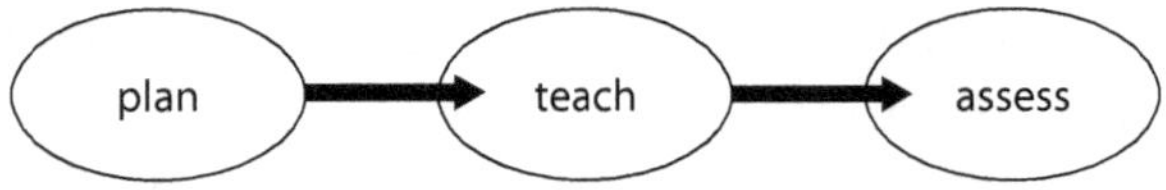

Linear models of assessment

In contrast, models of assessment should use assessment to inform teaching so that a more cyclical approach to planning, teaching and assessing is implemented. Such models see assessment as an integral component of good teaching. Using the information gained through assessment, good teachers are able to improve their understanding of the teaching/learning process, and then use this understanding to improve the learning environments for the students in their classes. Current practice in good teaching requires teachers to be reflective about their teaching, and incorporating what has been learned through assessment is an important part of that process.

If what has been learned through the assessment process is not fed back into the learning loop, then one needs to ask some serious questions about the purposes of assessment. For example, when teaching a particular concept or process in mathematics, teachers should be assessing what students know (and do not know) and then using this information to decide where to move next in their teaching. At the simplest level, if students understand what is being taught, the teaching can move forward. If students do not appear to understand, then learning needs to be consolidated or revised rather than moved forward. Hence, assessment is an integral component of the learning loop.

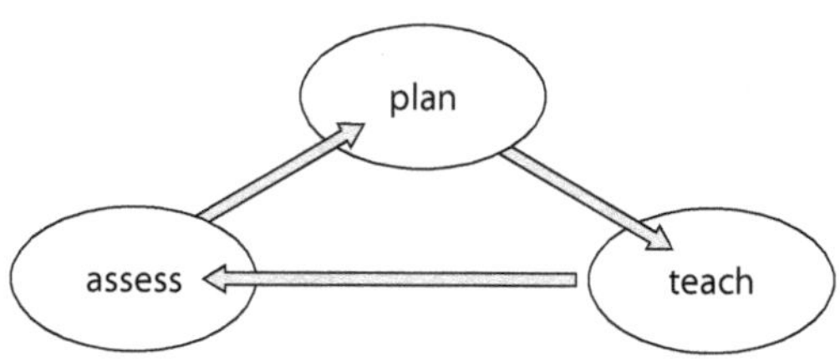

Cyclical models of assessment

Contemporary assessment practices have seen quite dramatic shifts both in how assessment is undertaken and the purposes of that assessment. Some of these shifts can be seen in the table below, where a reliance on Friday tests (usually developed by the teacher or taken from a resource book), on the basis of which students were graded using a scoring method, was replaced by teachers using a much wider range of assessment techniques built into their teaching practice.

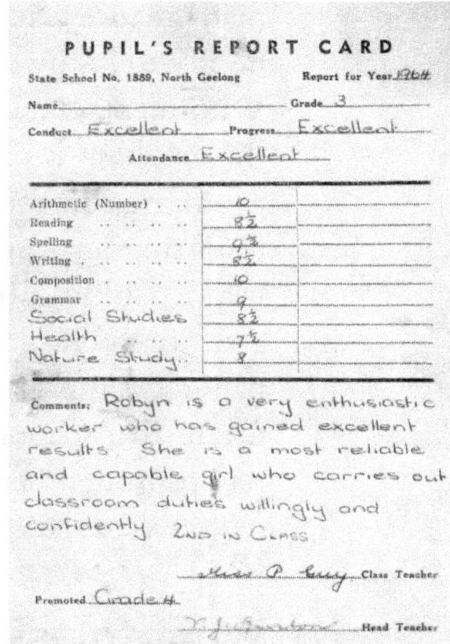

PUPIL'S REPORT CARD

State School No. 1889, North Geelong — Report for Year 1964

Name ______ Grade 3

Conduct Excellent Progress Excellent

Attendance Excellent

Arithmetic (Number)	10	
Reading	$8\frac{1}{2}$	
Spelling	$9\frac{1}{2}$	
Writing	$8\frac{1}{2}$	
Composition	10	
Grammar	9	
Social Studies	$8\frac{1}{2}$	
Health	$7\frac{1}{2}$	
Nature Study	8	

Comments: Robyn is a very enthusiastic worker who has gained excellent results She is a most reliable and capable girl who carries out classroom duties willingly and confidently 2ND IN CLASS

Class Teacher

Promoted Grade 4

Head Teacher

1960s-style report card

Changes in assessment practices can be seen in the report cards used in schools. Consider the 1960s-style report card shown in the photo, on which students were given a mark for each curriculum area and then a general comment. While the card may have been less time-consuming than the reports used in today's classrooms, it tells little about what the student knew, could do, needed support with, was confident in, and so on. In some cases, the raw score was entered against the class average so that parents could relate to where their child performed against the class, but this did little to convey the student's learning. However, this type of report is still seen by many parents and members of the wider community as valid assessment. They want to know students' 'scores' on tests. They want to know where the student is in relation to others in the classroom—that is, their ranking in the class—and there can still be considerable pressure on the teacher to conform to these expectations.

Examples of old and contemporary assessment practices

Old practices	Contemporary practices
Gather data periodically	Gather continuously
Assess as isolated facts or skills	Integrate assessment with teaching
Rely on tests and quizzes	Use a range of tools
Externals assess students	Students self-assess
Use a single source of evidence	Need multiple evidence
Indicate right/wrong answers	Continual feedback to students

Modern report cards tend to be more descriptive and to reflect students' strengths. Often couched in positive terminology, they can, however, be criticised for failing to identify areas of weakness. More recent practice has demanded that teachers report what the student knows or can do against curriculum requirements, but this is often couched in language that is beyond parents' comprehension. In modern society, the mathematics teacher needs to be aware of the pressures relating to, and be able to justify contemporary thinking about, assessment and reporting. Within contemporary practices, particularly outcomes-based education, the emphasis is on documenting learning rather than rankings.

Alongside the national testing systems, there is also mandatory reporting against which teachers must also report. This includes the use of rankings on an A–E scale. The government argues that this is a requirement demanded by parents. Education systems and teachers may resist this pressure, but, unfortunately, the pressure of government means that schools must report using these scalings. Many schools also report using other forms of reporting alongside the mandated reporting requirements.

What to assess

When thinking about assessment in mathematics, it is important to consider what is being assessed. Past testing practices often assessed relatively trivial aspects of mathematics—that is, knowledge and skills. More complex aspects—such as attitudes towards mathematics, problem-solving and mathematical thinking—are more difficult to assess through simple tools such as tests. Contemporary assessment in mathematics extends what should be assessed to include mathematical knowledge, using and applying mathematics, and dispositions towards mathematics. These aspects of mathematics need to be incorporated into assessment practices.

A common misperception of the testing process is that it is an objective tool for assessing students' mathematical understanding. For example, research into the language of mathematics points to some very critical points. Pimm (1991) holds that approximately 80–90 per cent of mathematical word problems can be solved using a key word strategy. Consider the two questions in this diagram.

1. My jug holds half as much as John's jug. If John's jug holds 4 litres, how much does mine hold?

2. My jug holds half as much as John's jug. If my jug holds 4 litres, how much does John's hold?

Posing these questions to students produces some interesting results (Zevenbergen et al. 2001). Using a key word strategy, students were likely to identify the key information as the '4 litres' and the 'half', and, since half generally means 'to divide by 2', they were likely to produce an answer of 2. This works for question 1, but not for question 2, which requires a doubling process. A correct response to question 1 may be interpreted as 'the student understands and is able to undertake the mathematics within the question'. This leads to an assumption that the student is working with the ideas, knowledge and skills needed for fractions and volume. The same key word strategy does not work on question 2, and the teacher may assume that the student is having difficulties with volumes and/or fractions. In fact, there is little evidence of a mathematical strategy or knowledge being used in either working. Thus, the responses to questions, and the questions themselves, need to be considered beyond what is often assumed to be the student's understanding.

Furthering this idea, research by Ellerton and Clements (1997) on the questions posed in tests has indicated that there is potential for a gap of 28 per cent between the student's performance and the student's understanding. They propose that such differences can be the result of students being able to give 'correct' responses without understanding the task, or, conversely, having full or partial understanding but giving an incorrect response. Extending this work, Zevenbergen (2000) analysed the questions posed on a state mathematics test for language barriers and showed how language impacted interpretations, creating difficulties in the testing process. These findings challenged what was being tested—mathematics, language or cultural background—and pose serious challenges to the 'objectivity' often associated with test questions. It is not safe to assume that test items test only mathematical knowledge and skills. Rather, other aspects of learning are also being applied and tested.

The limitations of pencil-and-paper testing have been well documented in terms of what is being assessed and how the evidence that is collected through the process can be severely limited and biased (Ellerton & Clements 1997). Thus, other forms of assessment may be more reliable in assessing what students know, can do and feel about mathematics. Good teachers use more than one method for assessing student learning; when various methods produce similar evidence, there is a greater probability that the students understand the mathematics.

Planning assessment

Much like a researcher or detective, a teacher needs to identify what information is needed from the assessment process, how to collect the information, how to make sense of the information and how to report the results. The assessment process is just like good detective or research work—and the target of the information gathering could be the student, the program or the teacher. The assessment process is divided into four components:

1. identifying what is going to be assessed and what will be the most appropriate tool for collecting the information
2. collecting the information
3. analysing or interpreting the information
4. using the results.

The ways in which the results will be used will depend on why the assessment was undertaken. For example, if the reason for assessment was to monitor student learning, then the results will be used to inform teacher action in order to create a learning environment that will promote learning. If the reason for assessment was to inform instructional decision-making, then the results will be used to determine whether or not the next lesson will consolidate learning or move on to new knowledge.

In the initial stages of planning assessment, the teacher needs to consider the actual purpose of assessment. This will inform what methods will be most appropriate for collecting the information needed. For example, if the teacher needs information about students' mis/conceptions about fractions, it becomes necessary to think about the best method for collecting such information—observing when working through tasks, listening to conversations with peers about the tasks, and a worksheet could all be useful tools for assessment.

Having collected the information, it is then necessary to think about how to make sense of what has been collected. This involves the data relating to individual students, to groups and to the whole class, as well as the students' overall responses. Analysing one student's responses can be useful for determining what strengths or problems that student may have with fractions. Collectively, the data may show that the whole class is making a common mistake—that there is something other than an individual misunderstanding—and so something in the teaching process may need to be considered.

Once the data have been interpreted, consideration of how the assessment results will be used must be undertaken. Individual results may be used to inform intervention strategies to support development of a student's knowledge. This process is often to referred to as *diagnostic teaching*. Class results may be used to plan whole-class teaching or programs so that revision of previous learnings may be implemented—for example, more work with fraction kits may be undertaken by the whole class.

The information-gathering processes used are an important consideration. Some assessment tools are better than others for the information being sought. Worksheets and tests may produce some information, but may not be detailed enough to provide the insights that are needed; in this case, discussions with the student may be more appropriate. If students are experiencing difficulties with the concept of area, for example, observing them as they work with covering activities can facilitate the teacher's understanding of their working processes. The ways in which students work with the blocks or talk with peers can provide valuable insights into their thinking, which can be further supplemented by asking questions as they work through their tasks.

When undertaking assessment, through whatever means have been decided, it is equally important to consider the criteria that will be seen as evidence for the desired learning. Observing students for what would constitute evidence of learning (or confusion) is a key consideration.

Making assessment criteria explicit allows the teacher to consider the expectations that should emerge from the teaching, as well as the common errors that might occur and the misconceptions that can develop. Being cognisant of what can and should be expected from a given teaching episode (a lesson through to units), teachers are able to assess more effectively. The teacher notes in the following diagram indicate a teacher's thinking about what will be observed in a place value lesson, what is expected if the students are understanding the intended learning of the lesson and what mistakes the students might make.

Observe: *blocks in right columns, using right language of place value, recording numerals in correct columns, correct naming of place value.*

Mistakes to expect: *recording 6 + 7 as 13 and leaving both digits in the ones column, not trading.*

Place value lesson

Having collected and interpreted the information, the next step is to consider how the results will be reported to others. Information must be conveyed to the intended audience (students, parents, school administrators) in a way that communicates the assessment's intended outcomes. This could relate to students' mathematical knowledge, their dispositions towards mathematics, the ways in which mathematics is being used, and students' ability to think mathematically. Readily understood formats for communicating this information are needed—descriptive narratives, profiles where grids indicate the goals of the assessment, grades, and so on.

Assessment tools

Even before considering which tools will be used, the first and most important consideration is to determine what is going to be assessed. If teachers want to know whether their students are able to communicate mathematical understandings or ideas, they need to select an assessment activity that will enable students to do this. Depending on the type of communication desired, this could require students to undertake an oral presentation, participate in a discussion during the lesson, prepare a written report or journal reflection, or create an advertising brochure. These formats require different communication styles, and consideration of this is therefore also necessary. If the requirement is to assess students' engagement, persistence and/or motivation in mathematics, the teacher might decide to implement a problem-solving task and, through the students' final products, as well as observations of students working and/or of journal reflections, be able to assess whether or not the desired characteristics are apparent.

The means by which students are expected to demonstrate their mathematical knowledges, attributes and dispositions must be conducive to facilitating the intended outcomes. It is unreasonable to assess when students have not been given ample opportunity to develop understandings of what is being assessed.

Once a decision has been made about what will be assessed and how best to engage students in activities that will facilitate what is to be assessed, consideration needs to be given to how the data will be collected. For example, if the task is for students to demonstrate their understandings of subtraction with internal zeros, the teacher might ask them to send an email to a friend who is having trouble with the process. In this way, teachers can observe students working through tasks. They can also listen to students' conversation with peers or invite students to describe in their journals how they did the task. These activities will allow students considerable scope for demonstrating their understandings. The processes by which assessment will occur should be an integral part of the teaching episode.

Observation

Perhaps the most common assessment tool in the primary-school classroom is observation, a non-intrusive form of assessment whereby the teacher simply observes what students are doing and/or saying. This form of assessment is used a great deal in the early years of schooling, and becomes increasingly formal as students move through school. That is not to say that observation is not used in the upper-primary grades; rather, it is not seen to be so popular in the years where there is an increasing reliance on traditional forms of assessment. However, in everyday interactions, observation is a key tool for judgement, and many teachers' 'intuitive' knowledge about students comes through this means. In the upper years, teacher observation can often result in a challenge to formal results on pencil-and-paper tests when teachers 'know' that the results are counter to their expectations.

To fully utilise observation as an effective assessment tool, teachers need to articulate what they want to observe and how they will record their observations. Being aware of what to observe is central to good observations (Artzt et al. 2015). This applies equally to the use of technologies such as IWB or tablet devices, which are occupying more and more space in classroom practice. It is important to observe student interactions on these devices so that the devices are seen to be an integral part of the classroom repertoire.

Recording observations on Post-it® Notes

Identifying the key behaviours, language, and so on in the classroom is essential to the assessment process. The observations link with the intended outcomes of the teaching episode, the common errors that are likely to occur and other features of the lesson (such as cooperation in groups, sharing, and so on). The teacher should have a good knowledge of these, as will be evident in the planning process. The focus of the observations can vary from individuals through to groups, and can relate to any of the foci of the mathematics program.

One of the key features of good assessment through observation is keeping records of the observations. These typically come as anecdotal

	Counting: May 13
John	*Basic OK, but difficulties with facts*
Mark	*Having difficulties with all skip counting. Only counts in ones*
Tim	*Has shown competence with all counting*
Jo	*Can do all counting tasks easily. Can do counting in 10s!*
Meg	*Has problems with 5s. Wants to revert to 2s*
Ny	*Problems with language*

Anecdotal observations

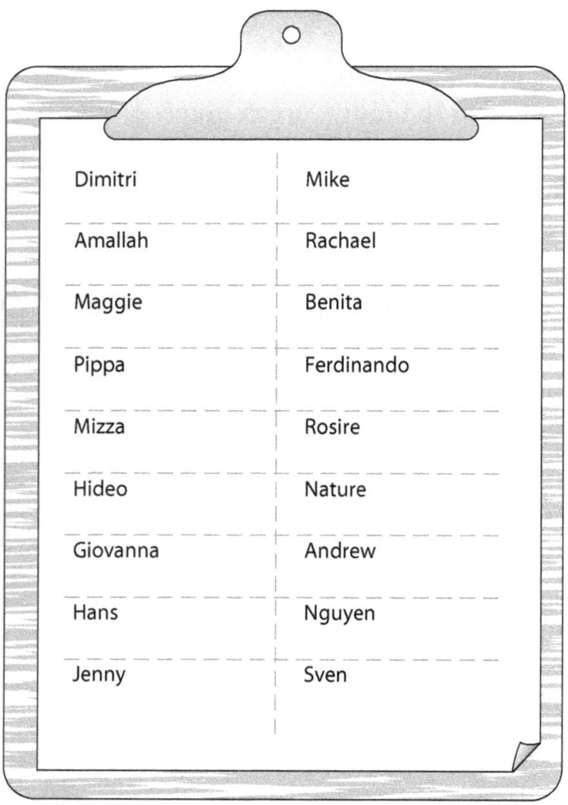

Anecdotal observations may be made using a clipboard

records. An effective method is the use of Post-it® Notes. These can be made for individual students and pasted into the individual's profile at the end of the lesson or day, enabling quick transfer of information. Alternatively, a flipboard can be made using file cards for each student. These are taped onto a clipboard, slightly overlapping, as shown in the diagram. Notes can be made on the cards, which when full are placed in the file-card system. Notes can also be made on handheld devices. The advantage of the digital approach is the capacity to take photographs of students' work and accompany these with notes. This makes for easy recordkeeping. Journals can also be used, and anecdotal notes taken as the teacher is working with the students. This system is easy to implement, but transferring information from the journal to individual student records can be time-consuming.

A further tool for observation is the use of running records, which can be recorded using a checklist or a similar class-list record. The records can be checked off, if an outcomes list or specific list of expected behaviours is being used, or through the use of anecdotal records against the student's name. For example, in the checklist illustrated in the diagram overleaf, the teacher has noted the expected outcomes for the teaching episode along the top of the checklist and the students' names down the side of the page. Using a 'traffic light' system for recording observations, the teacher uses red to indicate no understanding, yellow to indicate some understanding and green to indicate full understanding. This simple system can be used to record observations in a quick and manageable fashion. In some cases, teachers even note the date of the observation. This is particularly useful when students' understandings appear to be developing.

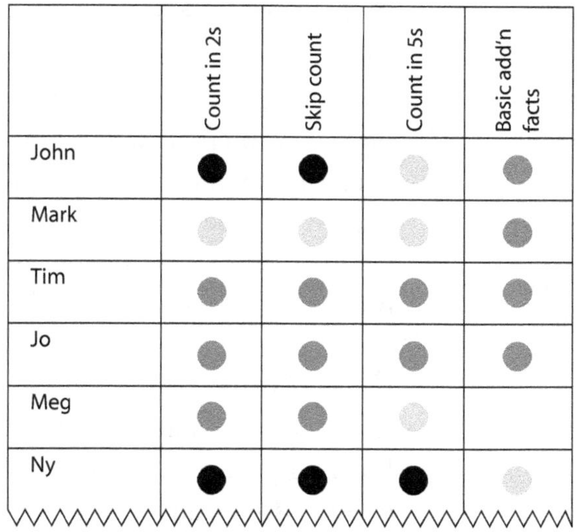

Checklists with outcomes

Another tool for observation is more open-ended, whereby the teacher has a blank class list; as behaviours are observed, the teacher records them against the students' names. This type of observation schedule allows for more descriptive accounts to be entered and for the inclusion of observations that may not have been possible using the more prescriptive checklist with nominated outcomes.

All five formats allow for teachers to record their observations, albeit in very different ways. The format selected can depend on the immediate need for the assessment (such as report cards), the teacher's personal style and what the data will be used for in teaching. For those teachers utilising digital technologies in their teaching, many of these formats can be supported with software or apps. For example, a classroom website can be created that allows students to submit work samples electronically that the teacher can assess. Teachers can also take photos of student work and add these to student samples. The app 'Explain Everything' is a useful addition to assessment procedures. It tracks student work as it is completed, and students can also record audio annotations explaining their work. This allows teachers an insight into the thought processes of students as they complete mathematics tasks.

Performance tasks

As students are engaged in working on mathematical tasks, many of the skills that they are learning and developing can be assessed through observing them at work. For example, the ability to use a protractor to construct or measure an angle can be assessed through observation. In many cases, performance tasks resemble the tasks that would be undertaken in the world beyond school—measuring up a pool, building a frame for a window or tendering correct change.

Consultation

Unlike observation, consultation involves some intrusion into the teaching/learning process. In most cases, the consultative process is simply posing questions to students and seeking their responses. It is often extended from observations, and it can be used to complement the observation process. For example, a teacher, while observing students working in small groups, may not understand an action, and so pose some questions in order to gain more insight into the students' thinking. Alternatively, when presenting instructions at the start of the lesson, quizzical looks on some students' faces may suggest they do not understand, so the teacher can interact with the students to clarify their concerns. Interactions with students allow the teacher to gain a better knowledge of, or insight into, the students' thinking, problems or concerns.

Focused analysis or questioning

Part of the consultative process is the posing of good questions, which takes time and effort. Sullivan and Lilburn (1997) argue that most of the questioning in mathematics classrooms falls within the lower regions of Bloom's taxonomy of questions (Bloom 1956), and hence involves knowledge, comprehension and application. In contrast, few questions (even in major mathematics texts) lie within the regions of analysis, synthesis or evaluation. Sullivan and Lilburn (1997) define good questions as having three aspects:

1. They require more than remembering a fact or reproducing a skill.
2. Pupils can learn by answering the questions, and the teacher learns about each pupil from the attempts.
3. There may be several acceptable answers (p. 2).

In defining what makes a good question, Sullivan and Lilburn argue that good questions are useful for teachers when it comes to accessing and assessing what students know, since they create greater opportunities for students to display their understandings—or, indeed, their lack of understanding. Such questions move beyond mere memorising, whereby students demonstrate understanding of concepts rather than rely on recall of a technique. A common task, such as calculating the mean of a given series of numbers, often allows students to calculate the answer correctly, since they may know the procedure for such calculations without necessarily understanding the concept of 'mean'. Such questions are closed—that is, they have only one acceptable answer, and they generally restrict methods for resolution. In contrast, asking a question such as, 'Over the past six matches, a soccer team has averaged a score of 3. What could be the numbers of goals scored in each of the matches?' allows for greater diversity in the possible responses and in the logic or methods that students will use to arrive at an answer. The opening up of questions in this way allows for greater flexibility for students, as well as greater diversity in the responses and methods used. It also allows teachers greater access to what students know and can do in a given topic. Further details regarding questioning are available in the section on 'Questioning' on page 93.

As we indicated in Chapter 5, many teachers use questioning rather than direct instruction as the basis for their teaching. Being able to do this requires considerable skill. Questioning enables students to be actively engaged in the teaching episode while simultaneously providing the teacher with valuable information about what they are thinking. When planning teaching episodes, consider the form of the key questions to be used. The questions should move from closed, low-level forms to include higher-order, open-ended questions. Posing questions that demand more reflective or analytical thinking may involve less interaction across the class. To ensure participation from a wide range of students, a useful strategy can be to use the 'think–pair–share' strategy, whereby students formulate a response, share with a peer in pairs, then move to bigger groups—which could include groupings of four, eight and/or the whole class. This strategy is less threatening and allows students the time and space to work on their responses. Sharing with peers can also help to finetune, adjust or rethink their responses and thinking strategies.

Open-ended questions

There are two main forms of questions—closed and open. A closed question provides some very focused questioning—such as, 'What is the mean of 5, 4 and 3?'—where there is only one answer, which students can get either correct or incorrect, and one identified method for solving the task. It yields little other information (other than using error analysis).

TEACHING IDEA

Open-ended questions

- I drew a shape with 6 sides. What might my shape look like?
- If you have 8 straws and 4 joiners, what sort of skeletons can you make?
- If a decimal number is rounded off to 6, what might the number be?
- I am thinking of fractions between 1 and 2, what might they be? Give at least 12 answers.

By contrast, Sullivan and Lilburn (1997) have provided numerous examples of open-ended questions in mathematics. They argue that open-ended questions typically do not have an obvious path of resolution, and instead have multiple answers. An open-ended question provides for greater diversity of responses across a range of levels of understanding. Consider the question, 'If 5 is the mean, what six numbers could make this mean?' Such a question allows students to provide a range of responses, which could include numbers that are all the same through to very large and small numbers, as well as common fractions and decimals. The range of responses offered by the students allows the teacher greater access to students' deep understanding of the concept of mean, as well as catering for the range of understandings across the classroom.

Examples of closed and open questions

Closed	Open
What is the area of a sandpit whose dimensions are 15 cm × 20 cm?	If a sandpit has an area of 300 cm^2, what might its dimensions be?
What is the sum of 7 and 8?	If the total of two numbers is 15, what might the two numbers be?

To change a closed question to an open one, a simple strategy is to look at the answer to a closed question—for example, 'the area is 300 cm^2—and then work backwards; given the answer is 300 cm^2, the students have to work out the dimensions. The goal of the activity (in this case, area) and the process (calculating using multiplication) are the key aspects of the task; these remain preserved and the task is inverted. There will be a number of correct and different answers to this question.

Open-ended questions provide a wide entry point, since students can respond to the questions in ways that allow for their particular learning levels. Consider the following example, in which students were given the following task: 'At 4.00 p.m. each day, the average number of people in each car on a busy suburban bridge is 2.5. At any one time, there will be 16 cars on the bridge. Draw what this might look like.' The drawings shown in the diagram illustrate the variety of responses offered by students in the upper-primary grades.

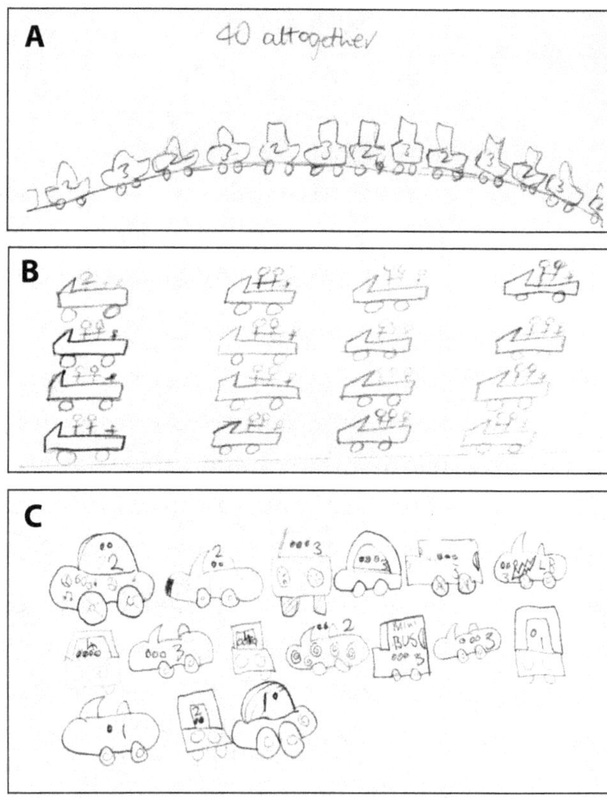

Student work examples of open-ended responses

The range of responses shows that some students have complex understandings of how to represent a mean of 2.5; in diagram C, a variety of cars is drawn, with a diverse range of people in each car. Other students (diagram A) showed that they knew the total had to be 40 people, representing this as a simple group of cars containing two and three people. Other students took the question literally and drew two people and half a person, or represented the half person as a smaller person (diagram B). These diagrams represent a range of different understandings of the concept of mean that would not be accessible through closed questions. All students have been able to participate and engage in the task, and to represent their understandings. These responses allow the teacher greater access to the ways in which students are thinking about the 2.5 as a mean. Unlike the conventional task, which would have been, 'If there are 16 cars on the bridge and 40 people, what is the mean number of people per car?', the open-ended task allows students to demonstrate their understanding of the concept of mean, the ways in which they would calculate it and how to represent it. It allows the teacher greater access to the knowledge that students have about mean, which would not be accessible if the task were of the standard format. It allows for students to respond in ways that are appropriate for them, thereby allowing almost all students to respond, and to demonstrate their current levels of understanding.

Self-assessment and peer assessment

Most students have a good sense of what they know, what they have problems with and what they enjoy. They often also know why this is the case. Yet teachers spend inordinate amounts of time and energy trying to access this information through a range of assessment instruments. However, students are given little opportunity to self-assess. Self-assessment allows students to disclose their mathematical understandings (or lack thereof). Developing a supportive ethos within the classroom is central to developing this aspect of assessment. This may take some time, since students have to learn to self-disclose; for most students, the practice of assessment in mathematics is to hide what you don't know, rather than to show what you do and don't know.

Journal writing

Asking students to write in a personal journal about their mathematics learning, lessons and experiences can provide the teacher with valuable insights that would otherwise be inaccessible. Catalysts that help students write about the focus of assessment are an important consideration. Questions that can be posed include:

- What did I learn today/this week?
- What do I need more help with?
- What don't I understand?
- What did I enjoy about this week's mathematics?
- What would I like to learn more about in mathematics?
- What can the teacher do to help me more?

- What did I think the teacher did well (or poorly) in teaching mathematics?

These types of questions are useful because they allow students to discuss freely the issues they are confronting in the mathematics classroom. An app that is useful for students is AudioNote. It allows students to record their thoughts about a mathematics lesson in either written or audio form. This may be appropriate for young students or for students with literacy limitations.

A particular ethos needs to be developed in a classroom when undertaking journal writing. Depending on the ethos that develops, various forms of journal writing can be nurtured. For example, a teacher may decide that journals are a personal communication between student and teacher only, so that quite personal information may be included. Alternatively, the journals may be a more open document so that other information will be included (or excluded) depending on who will read the documents.

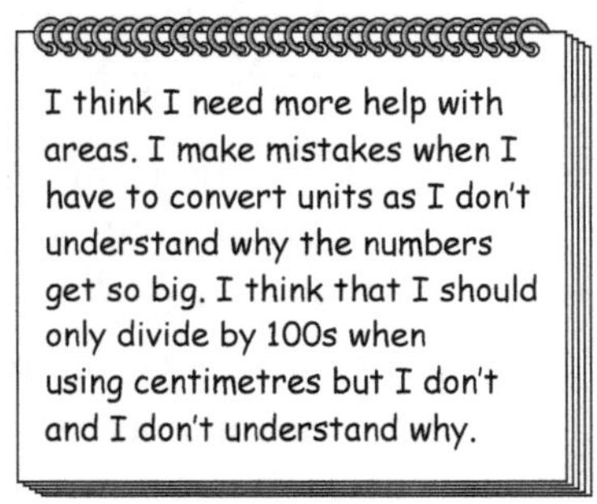

Example of journal writing in mathematics

Reflection

Most of the work of self-assessment requires students to be self-reflective, and to pose questions to and of themselves: 'Could I do this in a better way?' 'What have I learned from doing this work?' 'Why am I doing this this way?' Reflecting on what they do in mathematics and how they do it, how they feel about mathematics, and what they need to know more about and why can encourage students to recognise their strengths and weaknesses in mathematics.

Work samples and portfolios

The range of work undertaken by students in the mathematics classroom can be seen as representing their understandings of and/or dispositions towards mathematics. This work can be collected and collated into a systematic filing system so that it is seen to constitute pieces of evidence about students' learning. The work samples should show clearly how they demonstrate learning. The pieces can include reports, diagrams,

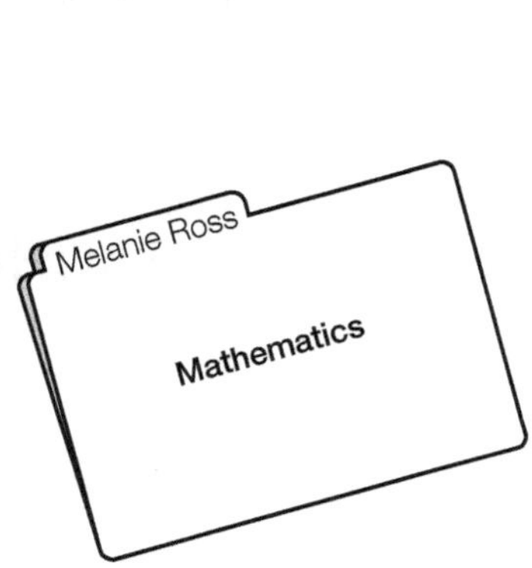

Example of student portfolio

worksheets, problem-solving activities and the like. Some pieces may need to be annotated in order to clarify the reason for their inclusion.

A number of issues need to be considered in order to develop such a portfolio:

- Is the purpose of the portfolio to document mathematics learning?
- What will constitute evidence, and who decides what will be included?
- Should the evidence represent full understanding, best work, developing ideas, drafts, etc.?
- What is the best way to store and keep the portfolio?
- Will it be used for a particular purpose, or be ongoing?

The traditional format of a portfolio consists of a manila (or similar) folder holding pieces of work and a page that indexes the contents and states the point of the documentation. Other formats can also be used, such as computer documents or digital portfolios that can include video of students' work, which can be shared easily with parents or guardians.

Student writing

One of the less commonly used formats for assessment is student writing. This can be in the form of a report or a similar text. A catalyst can be used (such as those in the mathematics and curriculum teaching program packages in Lovitt and Clarke 1992), and the student writing used as evidence of learning. Consider the example in the following Teaching Idea. Students are asked to write a story about the number of jelly beans in the jar as represented by the graph. The task requires students to interpret information from the graph and to communicate this to colleagues and peers. This demands a high level of communicative skill.

TEACHING IDEA

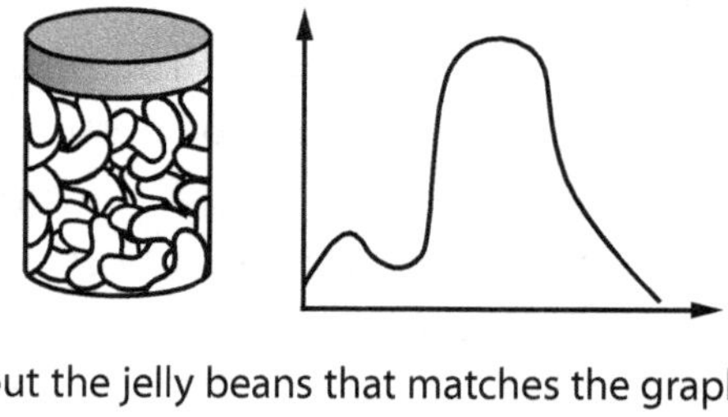

Write a story about the jelly beans that matches the graph.

Another strategy is to ask students to write a letter to a friend who was not at school and explain what they learned in mathematics that day. This encourages reflective thinking and aids in the development of a language of explanation.

Tests

Tests do have a place in assessment, albeit a small one. Tests have limited value in terms of assessing what students know, feel and believe. Not only are tests difficult to construct well, but they can be notoriously unreliable in testing what they are thought to be testing. A good assessment tool can be a student-constructed test. Such items serve two purposes: the construction of the test items by students gives the teacher insights into what students know or can do, and the final test provides teachers with information about students' performance. These tests are useful for classrooms because, since all students have had some input into the construction of the items, the results should represent the diversity of understandings within a particular classroom. With this diversity represented in the test, all students should have success in answering questions. The tests are also useful in that they are owned by the students and will be less stressful since they themselves constructed the test.

Rubrics

0	Shows little or no understanding of place value. Can't trade. Subtraction incorrect.
1	Places numbers in right columns, trades but can only do in one column. Subtraction incorrect due to incorrect recording.
2	Place value recorded OK. Subtraction in tens but does not transfer to ones column.
3	Place value OK. Trading done correctly but records ones place value incorrectly when trading the 10 to 1.
4	Place value, trading and subtraction all completed correctly. Justifies response.

Specific marking rubric

One way in which you can document students' understandings is through the use of rubrics. These tools have a number of very good functions for teachers. By requiring the teacher to consider what will constitute evidence of learning, and of degree of learning, the rubric documents what the teacher sees as the critical elements of the teaching outcomes. The rubric should specifically target the key learnings for the teaching/learning process. This requires the teacher to be aware of the common errors that could be

A	Little or no understanding. Cannot represent ½, ¼ or ⅕ numerically or pictorially.
B	Some understanding. Can divide the unit into the number of nominated parts but does not understand the notion of equal parts.
C	Full understanding. Can represent a range of fractions, can represent part–whole relationships, understands equal parts.

Simple rubric

made, and what is expected if full understanding is evident.

The rubric in the diagram shows what the teacher is expecting to observe and how it will be recorded. The rubric rating scales are nominal scales. The scales are used to indicate the types of response given. Performance is scale-marked along a continuum so that the overall performance is rated. This rubric is quite specific, and is targeted at the intended learning outcomes of a highly specific teaching episode.

Rubrics can be used in a more holistic way, be very detailed and contain up to ten rating scales, or be quite simple and involve only three rating scales. The scales are generally designed to describe the outcomes rather than to act as a scoring system. The detail provided in the rubric is designed to support the intended outcomes for the teaching episode, thus reinforcing the overall impetus for the lesson or unit. Teachers need to be very clear about the intended learning outcomes for the teaching episode in order to develop a marking rubric. At the simple level illustrated in the diagram, a rubric could be used for students' fraction knowledge to indicate little or no understanding, a developing understanding and full understanding. What constitutes these categories is the basis for the rubric, so the teacher needs to be fully cognisant of what is expected from students.

Opening up assessment

The most common relationship in assessment practice is that it is undertaken by the teacher on the students for the purpose of reporting to parents, usually through testing procedures. Clarke (1992) has strongly advocated for assessment networks to be opened up to create better assessment practices, in pathways represented in the following diagram.

Assessment of	Assessment by	Assessment for	Assessment using
Students	Teachers	Parents	Tests
Teachers	Parents	Students	Journals
Programs	Students	Teachers	Projects
		Systems	Investigations
			Observations
			Consultations

TEACHING TIP

Each student creates a question on the current topic. This can be done individually or in pairs. Collate the questions and create a 'test' using the questions. Distribute in the class. Discuss the answers and the questions with the class.

Monitoring and recordkeeping

Keeping accurate records of student performance is a key task in assessment. Teachers need to develop techniques that can be incorporated easily into their work programs and styles of working. Some strategies have already been listed in the preceding sections, including a range of checklist formats and anecdotal records that can be noted in a general list or through a file-card system. Keeping track of anecdotal records jotted while observing students at work can be difficult. A system that can work is the use of a folder where pages can be inserted (so that more pages can be added at any time). The teacher can use the document to scribble notes about the student, as a collection point for the Post-it® Notes (noted in an earlier section) or for

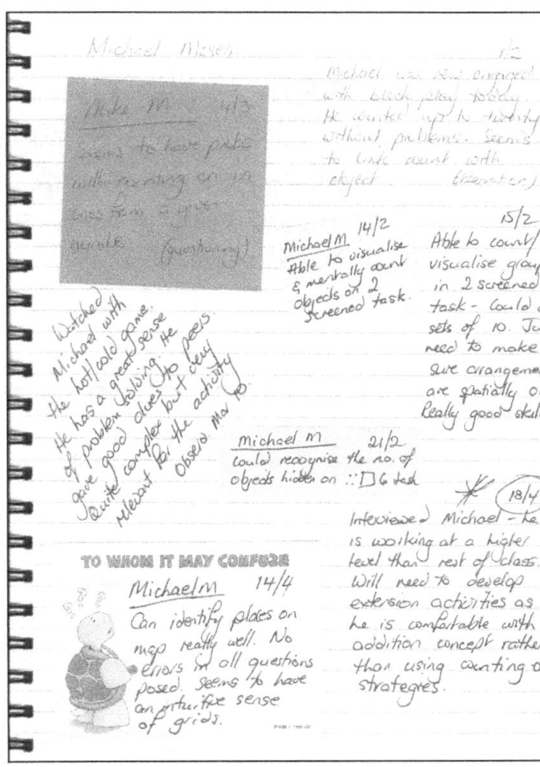

Teacher notebook

		Test/16	Project/20	Test/20
		Feb-10	Mar-30	Apr-06
Jade	Scott	13	18	12
Melanie	Brown	11	17	18
Martin	Cumolov	12	19	18
Harley	Downer	7	11	19
Terri	Dunston	13	18	18
Paulette	Winston	12	16	17

Using a spreadsheet for recordkeeping

storing other information collected about the student. Having tabs at the side of the page allows the teacher to access any student very quickly. Such a document may not necessarily be neat, but it provides a common collection point for the various items that will be included.

Technology can also be a useful tool for recordkeeping, particularly when using a quantitative program. Recording scores into this format can be simple and, once entered, scores can be easily calculated, plotted and made into reports for authorities and parents. As indicated earlier, many schools are now using electronic portfolios (ePortfolios) to record and communicate rich information (written, audio, video, etc.) about student learning.

TEACHING TIP

Provide experiences for students to give feedback to their peers. Modelling feedback is a good strategy. Creating resources such as posters that have feedback stems can prompt students when giving feedback (e.g. 'It is good how you have . . .').

Communicating results

Results of assessment need to be communicated to students, teachers, parents, administrators and, increasingly, to authorities. Test results tell any of these parties in a student's education what the student is capable of doing. They may be useful for monitoring large-scale systems but have little impact on learning.

Students can be informed about what they know, what they need to know and how they can improve their understanding. Qualitative information specifically targeted to the goals of the learning can support this. Communication of results to parents and caregivers, however, is often constrained by school policies and reporting formats. Most schools have some form of parent/teacher evening where individuals can discuss their child's progress with the teacher. This is often supplemented by a report card or ePortfolio, usually at the end of each semester. Results are usually passed on to the teacher who will next work with the student—within the same school setting or in another school. Part of the rationale behind the outcomes-based education that is dominating the considerable reform in Western mathematics is the need for a standard language and format for monitoring and reporting student learning. Being able to state that a student is at a particular level in number, for example, and working with particular concepts within the number sub-strands, enables teachers to have a better understanding of what the student is able to do and understand. These results are often supplemented by qualitative and other data from the ePortfolio that moves with individual students as they progress through school. As the student moves through the mathematics curriculum, pieces of evidence are inserted and removed to show growth in learning and to provide evidence of how that growth has been demonstrated.

In terms of reporting to parents, developments within outcomes-based reporting include the use of software packages to support teachers in writing report cards. In some cases, it is simply a case of using a cut-and-paste mechanism whereby outcome statements and subsets of those outcomes—elaborations or indicators, as determined by the jargon of the initiative—can be taken from a database of the curriculum and inserted into the student's report card. In other cases, parents will be provided access to the ePortfolio so that they can see their child's progress for themselves. These processes document achievement levels and move away from the qualitative comments found on many report cards.

Trends in reporting

Over the past few decades, changes have occurred in the ways in which teachers report students' results. Sometimes teachers will be asked to present student information in particular ways due to parental (or employer) expectations. By understanding where interest groups are coming from and why they have particular expectations of teachers, teachers are able to justify their approaches to reporting. Changes in reporting over the past 30 years are summarised in the following table.

Changes in reporting

Period	Type of reporting	Strengths	Weaknesses
Up to 1970s	Recorded scores—e.g. 6/10 or 60%. Position in class	Gave rankings or position in class	Did not indicate what was known or not known
1970s to mid-1990s	Descriptive text—e.g. participates actively in discussions	Indicated strengths (and sometimes weaknesses) of the student	Did not indicate learning—what students did know
Current	Learning outcomes—e.g. operates on two-digit numbers	Describes what students have learned in terms of outcomes stated in syllabus	Emphasis on learnt outcomes, not on attitudes or areas of growth

As this table indicates, there are pros and cons to most approaches to reporting. As with all teaching, mathematics teachers need to identify the reasons for reporting in particular ways. For example, while contemporary education focuses on reporting against learning outcomes, there are cases in which test scores are featured. This is evident in the massive push by authorities to test students through state/territory-wide (or national) test schemes.

Data-driven pedagogy

The importance of making informed decisions about pedagogy and curriculum cannot be understated. As this chapter has stressed, there are many ways to assess learners and learning. But these assessments need to inform subsequent decisions and actions. Performance data of students can be collected, collated and analysed in many ways. Teachers are increasingly required to have good data on students that documents their learning and progression, and that is mapped against national (or state/territory) standards. There can be a tendency to over-assess learners, whereby too much data are collected, but are not then used to inform subsequent actions. There is a sense that the collection of data can hinder good teaching if it becomes the driver of education (Roberts-Holmes 2015). Teachers need to make decisions based on evidence, so it is prudent to assess learners and then decide based on empirical evidence. Systems are requiring teachers to make greater use of data to inform their work.

TEACHING TIP

Data walls can be made that document students' levels of learning as well as their progressions. A data wall can be created in the classroom so that students can see progress (rather than levels), and these data are also available for families to view and celebrate (or question) students' progress.

Review questions

6.1 Explain the relationship between teaching, assessing and learning, using an example to illustrate the points being made.

6.2 Outline and discuss three approaches to assessing learning other than pencil-and-paper testing.

6.3 What are the strengths and weaknesses of the approaches outlined in the previous question?

6.4 Develop and implement an open-ended question for a classroom. Bring examples of student work into class and discuss what you have learned about students' learning.

6.5 What are some of the ways in which teachers can report student learning?

6.6 Visit the ACARA My School website and see if you can identify a relationship between ICSEA scores and numeracy scores.

Further reading

Clarke, D.J. (1992). Activating assessment alternatives in mathematics. *Arithmetic Teacher, 39*(6), 24–9.

Cole, K. (1999). Walking around: Getting more from informal assessment. *Mathematics Teaching in the Middle School, 4*(4), 224–47.

Ensign, J. (1998). Parents, portfolios, and personal mathematics. *Teaching Children Mathematics, 4*(6), 346–51.

Flores, A. (2010). Learning mathematics, doing mathematics: Deductive thinking and construction tasks with the Geometer's Sketchpad. *Romanian Journal of Education, 1*(4), 7–14.

Lee, C. (2001). Using assessment for effective learning. *Mathematics Teaching, June,* 40–3.

Mertler, C.A. (2016). *Classroom assessment: A practical guide for educators.* London: Routledge.

Novak, J.D. (2010). *Learning, creating, and using knowledge: Concept maps as facilitative tools in schools and corporations*. London: Routledge.

Parker, D.L. & Picard, A.J. (1997). Portraits of Susie: Matching curriculum, instruction, and assessment. *Teaching Children Mathematics, 3*(37), 376–82.

Schloemer, C.G. (1997). Some practical possibilities for alternative assessment. *Mathematics Teacher, 90*(1), 46–9.

Sumsion, J., Barnes, S., Cheeseman, S., Harrison, L., Kennedy, A. & Stonehouse, A. (2009). Insider perspectives on developing belonging, being & becoming: The early years learning framework for Australia. *Australasian Journal of Early Childhood, 34*(4), 4–14.

Thompson, D., Thompson, K. & Else, N. (2000). Alternative assessment. *Australian Primary Mathematics Classroom, 5*(4), 29–32.

Tonack, D.A. (1996). A teacher's views on classroom assessment: What and how. *Mathematics Teaching in the Middle School, 2*(2), 70–3.

Watt, H.M. (2005). Attitudes to the use of alternative assessment methods in mathematics: A study with secondary mathematics teachers in Sydney, Australia. *Educational Studies in Mathematics, 58*(1), 21–44.

Wright, B. & Stewart, R. (1999). Can teachers know too much? *Australian Primary Mathematics Classroom, 4*(2), 4–7.

References

Artzt, A.F., Armour-Thomas, E., Curcio, F.R. & Gurl, T.J. (2015). *Becoming a reflective mathematics teacher: A guide for observations and self-assessment*. Routledge.

Australian Curriculum, Assessment and Reporting Authority (2015). *NAPLAN*. Retrieved from https://nap.edu.au/.

Bloom, B.S. (1956). *Taxonomy of educational objectives: The classification of educational goals: Handbook 1, Cognitive domain*. New York, NY: David McKay.

Clarke, D. (1992). *Assessment*. Melbourne: Curriculum Corporation.

Department of Education and Early Childhood Development Victoria (2006). *Early Numeracy Research Project*. Retrieved from www.education.vic.gov.au/studentlearning/teachingresources/maths/enrp/default.htm.

Department of Education and Training Western Australia (2004). *First steps in mathematics*. Melbourne: Harcourt.

Ellerton, N.F. & Clements, M.A. (1997). Pencil-and-paper mathematics tests under the microscope. In K. Carr (ed.), *People, people, people: Proceedings of the 20th Annual Conference of the Mathematics Education Research Group of Australasia* (vol. 2, pp. 155–62). Rotorua: MERGA.

Pimm, D. (1991). Communicating mathematically. In B. Shire (ed.), *Language in mathematical education: Research and practice* (pp. 17–24). Philadelphia, PA: Open University Press.

Roberts-Holmes, G. (2015). The 'datafication' of early years pedagogy: 'if the teaching is good, the data should be good and if there's bad teaching, there is bad data'. *Journal of Education Policy, 30*(3), 302–15. doi:10.1080/02680939.2014.924561.

Sullivan, P. & Lilburn, P. (1997). *Open-ended maths activities: Using 'good' questions to enhance learning*. Melbourne: Oxford University Press.

Wright, R.J., Martland, J. & Stafford, A.K. (2000). *Early numeracy: Assessment for teaching and intervention.* London: SAGE Publications.

Zevenbergen, R. (2000). 'Cracking the code' of mathematics: School success as a function of linguistic, social and cultural background. In J. Boaler (ed.), *Multiple perspectives on mathematics teaching and learning.* New York, NY: JAI/Ablex.

Zevenbergen, R., Hyde, M. & Power, D. (2001). Language, arithmetic word problems and deaf students: Linguistic strategies used by deaf students to solve tasks. *Mathematics Education Research Journal, 13*(3), 204–18.

CHAPTER 7

WORKING MATHEMATICALLY

One of the most common descriptions of mathematics is that it is about doing pencil-and-paper calculations. This view is under challenge in current reforms in mathematics education. Contemporary mathematics education focuses on thinking and working mathematically so that mathematics becomes a way of seeing and acting in the world, rather than an activity in and of itself. Encouraging students to think and work mathematically requires them to engage in a range of tasks, problems and investigations (Schoenfeld 2017). The aim of this chapter is to present the notion of working mathematically in relation to problem-solving, connecting mathematics beyond the mathematics classroom and communicating mathematics through multiliteracies.

Working and thinking mathematically

In her study of mathematicians, Burton (2004) found that for research mathematicians working with mathematical concepts and processes, working mathematically is vastly different from the practices found in school mathematics. Burton reported that mathematicians saw mathematics as something related to aesthetics, coherence, intuition, messiness, insight and fun. She argues very strongly that the practices used by mathematicians should be fostered in school mathematics so that students can learn how to work mathematically. Rather than rely on teacher direction, individual work and rote learning, Burton argues that a stronger focus must be placed on working collaboratively and talking about mathematical ideas. She maintains that arguing and talking about ideas, use of intuition and passion, and regularly going back to known knowledge to extend and build new knowledge are key processes that should be featured in contemporary classrooms.

Using many of the principles identified by Burton, Boaler (2008) similarly reports that quality teaching in mathematics classrooms included the use of group work so that students could talk, discuss, support and challenge ideas in ways that were far less intimidating than in whole-class teaching. Boaler also suggests that teachers need to set tasks that create opportunities for high levels of mathematical understanding, that the teacher should take a role that enhances and supports learning (rather than directs learning) and that students need to have an opportunity at the end of the lesson to share their learnings with their peers. This final part of a lesson is one that encourages debate among peers, but within a supportive environment. In the development of this new model of teaching mathematics, some time is required for students to learn the new protocols, but when they do the results are strong. Boaler's work shows that there are strong outcomes when the model of teaching moves from direct instruction to one that fosters deep mathematical learning by creating new spaces for working mathematically.

Working mathematically is an important life skill for effective living in the world beyond school. In order to be competent and effective citizens, students need to exit school with the dispositions and competencies in mathematics/numeracy that will allow them to participate fully in the activities they undertake. This means being able to use and work with mathematics in a way that empowers them in their everyday lives. Working mathematically is enacted in the Australian Curriculum via four proficiencies that underpin each of the strands of Number and Algebra, Measurement and Geometry, and Statistics and Probability. The four proficiencies are Understanding, Fluency, Reasoning and Problem-solving. Further details about the proficiencies in the Australian Curriculum can be found at www.australiancurriculum.edu.au/f-10-curriculum/mathematics/key-ideas/. Most mathematics curricula include something similar to these proficiencies.

Contemporary approaches to mathematics education encourage students to think mathematically. *Thinking mathematically* refers to a disposition to use mathematics to solve problems and tasks in a manner that is logical and based on mathematical principles. Mathematics is a very useful tool to work through problems. Thinking mathematically is encouraged through the curriculum by the posing of problems that adopt and apply mathematical knowledge, skills and processes. The linking of realistic examples to mathematics is a key feature of teaching, since students need to see the relevance and purposefulness of mathematics to real-world situations.

Problem-solving

The idea of the importance of problem-solving for fostering mathematical thinking gained prominence in the 1970s. It has been used in mathematics classrooms, with greater or lesser effect, ever since. For problem-solving to have greatest effect and benefit, it needs to be an integral part of the classroom ethos rather than simply an activity (Clarke et al. 2007). Problem-posing has been emphasised in addition to problem-solving (Singer et al. 2015). Digital tools, when chosen carefully, assist in the process of problem-solving (Santos-Trigo & Moreno-Armella 2016)

A problem has been defined by Mason and Davis (1991) as something that gets inside the head of the learners so that they become motivated and challenged by the task or question. It is a question or task that does not have an obvious answer or a clear path to resolution. Problem-solving is more than solving word problems such as, 'Three birds were sitting on a fence; one flew away. How many are left?' For students who cannot determine that this is a subtraction task, this is a problem; but for most students, it is little more than a routine task embedded in a word problem. Fostering mathematical thinking requires more motivating problems.

Teaching problem-solving requires teaching both *through* problem-solving and *about* problem-solving. Teaching through problem-solving means immersing students in a variety of novel, challenging and motivating problems as a natural part of the mathematics program. Teaching about problem-solving scaffolds learning through the provision of relevant strategies. However, students also need to have an adequate understanding of the mathematics involved in the problem before embarking on tasks (Pengally 1989).

Polya's (1973) four-stage model for problem-solving is a generic strategy that underpins much teaching with problem-solving:

1. Understand the problem—what is the problem asking?
2. Devise a plan for solving the problem.
3. Carry out the plan.
4. Look back and reflect on the solution obtained.

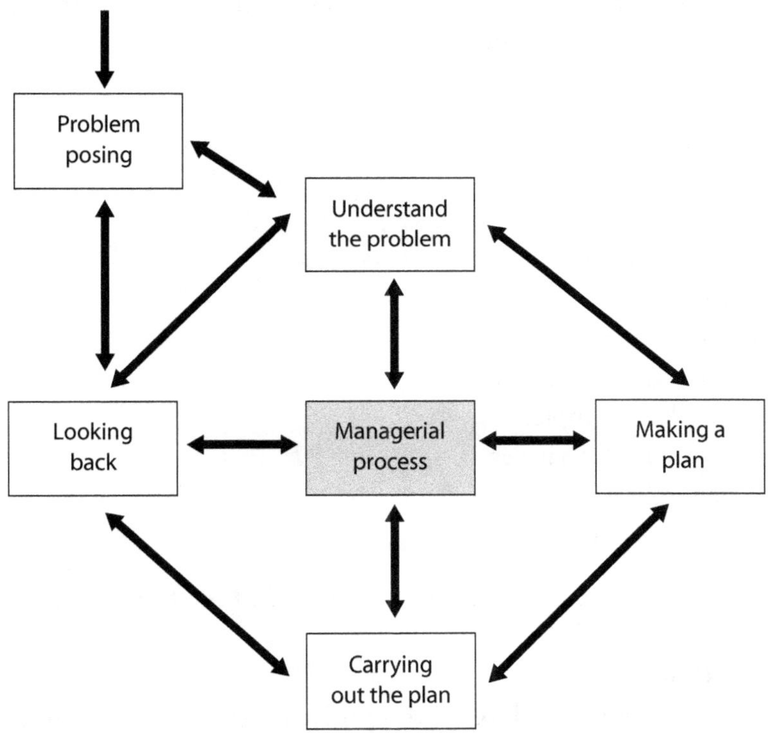

In classrooms, this model is often referred to as *see, plan, do, check.*

While the four steps are proposed as isolated units, Mason and Davis (1991) have suggested that this is not the case, and that the steps in fact tend to blur. Students can move backwards and forwards between the steps as they attempt to work out ways to solve the problem. It is possible for students to become caught up in an endless cycle of reading the problem, thinking about ways to solve it and then rereading, to the point that they eventually give up without even attempting the problem. Students need to be supported to move beyond the blueprint of problem-solving to enable them to develop specific strategies for solving problems. A slightly different model has been proposed by Fernandez and colleagues (1994), which adds the additional element of problem-posing to the model as well as representing the process as cyclical.

Problem-solving is enabled by having a wide repertoire of strategies from which to select when encountering different problems. Particular strategies suit particular problems. Knowledge of strategies is a product of experience and engagement in solving a variety of problems. The teacher's role is to provide such experiences. Common problem-solving strategies include the following:

- Create a table.
- Make a drawing.
- Think aloud.
- Act it out.
- Look for a pattern.
- Guess and check.
- Work backwards.
- Identify unwanted information.
- Use a simpler example.
- Identify other alternatives.
- Make generalisations.
- Check the answer.

These strategies are discussed in more detail in the following section.

Create a table

- A carpenter constructs desks with four legs and stools with three legs. At the end of the day, she has used 43 legs. How many desks and stools have been made?

1 desk	2 desks	3 desks	4 desks	7 desks	10 desks
43 – 4 = 39	43 – 8 = 35	43 – 12 = 31	43 – 16 = 27	43 – 28 = 15	43 – 40 = 3
39 ÷ 3 = 13	35 ÷ 3 = no	31 ÷ 3 = no	27 ÷ 3 = 9	15 ÷ 3 = 5	3 ÷ 3 = 1
13 stools	no	no	9 stools	5 stools	1 stool

Using a table to organise information helps students to see patterns emerging and can help to identify any missing information. After the first four attempts, a pattern seems to emerge whereby every time the desk number increases by 3, there will be a remainder divisible by 3, which can result in the stool being constructed. Using a table allows students to see a pattern beginning to emerge (every third time) and to then trial the idea (guessing and checking).

Make a drawing

- If there are five people in a room and they all shake hands with each other, how many handshakes are there altogether?
- A frog in a well hops up 3 metres each day and slips back 2 metres each night. If the well is 10 metres deep, how long will it take the frog to hop out?

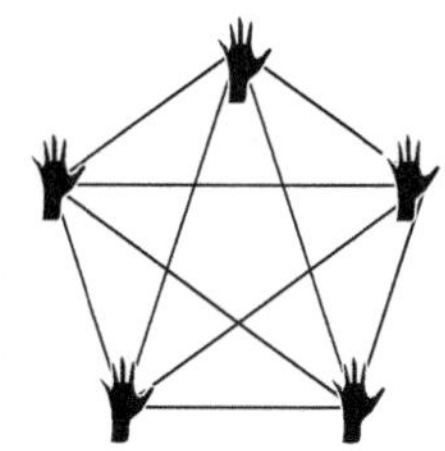

Using drawings allows students to construct the problem visually. Some problems, including spatial problems (such as maps and paths), are best solved using this strategy. It allows students to 'see' relationships. When using this strategy, it is important to stress to students that they should not be spending time on details in the drawings, but only providing sufficient information to demonstrate the problem. Doing this also supports students in learning what the key information in a problem is and ignoring redundant or irrelevant information.

Think aloud

- As I was going to school, I met a teacher who had 24 students in her class. Each student had two siblings. Each sibling had two pets. How many teachers did I meet?

By encouraging students to think aloud, they are able to hear their verbalisations. This allows two processes—thinking and talking—to support students' problem-solving. Speaking their thinking processes supports communication as well as encouraging metacognitive processes.

Act it out

- A party of three people goes to a restaurant. The dinner is $10 a head, so it costs $30. Since they are regulars at the restaurant, the manager decides they can have a $5 discount. The waiter decides it is too difficult to share $5, and so gives each person a discount of $1 and keeps the other $2. The meals are now $9, which means that the dinner cost a total of $27. The waiter kept the other $2, which makes for a total of $29. Where did the other dollar go?

In this case, students would be able to act out the process of the bill sharing to see where the missing dollar went. By modelling the process, they will be able to see the problem.

Look for a pattern

- Square numbers are so called because of the pattern they form as the square 'grows'. How many squares are there when the square has a side of 5, or of 10? How long will a side be when there are 81 squares in the square?
- A king has decided to reward his servant for his good deeds by offering him a choice of whether he would like to take the wheat spread in each square of a chessboard multiplied by 10, or whether he would prefer to place a grain of wheat on a chessboard square and double the number in each subsequent square. Which option would you take, and why?

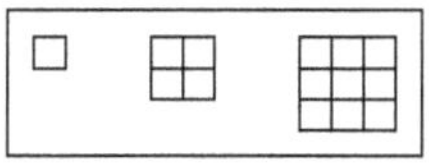

Early mathematics involves considerable work on patterning, both in number and spatially. This is a very important skill (and attitude) for competent thinking in mathematics, and hence should be continually encouraged and reinforced. Many mathematical discoveries have been made through people identifying patterns (such as Fibonacci's numbers). Often, the pattern can be seen when the data are entered into a table; in other cases, such as spatial patterns, the patterning needs to be seen spatially.

Guess and check

- Place the numbers 1 to 9 in the grid so that the numbers add up to 15 in any direction.

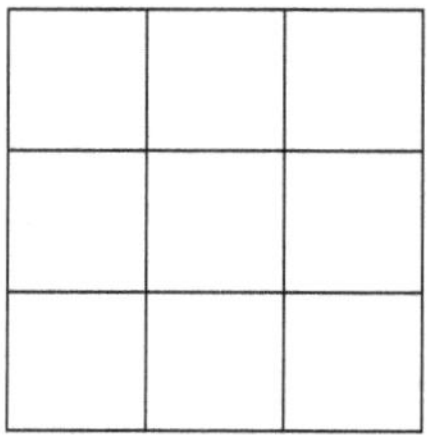

The guess-and-check strategy is often frowned upon in mathematics, as if it were an inferior process. However, it can be useful and should be encouraged as one of many possible strategies. When using an educated guess, students need to rely on identifying the key information and some strategies for the resolution of the problem. Students should be encouraged to develop guessing strategies that are based on some knowledge and experiences rather than indulge in blind or wild guessing.

Work backwards

- You spent \$21 at the supermarket. The fruit cost twice as much as the meat, which cost twice as much as the chocolates. How much did the meat cost?
- A length of rope was cut in half to share between two landowners. One owner needed to use ⅔ of it for tying his dog to the shed. The piece that is left is 1 metre long. How long was the original rope?

The working backwards strategy encourages the student to see what the answer is and then systematically work from that point to account for what has happened earlier.

Identify unwanted information

- A new car's purchase price is listed at \$19 760. The fuel consumption is listed as 100 km on 10 L. If the new owner drives an average of 1500 kilometres per month, how many litres of petrol would have been consumed in a year?

Students are often given mathematics tasks and problems in which they need to work with the numbers provided. However, in the world beyond school, such examples rarely exist, so there is a need to recognise what is important and relevant information to solve a problem. Examples should be given to students that encourage this disposition to develop.

Use a simpler example

- How many squares are there on a chessboard?

Often, problems can be complicated by the size of the numbers, or the nature of the patterns, being used. By making the task smaller or simpler, reducing the size of the pattern or breaking the task into smaller components, the task can come to be seen as manageable. With the example of a chessboard, by reducing the original chessboard pattern into a smaller task—such as a 3 × 3 grid—the inherent components of the task remain unaltered and the student is able to manage the task. By gaining insight into solving the smaller task, students can extend this learning to the larger or more complex task.

Identify other alternatives

- The monkey needs to get to the bananas. How many ways can it get there? What is the shortest route to the bananas?

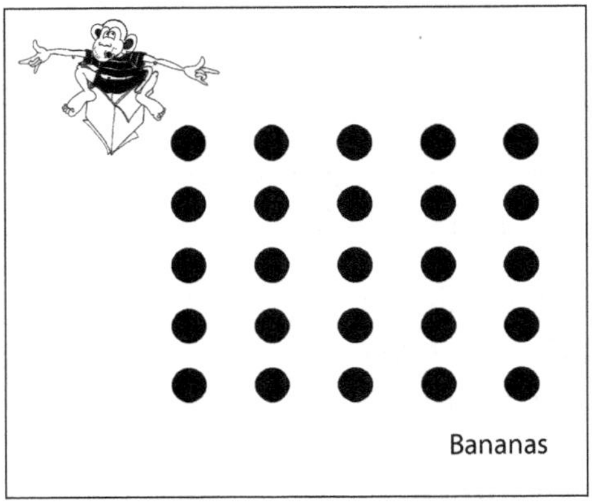

Encouraging students to seek alternatives—either in the resolution of the answer or in the process—allows them to generate new ways of seeing the problem and of how to solve it. A useful process for developing these outcomes is to promote group discussions—within a group and between groups. Allowing students to hear how others have solved the problem and the answers they have generated enables them to see alternative modes and responses to the same task. Once this ethos is a part of the classroom culture, it is useful for students to evaluate the responses their peers propose. Such a process should be undertaken with care so as to nurture a constructive process and ensure that a negative ethos does not develop. One way this can be achieved is through the *SWOT* process, whereby students identify *strengths, weaknesses, opportunities and threats* to the processes being used. Fostering this metacognition is an important aspect of working mathematically.

Make generalisations

- When three consecutive numbers are added together, their total is 3. What are the numbers?
- A rectangle has sides that are twice as long as its height. How long might the sides of a rectangle be?

When using this type of question, the way in which it is resolved can lead to a general rule for this type of problem. Focusing on the broader features of the problem can result in students gaining insights of greater significance than when they focus on the specifics of the task. Once students have gained proficiency in problems with numbers, they can start to generate general principles, leading to algebraic thinking about problems that have no numbers.

Check the answer

A most important strategy for students to learn is checking their answers. This enables them to identify errors—in the answer as well as in the process. Using a strategy of checking the reasonableness of the answer, students can be encouraged to see whether their answer is plausible. Estimating before working through a problem can be a useful process in that it allows students to see whether the answer they arrive at is reasonable.

Problem-posing

In contrast to problem-solving, problem-posing encourages students to create their own problems. Using the principles and strategies discussed above, problem-posing is seen to create significant opportunities for students to engage in mathematical ideas, with problem-posing offering better transfer of learning than what is achieved in problem-solving (Kapur 2015)

Moses and colleagues (1990) propose four principles to support students with problem-posing:

1. Focus students' attention on the key information to be contained in the problem.
2. Commence with familiar concepts or ideas.

3. Encourage the use of ambiguity when designing problems; this opens up the problem.
4. Encourage students to set restrictions (or domains) to their problems.

The problem-posing approach is gaining a strong foothold in classrooms, where teachers are attempting to make teaching more authentic. The use of digital resources in problem-posing has been seen to offer powerful ways for students to consider, identify and evaluate different concepts and processes in problem-posing (Santos-Trigo & Moreno-Armella 2016).

Making connections

For too many students, the experience of school mathematics is not positive—they perceive it to be a difficult and irrelevant subject. The task of the teacher is to create meaningful and purposeful connections between mathematics and other spheres of life and school. That is, mathematics needs to be *transdisciplinary*—it must link realistically to other curriculum areas—for students to understand why they study it. That mathematics informs many other areas is a connection that is rarely made explicit to students. It is very limiting to undertake social studies without knowledge of number, area, graphing, location or statistics; likewise, studying science is difficult without knowledge of number, exponentials, measurement, or recording and interpreting data.

Mathematics applies to all areas of life. Without a deep understanding of mathematics, people's lives are impoverished, and in many cases severely restricted. Not being able to undertake calculations can limit a person's understanding of salaries, budgets and banking; not being able to measure accurately can affect such activities as cooking, gardening, handy work, and so on. Mathematics is an integral part of a quality life, yet considerable numbers of school leavers exit with very negative experiences of mathematics and self-concepts of themselves as users of mathematics.

Teachers need to make connections between mathematics and other curriculum areas and activities beyond the school in ways that are not tokenistic but realistic. In many early-year classrooms, a common activity is that of the class shop, where students practise shopping through purchasing and operating with money. Students see that these are not real activities but pseudo-activities, where the shopping activity is a mask to justify the mathematics. Making connections so that students can see the value of mathematics demands that realistic activities are undertaken.

The value of activities such as purchasing the items on a shopping list becomes meaningful, relevant and purposeful when the items are realistically priced and there is some culminating activity—such as going shopping to purchase goods, or making a class party for which foods are purchased and/or cooked. When the classroom activity resembles the activity undertaken in the world beyond school, there is every opportunity for students to make the connections between school mathematics and the wider world. They can see why they need to study mathematics and that mathematics is not an irrelevant area of study.

Making connections happens on three levels:

1. *Within mathematics*—where links are made between the various strands of the curriculum. It is not possible for number study to exist in isolation from other areas. For example, determining the area of regular shapes relies on a knowledge of multiplication or repeated addition. These links need to be developed.
2. *With other discipline areas*—where connections are made with other curriculum areas, such as science, health and physical education, and social science.
3. *In worlds beyond schools*—where the mathematics is placed in realistic contexts. These contexts demonstrate how mathematics can be used to support and enhance problem-solving.

Thinking mathematically

Schoenfeld (2017) argues that thinking mathematically means that a person needs to do the following:

- Develop a mathematical viewpoint so that they value the processes associated with mathematising a problem, and then have the desire to use both the mathematics and the abstraction.
- Competently and confidently use the tools available in and through mathematics to better understand the underlying patterns and structures of mathematics to make sense of the task or problem.

There are many ways to think mathematically. Imagine a three-dimensional object that is to be rotated. The object must be visualised, and then students might either draw their own image or identify an image from a collection. The task can be solved in a number of ways. One approach could be using visual imagery, where the object is rotated in the mind; another might be to solve it logically through identification of particular vertices, and then working out the orientation of the rotated object. Both of these strategies rely on thinking mathematically. If a student simply rotated the image, little to no deep mathematical thinking would be used to solve the problem. The disposition to solve something mathematically needs to be nurtured and developed by teachers.

TEACHING IDEA

Use your knowledge to work out how you can create a formula for the sum of the internal angles of a polygon. What knowledge did you draw on? How did your group work out the formula? How did other groups work it out? What are the strategies used? Are some better than others? Why?

Communicating mathematically

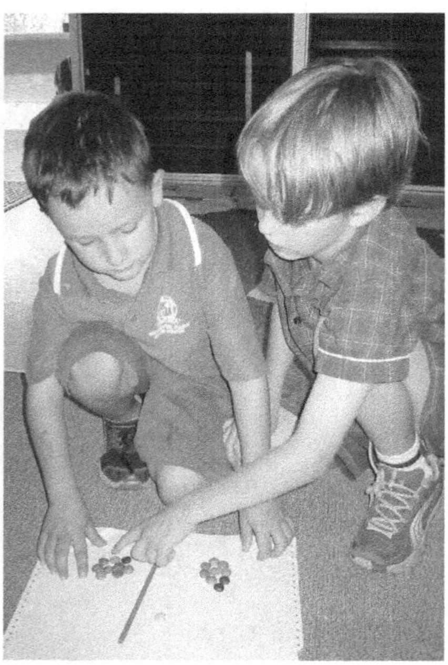

Students talking about representations

Current thinking in the area of literacy involves four different roles of the reader (Freebody 1992). This model of multiliteracies provides a strong link to the roles of the numerate student. More than simply considering mathematics teaching as linking language to concrete and symbolic representations, multiliteracies encompass mathematical reading, representing, recording and communicating mathematically. The role of the reader is not passive, but integral to the construction and interpretation of mathematical 'texts', including concrete materials, photographs, pictures or books, video or non-static representations, teaching episodes where there is considerable interpretation of teacher actions by the students, and the traditional written formats of textbooks, worksheets and board work.

Consider the photograph opposite. A number of interpretations can be made of what is happening here. Is one boy helping the other? Is he telling the other he is incorrect? Correct? Many interpretations are possible.

In terms of multiliteracies, the mathematics classroom is itself a text about which students will make interpretations (or readings). When teaching is seen in this way, it becomes possible to understand the learner as a much more active participant in the classroom, and doing so allows the teacher to realise that students can construct very different interpretations of what has been said or done. This moves the emphasis away from seeing students as giving right or wrong answers to one where the role of the teacher is to understand why students construct responses and understandings in the ways they do. Not only are the communications related to mathematics, but so are the texts within which the mathematics is being conveyed to the students. Meaning making becomes multidimensional.

By applying the multiliteracies perspective to mathematics, communicating mathematically becomes more than simply recording with pen and paper. Communicating mathematically encourages several different types of communication:

- *Oral communication*—where students 'talk' mathematically. This can be in open discussions or in small groups. The teacher sets tasks that require students to talk about the mathematics they are using, the ways in which they are working and the answers they are constructing. This form of communication can include plays and other performances.
- *Visual communication*—this can be in the form of displays that may be flat, two-dimensional representations such as written projects, or three-dimensional displays such as working projects (e.g. constructions).
- *Digital communication*—creating displays using digital technology. This could include the use of specific tools such as video of students talking about representations, spreadsheets to show patterning (or pre-algebraic thinking), PowerPoint presentations to create a visual display that documents learning, or iPad apps to document and communicate student learning.
- *Textual communication*—the traditional format in which students write mathematically. This can be useful for explanations, justifications, refutations, conjectures, and so on. This form of communication is particularly important in developing students' metalanguage of

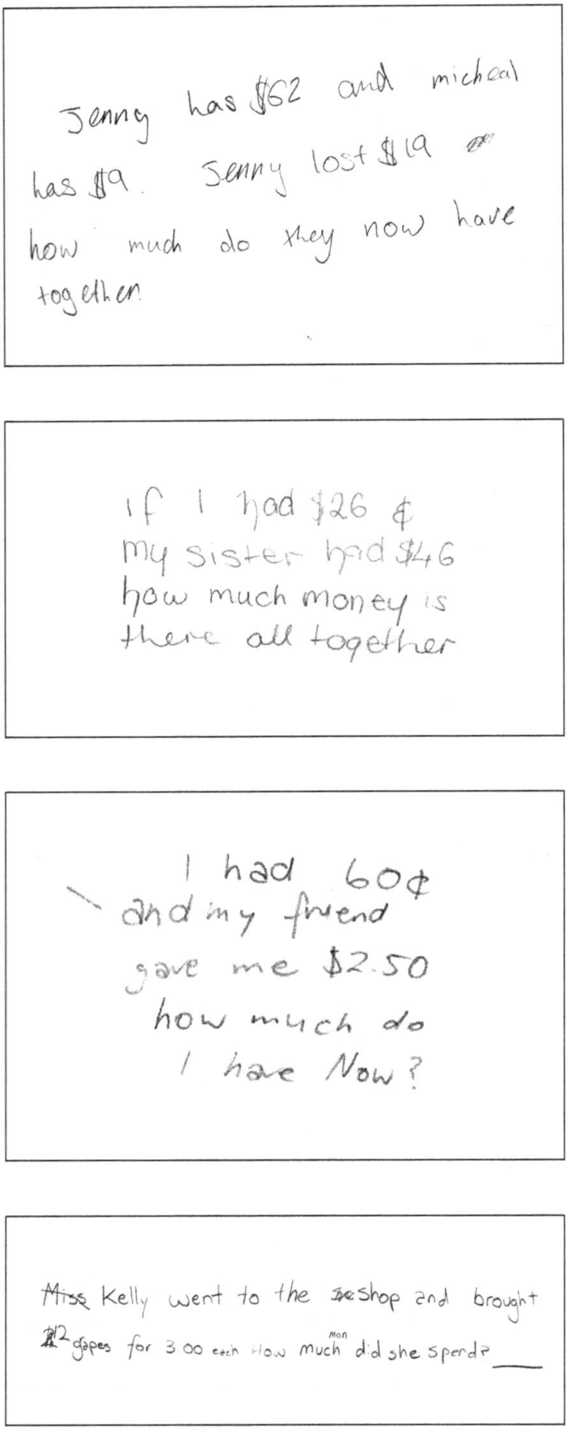

Examples of students' own mathematics questions

mathematics and encouraging them to think mathematically. Students can be asked to write their own mathematics questions, to reflect in journals about what they have learned or had difficulty with in mathematics or to write a letter to a friend explaining what they have learned. A commonly used example has students writing a story about a graph depicting the height of water in a bath. The students' stories indicate their understanding of graphs and their interpretation of data, slope, gradient, and so on.

- *Symbolic communication*—a unique feature of mathematics is its symbolic form. To communicate in this truncated language, students need to know the specific language of symbols. The use of symbols allows the development of a form of communication particular to mathematics. For example, students are introduced to $+$, $-$, $\times$, $\div$ and $=$; these are followed in later years by $<$, $>$, $\leq$ and $\geq$, among others. While these symbols are an integral feature of mathematics, they are also a key to a specific language. Common usage of these signs can result in misconceptions developing. For example, early addition experiences focus on language, such as '3 apples and 4 apples makes 7 apples', so that the equals sign comes to be seen as equivalent to 'makes'—that is, 4 and 3 is equivalent to 7. The language becomes inappropriate, though, in the case of '1 and 1 makes 2', since 1 and 1 can make 11 when the two numbers are placed side by side, or 7 if one is placed horizontally and the other at an angle to the horizontal.

Review questions

7.1 Why is working mathematically a strong emphasis in current practice in mathematics teaching and learning?

7.2 Identify six strategies for problem-solving. Briefly outline each, and discuss when each strategy is most useful.

7.3 Identify three modes of communicating mathematically. Provide an example of each mode and of how it could be used in the classroom.

7.4 Why is thinking mathematically an important skill for students to develop?

Further reading

English, L.D. (1997). Promoting a problem-posing classroom. *Teaching Children Mathematics*, *4*(3), 172–9.

Jitendra, A. (2003). Teaching students math problem solving through graphic representations. *Teaching Exceptional Children*, *34*(4), 34–9.

Lovitt, C. & Clarke, D. (1990). *Mathematics curriculum teachers package 1*. Melbourne: Curriculum Corporation.

Lovitt, C. & Clarke, D. (1990). *Mathematics curriculum teachers package 2*. Melbourne: Curriculum Corporation.

References

Boaler, J. (2008). Promoting 'relational equity' and high mathematics achievement through an innovative mixed-ability approach. *British Educational Research Journal*, *34*(2), 167–94.

Burton, L. (2004). *Mathematicians as enquirers*. Dordrecht: Springer.

Clarke, D., Goos, M. & Morony, W. (2007). Problem solving and working mathematically: An Australian perspective. *ZDM: The International Journal on Mathematics Education, 39*(5), 475–90. doi:10.1007/s11858-007-0045-0.

Fernandez, M., Hadaway, N. & Wilson, J. (1994). Problem solving: managing it all. *The Mathematics Teacher, 87*(3), 195–8.

Freebody, P. (1992). A socio-cultural approach: Resourcing the four roles of the literacy learner. In A.J. Watson and A.M. Badenthorp (eds), *Preventing reading failure*. Sydney: Ashton Scholastic.

Kapur, M. (2015). The preparatory effects of problem solving versus problem posing on learning from instruction. *Learning and Instruction, 39*, 23–31. https://doi.org/10.1016/j.learninstruc.2015.05.004

Mason, J. & Davis, J. (1991). *Fostering and sustaining mathematics thinking through problem solving*. Geelong: Deakin University Press.

Moses, B., Bjork, E. & Goldenberg, E.P. (1990). Beyond problem solving: Problem posing. In T.J. Cooney (ed.), *Teaching and learning mathematics in the 1990s (1990 Yearbook)* (pp. 82–91). Reston, VA: NCTM.

Pengally, H. (1989). *Becoming mathematical problem solvers.* In B. Doig (ed.), *Everyone counts* (pp. 1–5). Melbourne: Mathematics Association of Victoria.

Polya, G. (1973). *How to solve it*. Princeton, NJ: Princeton University Press.

Santos-Trigo, M. & Moreno-Armella, L. (2016). The use of digital technology to frame and foster learners' problem-solving experiences. In P. Felmer, E. Pehkonen, & J. Kilpatrick (eds), *Posing problems and solving mathematical problems: Advances and new perspectives* (pp. 189–207). Switzerland: Springer.

Schoenfeld, A.H. (2017). Learning to think mathematically: Problem solving, metacognition and sense making in mathematics. *Journal of Education, 196*(2), 1–38. doi:10.1177/002205741619600202.

Singer, F.M., Ellerton, N.F. & Cai, J. (2015). *Mathematical problem posing*. Dordrecht: Springer.

CHAPTER 8

NUMERACY

'Numeracy' is a term that has gained a strong foothold in school education. The term 'numeracy' was first coined in 1959 in the United Kingdom in the Crowther Report (Ministry of Education 1959), in which it was described as the mirror image of literacy. This term was revisited and elaborated in 1982, when it was described as encompassing scientific approaches to understanding phenomena, thinking quantitatively and combating statistical fallacies (Cockcroft 1982). During the term's infancy, there was a concerted effort by many policymakers and educators to put literacy and numeracy in the same basket, as if numeracy were a part of literacy. While it can be argued that many of the principles adopted by literacy educators can be seen to resonate with numeracy, there are distinct differences. For example, being able to read and interpret a text is a feature of literacy, and a needed skill in numeracy. However, applying mathematical processes (such as number operations) to check the accuracy of something reported in that text is a skill quite apart, and different from, literacy. Numeracy is a much-expanded version of literacy.

In education settings, there have been cases in which numeracy would be scheduled as a subject, in place of mathematics or as an alternative option. This approach was often established from a deficit model of learning; it was the option for those who could not 'do' mathematics. Such classes often took a 'back to basics' approach, and were designed under the premise that students who experienced difficulties learning mathematics would be better served by a curriculum that built basic skills that would then be applied to solve contrived 'real-world' problems. Other numeracy programs might aim to engage students in extended, authentic problem situations (e.g. planning and running the school disco). While such activities engage students in the application of mathematics in real contexts, they may not build and extend the mathematics knowledge needed by students so that they can continue the study of mathematics in higher grades. Other numeracy programs have a predominant focus on building number knowledge at the expense of other strands of the mathematics curriculum.

Many people find the application of mathematics in their everyday lives a daunting task and will declare often that numeracy is not their strong suit. Even many highly educated people will readily declare that they were 'never any good' at mathematics, and may even refer to themselves as innumerate. Yet very few people will ever declare that they are illiterate. Why do adults readily declare that they are innumerate rather than that they are numerate? Could the self-effacing label of innumeracy be more palatable than declaring yourself numerate, and therefore having the capacity to work with various aspects of mathematics in the everyday context?

Numeracy is a contested term, and it has many meanings. It is sometimes seen as the study of number, whereas other definitions regard it as encompassing more than mathematics. People may use the term in a way that is different from those with whom they are conversing, so it is important to establish a deep sense of what the term means so that positive communication can occur. Numeracy, in the contemporary sense, is bigger than mathematics per se, and also envelops mathematics. Mathematics is the knowledge that exists within the domain of mathematics and has meaning within that setting. Numeracy, on the other hand, is the application of mathematical thinking, knowledge and processes in both mathematics and beyond mathematics, and is context-bound. The relationship between mathematics and numeracy can be represented in the following way.

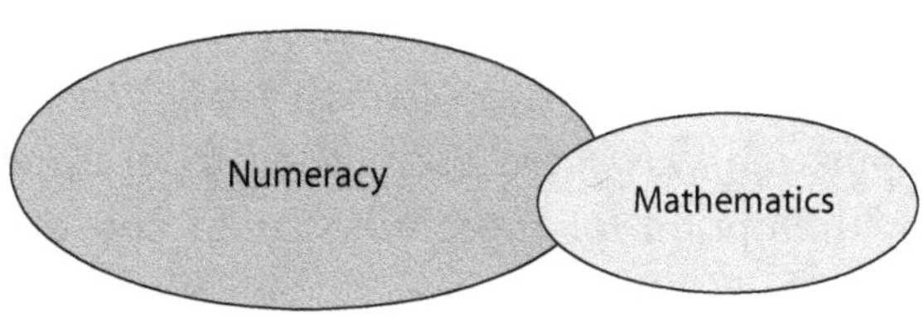

In this conceptualisation of numeracy, mathematics is a corpus of knowledge that exists in its own domain. Numeracy is the application of that knowledge in non-mathematical contexts. All numeracy is underpinned by mathematics, and the purpose of school mathematics is to introduce and develop mathematical thinking and knowledge. However, while a role of school mathematics is to develop numerate people and upskill them with the necessary skills and knowledge, this is not its only purpose. It is important that school mathematics fosters creativity, a love for mathematics, and an appreciation of the beauty and structure of mathematics as a discipline in its own right.

TEACHING IDEA

Engineers and tradespeople

There is a good story about the engineer who has done all the calculations, design and computer checking on a project. Yet when he goes to implement the project, the tradesperson is forthright and tells the engineer that it will not work. The engineer has the strong mathematics to create the model or design, while the tradesperson has the experience (and numeracy) to know that it won't work.

An interesting conundrum in the relationship between numeracy and school mathematics is that, while a student may gain the 'secret knowledge' of mathematics—that is, mathematical knowledge, skills and processes within the classroom context—this may not be enough to make a person numerate. The student may perform well in mathematics classrooms but not be able to transfer that knowledge and those ways of knowing into out-of-school contexts. This, by implication, may mean that the learner is innumerate. For example, a student may use his or her knowledge of ratio and proportion on school-based tasks, but, when deciding best buys at the supermarket, not be able to make informed decisions using their mathematical knowledge.

In today's curriculum, the main differences between mathematics and numeracy are that numeracy:

- does not exist solely in mathematics, since it is applied across all curriculum areas and to the world beyond schools
- is the application of mathematics to non-mathematical contexts
- is a disposition to use mathematics to solve problems in everyday contexts.

The numerate person

Part of the role of schooling is to create people who will live productive adult lives and make informed decisions. Being numerate is a life skill that is needed to function well in modern society.

Being innumerate is similar to being illiterate—people are unable to function fully if they do not have the skills, knowledge, dispositions and tools to solve everyday problems. Being unable to understand everyday finances and transactions severely impacts the everyday lives of everyday people. However, unlike illiteracy, which is not a trait that people announce proudly in their everyday lives, there is a sense that it is fine to be innumerate. Many people are quite happy to admit their limited skills in mathematics.

It is recognised that a significant portion of the adult community leaves school with less than favourable memories of their school mathematics. They often see themselves as unable to do mathematics—but they have framed mathematics as the work they did in school. In contexts outside of school, many decisions are made that require mathematics—shopping; banking; paying off loans; working out how much paint is needed for a room or how far a car will travel on the remaining fuel in the tank; making decisions based on the weather; time management; critically analysing news reports. These tasks all have some mathematics as their base, but people often do not see these as mathematics.

Statistics, statistics and lies . . .

Consider the common fallacy often reported in the media about the prices of houses. A common message is that the average price in a particular suburb has increased dramatically. Consider the table overleaf, which shows three different possible interpretations of a newspaper headline. These different forms of average are used to suit the purposes of those using the measure of central tendency.

Newspaper headline: The average price of houses in Redcliffe has gone up $300 000 in the past month.

Mean method interpretation	Mode method interpretation	Median method interpretation
Most homes in the area had gone up in price by about $300 000	A lot of houses had been sold recently that were $300 000 more expensive than the past mean	Only one very expensive home or property was sold and this pushed the average price up by $300 000

A numerate person can critically appraise the statement by considering what measure of central tendency has been used and how mathematics (i.e. statistics) has been used to the advantage of a particular interest group. A less numerate person may consider this message a truth (albeit a false truth) because it is based on mathematics. 'Answering back' to statistical information is regarded as an integral element of numeracy, and is referred to as having a 'critical orientation' (Goos et al. 2019). Opportunities for assisting children in developing a critical orientation, and therefore in developing critical numeracy, occur most naturally when teaching statistics and probability (see Chapter 14). But this should not be the only place for teaching critical numeracy. All real-world contexts for developing numeracy can be used to foster a critical orientation to ensure that children develop skills to recognise how mathematics may be used to distort, persuade or argue.

Being numerate has personal and social benefits. There is a much bigger picture associated with poor numeracy and the general population. A number of studies (e.g. National Numeracy for Everyone for Life 2014) have highlighted the costs of innumeracy in society, as summarised below:

- Poor numeracy skills make a person twice as likely to be unemployed.
- There is a direct correlation between numeracy skills and wages.
- There is a direct correlation between numeracy and health; poor numeracy is associated with depression.
- Students with low numeracy are more likely to face exclusion from school.
- Children with poor numeracy are twice as likely to be truant from school.
- A quarter of young people in custody have numeracy skills equivalent to (or below) what would be expected of a seven-year-old.

- Sixty-five per cent of adult prisoners have numeracy skills at or below what would be expected of an eleven-year-old.

While correlation does not mean causation, the links between poor numeracy and life chances suggest that numeracy has a major impact on quality of life for many people. Building numeracy is therefore a priority for the quality of life of both individuals and the nation.

TEACHING TIP

'Mum gives you $10 to buy a Mars Bar. If the Mars Bar costs $3.45, how much change would you get?'

How would you solve this? What is the process you would expect to follow in solving it at school? Is that process different from or the same as the one you would use outside the school context?

School mathematics	Numeracy

Defining numeracy

The terms 'numeracy' and 'mathematics' are often used interchangeably; however, they actually refer to very different concepts. Members of the wider community often bemoan the 'poor numeracy levels' of contemporary youth, but what they are actually suggesting is an inability to calculate. In this view, numeracy is seen as being about arithmetic. Some curriculum developers view numeracy as being related to the study of number only. Many curriculum documents focus on a set of skills and dispositions that are envisaged as part of teaching mathematics. Many of these, particularly those in strands related to working mathematically, align strongly with the prevailing and broad definitions of numeracy. Typically, aims include topics such as the following:

- Develop a disposition to use mathematics to help understanding and solve problems.

- Develop a positive attitude towards mathematics and engage with mathematics.
- Recognise that mathematics enables people to explain and influence aspects of one's life.
- Develop a repertoire of strategies to support working mathematically.
- Apply mathematical ideas to everyday contexts—through processes of interpretation, application and analysis.

Numeracy as a basic skill

Numeracy has a long history in adult education (Yasakawa & Johnson 2018). Knowing that many people exit school with low levels of mathematics, which impacts employment, employability and adult life, many tertiary colleges offer courses in basic numeracy. The courses focus on school mathematics as well as applications to real-world contexts. The courses build mathematical understandings and confidence in the adults to enable them to use mathematics and apply it to their worlds.

Numeracy as the study of number

The approach taken in the United Kingdom sees numeracy as the study of number. Much of the numeracy research and policy coming out of the United Kingdom focuses on the study of number.

Numeracy as a life skill

Poor numeracy is a drain on the economy; it affects not only individuals but also employers and governments. In fact, it has been estimated that poor numeracy costs the UK economy £20.2 billion each year! (National Numeracy for Everyone for Life 2014) If a country wants to improve its standard of living, increasing the population's numeracy level is a recognised tool for improvement.

Numeracy in early-years mathematics

The work undertaken in the early years is seen as foundational to later, more important mathematics, and so numeracy is seen as the precursor to mathematics. The view in many preschool settings encompasses the strands of mathematics, bringing them into conceptions of numeracy (Chigeza & Sorin 2016); here, there is little differentiation between mathematics and numeracy. Since early-years settings are predominantly play-based, the formality of school mathematics has not been established, so the use of numeracy is appropriate.

Numeracy as applied mathematics

As early as 1997, the term 'numeracy' was appearing in mathematics education discourse where the Australian Association of Mathematics Teachers defined it as the knowledge and skills needed for informed participation and decision-making in the world beyond schools (Australian Association of Mathematics Teachers 1997). The application of mathematics and the disposition to use mathematics in non-mathematical settings appears to be the overriding feature of numeracy in contemporary education.

More recently, Goos and colleagues (2015) have promoted the term numeracy to mean the successful application of mathematics in non-mathematical contexts. Their approach to numeracy is to recognise four key elements: contexts, tools, dispositions and mathematical knowledge. Collectively, these elements make up what they define as numeracy.

Numeracy across the curriculum

Within the current education context, the need to recognise that mathematics is not solely the domain of the mathematics curriculum is embedded in curriculum reform (Goos, Dole & Geiger, 2012). Different curriculum areas—such as English, social studies, science, physical education, and health—all have inherent numeracy demands. Teachers need to build and apply numeracy in all curriculum areas, including mathematics. The strength of a 'numeracy across the curriculum' approach is that it enables students to see that mathematics is useful in other contexts, and that mathematics is a very useful and applied body of knowledge.

Consider social studies and the applied mathematics (numeracy) that is used in this curriculum area. Some of the numeracy skills needed in this area include those related to:

- population charts/tables/graphs (creating and interpreting graphs and other graphical information)
- maps (with longitude and latitude)
- number sense, to comprehend the numbers involved (e.g. what does the population of India or China look like in comparison to Australia's?)

- density of populations—across a diverse country such as Australia in comparison to other countries such as Nepal, China or Brazil
- proportion of people in a population based on certain characteristics—religion, geolocation, ethnicity, gender (these figures may be presented in graphs, charts, etc.)
- length—distances between locations; coastlines
- time—world times.

There are two different approaches that can be taken towards numeracy across the curriculum. The first is to understand how different disciplines represent their numeracies, which may be different from how mathematics does so; the second is to impose the mathematics onto the discipline.

TEACHING TIP

Why do I need mathematics?

Create a chart listing different curriculum areas and the numeracy that is used in that curriculum area.

Assessing numeracy across the curriculum

As with all assessment in mathematics, assessing numeracy across the curriculum is complex. Informed teacher judgement is a critical component of assessment of students' work, and should include a combination of a range of strategies including, but not limited to, observations, conferencing and student work samples. Since numeracy is an applied concept, evidence should be collected over time and across contexts. The different ways in which numeracy is enacted in the different curriculum areas may mean that students can show understandings of a concept, such as graphing, in one curriculum area but be less confident in another curriculum area. The teacher needs to make an informed decision about whether the student understands the concept or not. Using digital tools to help with assessment such as portfolios, photographs and the like can help to manage the wide range of data that may need to be collected and collated.

Numeracy is everyone's business

Numeracy is no longer the domain solely of mathematics and mathematics teachers. With many schools now opting to employ specialist mathematics teachers, including in the primary years, there is a need for all teachers to take responsibility for numeracy. Teaching numeracy as part of mathematics is limiting; numeracy needs to be taught at as many opportune moments as possible. The generalist primary teacher can use moments in other curriculum areas to reinforce or to apply mathematics to the content being covered. For example, a science lesson in which temperatures are being explored makes for a teachable moment about negative numbers. In a health lesson on heartbeat, rate can be explored with different heartbeats after exercising. In a social studies lesson, large numbers can be explored when talking about the length of time for which Australian Aboriginal people have lived in Australia, for example, scale models can be created of events that occurred during that 60 000-year period.

Numeracy, families and community

Knowing the community from which students come is an essential part of good teaching. Knowing the activities that they undertake on the weekends or after school can create experiences for numeracy learning and application. Examples include:

- sporting groups (such as football clubs) can be used to show the power of numeracy through the keeping of scores (in Australian Rules, a goal is worth 6 points; rugby league gives 4 points for a try, 2 for a conversion and 1 for a field goal)
- living in tropical areas, particularly in remote communities, can provide experiences relating to hunting, location, the seasons and weather
- many families own small businesses, meaning some students will bring related knowledge to school
- community festivals provide opportunities for experiences related to the festival and its supporting facilities (stalls, food vending, car parking, etc.)
- families can share their work, experiences and histories with the class, with the information shared being used for numeracy work.

Building numeracy knowledge in the prior-to-school years

Increasingly, governments have given greater recognition to the formative years prior to school and their impact on schooling. There is a concerted effort to close the educational gap between those families who have traditionally been excluded from success and those who experience success. This includes marginalised groups such as Indigenous students, students with a low socio-economic status, students for whom English is a second language, rural/remote students and some girls. It is estimated that if 'the gap' can be reduced, significant social, educational and financial gains can be made. Considerable resources have been invested into the years prior to school and the first few years of schooling (up to eight years of age) so that learners can have enhanced experiences of numeracy (and literacy), with the aim of reducing the gap between learners.

Working with families in early-years settings, and taking numeracy experiences into the home, have been positive steps in mathematics education; games and play-based materials with support work for the families have been a further step in this direction. A federally funded project, Families as First Teachers (FaFT), was implemented across remote Indigenous communities to build positive early learning experiences that could be undertaken by families outside of school settings. Teacher-run playgroups provided experiences for families to learn through play with their children. Many foundational numeracy concepts were learned, and these helped with the transition into the early years of education. Families and children benefited from this and other similar programs. (Further details about the FaFT program can be found at https://education.nt.gov.au/education/support-for-teachers/faft.)

Building teachers' knowledge of numeracy

Teachable moments arise incidentally during the course of a lesson or over a day or a week. There is a need for teachers to recognise the numeracy that is found across the curriculum and to exploit the relevant opportunities when they arise (Bennison 2015). It has been found that teachers often fail to see teachable moments of numeracy (Goos et al. 2015). Incidents arise throughout the course of the day that can provide a moment of numeracy teaching, and recognising the opportunity and then exploiting it is good teaching. The diversion may be brief but still capture the application of mathematics to a real context; or it may become a much larger project, if it attracts students' interest (and aligns with teaching goals). Incidental teaching can provide students with very rich learning experiences.

TEACHING IDEA

Mel recounts the story of her family going fishing on the weekend and her uncle making her throw her fish back. This moment could be used to talk about the size of fish, bag limits, diminishing fish populations, the breeding size of female fish, the cost of fish and the risks commercial fishermen take to catch fish.

Digital technologies and numeracy

As we indicated in this text's introduction, we have been incorporating the use of digital technologies throughout the text; therefore, since many of these technologies also support numeracy, this section will merely provide a brief overview of the role of digital technologies in relation to numeracy. We primarily reference the work of Australian researchers Geiger and colleagues (2015) and Geiger and colleagues (2016), but we will also include other national and international research where appropriate.

Recent research has highlighted a number of roles for digital technology in the development of numeracy, including through meaning making in daily contexts; reasoning and problem-solving; representing mathematical thinking; using digital tools in daily numeracy contexts; and interconnectedness with other digital literacy skills (Geiger et al. 2015). Other researchers have noted the role of digital technologies in the construction of models for understanding daily numeracy tasks, as well as in the provision of a scaffold for students to develop problem-solving strategies applicable to numeracy tasks (Villarreal et al. 2010).

One avenue researchers have followed in seeking to understand the role of digital technologies and numeracy is to position technology as an agent that provides opportunities for teachers to tailor tasks they set for students, organise their classrooms to take advantage of the technology, and modify their assessment to support numeracy development (Geiger et al. 2016).

Irrespective of the various approaches to the issue of digital technologies and numeracy, what is clear is that, given the increasing use of data in mathematics education (see Chapter 14), the capacity of teachers to use digital technologies thoughtfully in the development of numeracy is critical. Steen (2001) argues that citizens need to be quantitatively literate—that is, be able to think about common issues in a quantitative way—and this skill clearly requires familiarity with a range of digital technologies. We have outlined many of these technologies in other chapters of this text.

Numeracy and teacher standards

The teaching standards are now part of the practice of teaching. These are standards or benchmarks by which teachers are judged. Numeracy has been seen as a key standard for a teacher. The Australian Institute for Teaching and School Leadership (AITSL) has adopted a number of standards, one of which focuses on teachers' ability to teach numeracy (and literacy) across teaching areas: 'Know and understand literacy and numeracy teaching strategies and their application in teaching areas' (AITSL 2017).

Review questions

8.1 A parent comes to your classroom and says that his child needs to be given more pencil-and-paper work—'where they do lots of sums'. You want to convince him that making links to the world beyond school is more important than doing a lot of sums on paper. What points could you make to convince him?

8.2 What might you consider an effective way to promote critical numeracy, so that children can develop a critical orientation?

8.3 A colleague suggests to you that mathematics is much more important than numeracy. What might your response be?

8.4 Consider your current teaching practices across the curriculum areas. How might you incorporate numeracy into your teaching of all curriculum areas (including mathematics)?

Further reading

Geiger, V. (2016). Teachers as designers of effective numeracy tasks. In B. White, M. Chinnapan and S. Trenholm (eds), *Opening up mathematics education research. Proceedings of the 39th Annual Conference of the Mathematics Education Research Group of Australasia* (pp. 252–9). Adelaide: MERGA.

Geiger, V., Goos, M. & Dole, S. (2015). The role of digital technologies in numeracy teaching and learning. *International Journal of Science and Mathematics Education, 13*(5), 1115–37.

Jorgensen Zevenbergen, R. (2011). Young workers and their dispositions towards mathematics: Tensions of a mathematical habitus in the retail industry. *Educational Studies in Mathematics, 76*(1), 87–100.

Rosa, M. & Clark Orey, D. (2015). A trivium curriculum for mathematics based on literacy, matheracy, and technoracy: An ethnomathematics perspective. *ZDM: The International Journal on Mathematics Education Mathematics Education, 47*(4), 587–98.

Stanley, G. & Council of Australian Governments (2008). *National numeracy review report, May 2008.* Canberra: Council of Australian Governments.

Willis, S. (ed.). (1990). *Being numerate: What counts?* Hawthorn: Australian Council for Education Research (ACER).

References

Australian Association of Mathematics Teachers (AAMT). (1997). *Numeracy = Everyone's business.* Adelaide: AAMT.

Australian Institute for Teaching and School Leadership (AITSL). (2017). *We're here for teaching.* Retrieved from www.aitsl.edu.au/.

Bennison, A. (2015). Supporting teachers to embed numeracy across the curriculum: A sociocultural approach. *ZDM: The International Journal on Mathematics Education, 47*(4), 561–73. doi:10.1007/s11858-015-0706-3.

Chigeza, P. & Sorin, R. (2016). Kindergarten children demonstrating numeracy concepts through drawings and explanations: intentional teaching within play-based learning. *The Australian Journal of Teacher Education, 41*(5). doi:10.14221/ajte.2016v41n5.5.

Cockcroft, W.H. (1982). *Mathematics counts.* London: HMSO.

Crowther, G. (ed.) (1959). *The Crowther report (1959) 15 to 18: A report of the Central Advisory Council for Education (England).* London: Her Majesty's Stationery Office.

Geiger, V., Calder, N., Loong, E., Miller, J., Tan, H. & Larkin, K. (2016). Transformations of teaching and learning through digital technologies. In K. Makar (ed.), *Research in mathematics education in Australasia 2012–2015* (pp. 255–88). Netherlands: Springer.

Geiger, V., Goos, M. & Dole, S. (2015). The role of digital technologies in numeracy teaching and learning. *International Journal of Science and Mathematics Education, 13*(5), 1115–37. doi:10.1007/s10763-014-9530-4.

Goos, M., Dole, S. & Geiger, V. (2012). Numeracy across the curriculum. *Australian Mathematics Teacher, 68*(1), 3–7.

Goos, M., Geiger, V. & Bennison, A. (2015). Conceptualising and enacting numeracy across the curriculum. In K. Beswick, T. Muir and J. Wells (eds), *Proceedings of 39th Psychology of Mathematics Education Conference* (vol. 3, pp. 9–16). Hobart: PME.

Goos, M., Geiger, V., Dole, S., Forgasz, H. & Bennison, A. (2019). *Numeracy across the curriculum: Research-based strategies for enhancing teaching and learning*. Crows Nest: Allen & Unwin.

National Numeracy for Everyone for Life. (2014). Why is numeracy important? Retrieved from www.nationalnumeracy.org.uk/why-numeracy-important.

Steen, L. (2001). The case for quantitative literacy. In L. Steen (ed.), *Mathematics and democracy: The case for quantitative literacy* (pp. 1–22). Princeton: National Council on Education and the Disciplines.

Villarreal, M., Esteley, C. & Mina, M. (2010). Modeling empowered by information and communication technologies. *ZDM: The International Journal on Mathematics Education, 42*(3), 405–19.

Yasakawa, K. & Johnson, B. (2018). Numeracy: Negotiating the world through mathematics. In B. Atweh, H. Forgasz and B. Nebres (eds), *Sociocultural research in mathematics: An international perspective* (pp. 279–94) New York, NY: Routledge.

CHAPTER 9

EARLY NUMBER

Early number study has tended to take a sequential approach. Historically, in the first year of schooling, the focus was on the numbers 1–10, followed by exploration of the teens. Numbers 20–99 were studied through a place value approach. The simultaneous linking of linguistic, concrete and symbolic representation of the numbers studied underpinned all work in this approach. This sequence has been questioned because it may not capitalise on students' intuitive counting and number understandings. Research has continually highlighted the importance of developing number sense in the early years, and of building on the intuitive knowledge that students bring to school (Gervasoni & Perry 2013; Mason 2004; Perry & Dockett 2013). This chapter presents early number from a number sense perspective, using a contemporary approach to the teaching of this foundational topic.

Number sense

One of the aims of school mathematics is to develop students' number sense. Number sense relates to a level of comfort and familiarity with numbers—an 'at-homeness' with numbers. It is about understanding number meanings, knowing relationships between numbers, knowing the size of numbers and knowing the effects of operating on numbers

(Anghileri 2006). Number sense is never complete; developing number sense is a lifelong process that is promoted through many and varied experiences with using and applying numbers.

Before they start school, many young children exhibit sophisticated levels of number sense that often go unnoticed by teachers. Take the example of a preschool child attending a school-entry interview. The mathematics task presented to the child was to take a packet of twelve coloured pencils and make two groups. After much deliberation, the child separated the pencils into a group of seven and a group of five. The child was then prompted to make two equal groups. After further deliberation and much consternation, the child removed two pencils from the larger group and held them in her hand under the table. This was clearly not the expected response, and the child displayed a level of discomfort with her action. Upon questioning, her reasons for the indecision became apparent. The first separation of pencils was according to 'boy' colours (black, dark brown, dark blue, dark green, grey) and 'girl' colours (yellow, orange, pink, red, purple, light green, light blue). The second separation was the removal of light green and light blue from the 'girl' colours—but, because they were girl colours, they could not be included in the 'boy' colours group. The child's response shows the level of intuitive knowledge of division and the depth of cognition that this task invoked.

Becoming aware of the mathematics knowledge and number sense that young children bring with them when they commence formal schooling is vital to ensuring that students maintain an interest in, and eagerness towards, school mathematics. Researchers' investigations into early number development (e.g. Clarke et al. 2006; Gelman et al. 2009; Gervasoni & Perry 2013; Wright et al. 2006) have provided research-based evidence of the depth of young children's intuitive mathematics knowledge—that is, knowledge acquired without formal instruction.

Once in the school situation, the teacher's task becomes one of capitalising on the number knowledge that students bring to the learning situation. Teaching strategies include:

- focusing on counting activities
- using calculators to explore big numbers and patterns within numbers
- working flexibly with numbers
- verbalising students' thinking and solution strategies

- visualising and articulating patterns in numbers
- recognising group size without counting
- developing number benchmarks (fives and tens)
- learning basic facts
- relating the sizes of numbers to meaningful contexts
- creating an awareness of the effects of operating on numbers.

The priority for early number work is to build upon and promote number sense.

Pre-number

Many young children display an awareness of number: they know how old they are and how many siblings they have, they know their parents' ages, they know which television channel their favourite show is on and which floor their mother's office is on; some even know their telephone number. Many children can count beyond ten before they commence school. To promote understanding of number and the development of counting skills is to capitalise on children's early number knowledge and to provide further, explicit activities relating to the processes of classifying, patterning and subitising.

Classification

Being able to classify objects into groups is a prerequisite for number study. In order to be able to count objects in a group, students first need to identify the objects themselves. Consider the collection of shapes in the diagram.

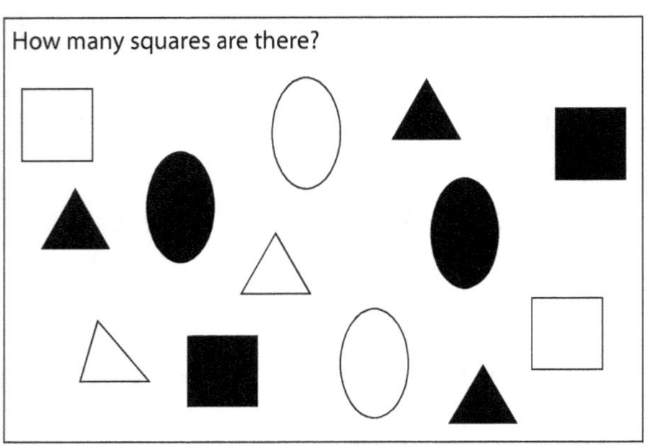

Asking students to count a particular item in a collection requires them to be able to identify the object to be counted. If they are unable to distinguish the circles from the squares, then counting the squares is not possible. Classification activities can arise from many collections of objects found in the classroom, the home, the playground and the environment. Buttons can be sorted according to colour, size, shape, texture, number of holes; toys can be sorted according to purpose, type, number of hands, colour of skin, and so on. Encourage creative thinking about classification of a collection. Classification skills relate not only to mathematics but also to other subjects, including science, art, music, social studies and environmental education.

Patterning

Patterning is a key aspect of mathematics. Providing students with quality experiences in patterning will mean they will be more likely to see patterns in problem situations, and be able to make generalisations about their patterns, which is the basis of algebraic thinking.

Patterning activities include the following:

- *copying patterns*—students are given a pattern and their task is to copy it. This can be a string of beads or an array of blocks, but the task for the students is to simply copy what is displayed.

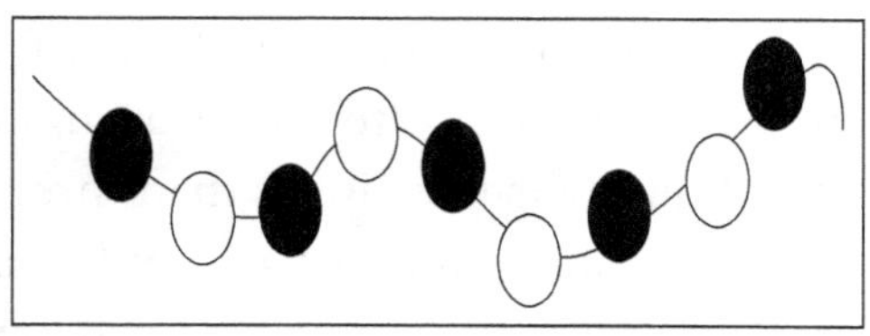

- *identifying the next item*—from a given pattern, students are asked to predict the next item(s).
- *extending patterns*—students are asked to extend the full pattern.
- *creating patterns*—this is creative pattern work where students construct their own patterns.
- *determining the missing element in a pattern*—a pattern is given, but one of the elements in the pattern is omitted. The student's task is to identify what is missing.
- *translating a pattern*—this is creative work where patterns created in one form (e.g. shapes) are represented in another form (e.g. movement). So, for example, triangle, square, circle becomes clap, skip, jump. This translation will be important in later years when students learn coding.

Patterns do not need to be limited to beads, blocks and counters; they can be created by using students themselves (one child standing, one child sitting, one child standing . . .). Patterns are also found in fabric (e.g. curtains, sheets, shirts, bedcovers). Encourage students to see patterns around them, so as to experience patterns in a variety of forms, before asking them to create their own. Students will also learn the difference between an everyday pattern and a mathematical pattern (with elements that repeat in a regular fashion). An app called Pattern Shapes (https://apps.mathlearningcenter.org/pattern-shapes/) can be a useful addition to a teacher's repertoire when teaching patterns.

Subitising: group recognition

Before they begin formal schooling, many young children have the capacity to subitise. Subitising is being able to identify the number in a collection or group without counting. This skill is very accurate for collections that number between one and five items. The skill of subitising is essential for the development of concepts of number (von Glasersfeld 1982).

Subitising is a useful skill for early number work because it saves time in counting and is often more accurate; it promotes number sense of the size of numbers (two is less than three is less than four), and assists simple mental addition and subtraction. It is important because it de-emphasises a counting-by-ones strategy for determining the number of objects in a collection. Instead, emphasis is on the automatic recognition of the number of items in the collection so as to free up cognitive space. This expedites subsequent work with operations and promotes flexible thinking about numbers and number combinations. Teachers of early-years classes need to determine the extent of their young students' subitising skills, and build upon them or promote their development.

Flashing small collections of counters on an overhead projector using flash cards, or showing small collections on boards, is useful for this purpose. The arrangement of collections for subitising must be considered in such activities. The app TouchCounts (www.touchcounts.ca) is a useful tool in assisting students to learn to subitise. Students use their fingers and an interactive screen to create collections. Later, these collections can be grouped and split to create larger and smaller collections for the students to subitise.

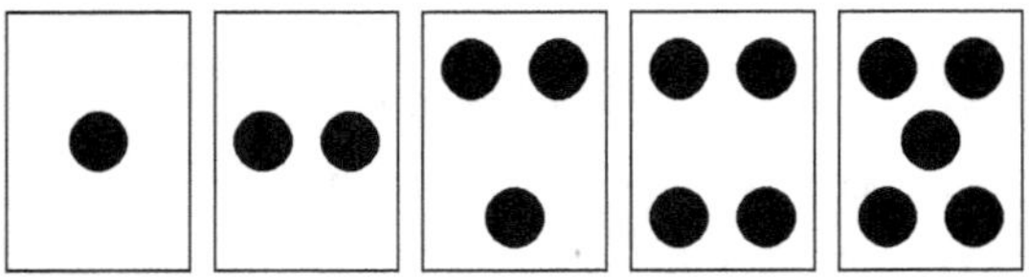

Many students rely on counting-by-ones strategies for addition and subtraction. Subitising reduces the need to rely on such strategies. For example, when students are asked, 'How many?' when presented with a group of three and a second group of four objects, through subitising they automatically recognise the collections of three and four—without counting—then mentally combine the groups to give the total.

Counting

Contemporary approaches to teaching early number recognise and emphasise the intuitive counting knowledge students bring to school. Students are often familiar with numbers beyond 10. Old approaches often saw the teen numbers as problematic, but contemporary research suggests that, through experience, young students know the teen numbers intuitively—often because of family members' birthdays and similar situations. Through playing with the language of numbers, students develop an aural sense of number names. Experimenting with number pattern names, students often count in tens—'sixty, seventy, eighty, ninety, tenty, eleventy'.

Difficulties often emerge with the study of numbers when formal study of place value is introduced. Recent approaches to teaching number shift the emphasis from place value to the intuitive counting strategies students bring to school and develop as they progress through school.

Many young children can 'count' to ten before they start school, but what they are really doing is merely repeating a sequence of words they have learned, just as if that sequence were a poem or a nursery rhyme. Counting is not just being able to say number names in order. To distinguish between, on the one hand, merely saying the number word sequence and, on the other hand, using counting in a problem-based situation, Wright and colleagues (2006) use the term 'counting' to refer to situations in which each number word is coordinated with an item. The items might be seen or might be visualised (i.e. imagined).

An important goal for the teacher is to determine the sophistication of a student's counting strategy beyond oral recitation of number names.

Counting stages

In the literature, there are two main approaches to describing children's development of counting strategies. The approach outlined by Steffe and Cobb (2012) has been applied extensively to classroom teaching, and also in intervention programs (Allen & Wright 2014), and is summarised in the following table.

Model for stages of early arithmetic learning

Stage 0: Emergent counting	Cannot count visible items. The child either does not know the number word sequence or cannot coordinate the number words with items.
Stage 1: Perceptual counting	Can count perceived items but not items that are screened (i.e. concealed). This might involve seeing, hearing or feeling items.
Stage 2: Figurative counting	Can count the items in a screen collection, but counting typically includes what adults might regard as reductant activity. For example, when presented with two screen collections, told how many in each collection, and asked how many counters in all, the child will count from 'one' instead of counting on.
Stage 3: Initial number sequence	Child used counting on rather than counting from one to solve addition or missing addend tasks (i.e. $6 + \times = 9$). The child may use a count-down-from strategy to solve removed items tasks (i.e. 17 – 3 as 16, 15, 14—the answer is 14) but not count-down-to strategies to solve missing subtrahend tasks (e.g. 17–14 as 16, 15, 14—the answer is 3).
Stage 4: Intermediate number sequence	The child counts-down-to, to solve missing subtrahend tasks (e.g. 17 – 14 as 16, 15, 14—the answer is 3). The child can choose the more efficient of the count-down-from and count-down-to strategies.
Stage 5: Facile number sequence	The child uses a range of what are referred to as non-count-by-ones strategies. These strategies involve procedures other than counting by ones, but may also involve some counting by ones. Thus, in additive and subtractive situations, the child uses strategies such as compensation, using a known result, adding to 10, commutativity, subtraction as the inverse of addition, aware of the '10' in a teen number.

Source: Wright et al. 2000, p. 26

In contrast, many curriculum documents identify a less in-depth approach, and nominate three distinct stages of counting that can be observed among young students. Each stage represents progressively more refined counting skills.

Rote counting

This is when students can say a string of number names, and this string is very stable. The string may include all numbers from 1 to 10 in the correct sequence, or some numbers in the sequence may be omitted (e.g. 1, 2, 3, 5, 7, 9, 10). Students at this stage appear to be counting, but they are not able to come to the correct amount of 'how many' objects are in a collection.

Point counting

At this stage, when given a collection of objects to count, students will make gestures of counting by touching objects in the collection, but they cannot tell how many objects there are. For a collection of five objects, students may count to 8 because they assign two numbers ('1, 2') when they point to the first object, and two more ('3, 4') when they point to the second object. At this stage, when asked again to state how many, students characteristically begin again at 1, stating there are '1, 2, 3, 4, 5, 6, 7, 8' objects.

Rational counting

Students are deemed to be rational counters when they demonstrate understanding of the following counting principles, as identified by Fuson (1988):

- Each item that is to be counted is assigned one (and only one) number name.
- The order in which items are counted must be undertaken in a specific and appropriate number order. That is, items must be counted in the order of 1, 2, 3, 4, 5, 6 . . .
- The order in which items are counted is irrelevant—it is possible to start at either end of a collection.
- The last number counted represents the number of items in a collection.
- Some things cannot be counted—e.g. water in a glass, air in a balloon, etc.

While not a curriculum document, the Early Numeracy Research Project (ENRP) (Clarke et al. 2000) identified six counting growth points, as follows:

1. rote counting
2. counting collections
3. counting by 1s (forward/backwards, including variable starting points; before/after)
4. counting from 0 by 2s, 5s and 10s
5. counting from x (x >0) by 2s, 5s and 10s
6. extending and applying counting skills.

One-to-one correspondence

The term *one-to-one correspondence* is often used to signal a student's progress in counting. One-to-one correspondence is evident when a student points to or touches an object and says a number. They may not be able to name the correct total number of objects in the collection, and they may use the incorrect sequence when 'counting'. But they will point to each object in turn and give it a different number name. This is a milestone in learning to count, in that the student knows that counting means that each object must be included and that when each object has been included, the final number tells how many. One-to-one correspondence is associated with the beginnings of rational counting, even though a student may not yet be demonstrating all features of rational counting. Some typical errors are when the student uses an incorrect string of numbers for the count but nevertheless allocates one number to each object (e.g., one, two, four, five). Another example is when the student uses the correct string for four objects (one, two, three, four) but then states a different number for the total (e.g., nineteen).

Conservation of number

The influence of Piaget on mathematics education was described in Chapter 2. Piaget's work on number provided us with the term *conservation* in relation to mathematics. From his research, Piaget suggested that the ability to see that the number of objects in a collection remains the same, regardless of how it is arranged and rearranged, depends on the level of cognitive development of the child.

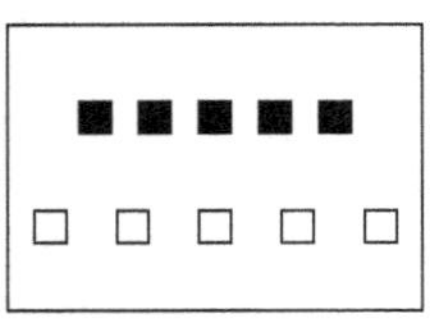

In the diagram, students may think that the bottom row holds more blocks than the top row. The inability to see that the number remains constant in spite of its spatial arrangement indicates that students are focusing on attributes other than 'how many' objects are in the collection. When students can recognise that the number of objects remains constant regardless of how the objects are organised, they are said to be able to *conserve* number. The implication of this is not that teachers should delay all counting activities until students can conserve number, but rather that they should provide students with many and varied counting activities in which collections are rearranged in a variety of patterns and arrangements, and encourage students to verbalise their thinking and mental strategies. Investigating the extent to which students can conserve number provides valuable information for building an early number program full of variety and with plenty of rich experiences and language.

Counting strategies

When students are rational counters, they need to be immersed in activities that promote the development of more effective and efficient strategies for counting. Three main strategies are counting on, counting back and skip counting. The effective use of the counting strategies is a precursor to the development of the four operations (addition, subtraction, multiplication and division), depending on the counting strategy. Students need to be immersed in a range of activities in which counting strategies can be applied so that they will develop flexible thinking when working with numbers, as well as utilise number sense.

Counting on

Counting on is where students begin with a given number and count on from this. In order to use this skill, students need to be able to 'break' the counting number string at any point by being able to determine what number comes after, or what number comes before, any given number. Being able to break the string means that the learner does not have to start counting at 1 to determine the next number forward from any given number. This is an important skill to apply to addition and subtraction problems.

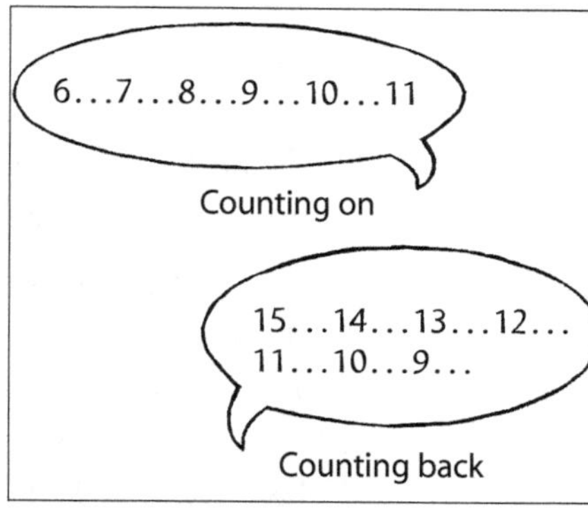

Two counting-on strategies are useful for simple addition:

1. *Counting-up-from* is where two numbers are given, the student identifies the larger amount and counts up the smaller amount starting from the larger number—e.g. 3 + 8: 'Begin with 8, and then count 9, 10, 11. So the result is 11.' Note that the student will also notice that 3 + 8 results in the same total as 8 + 3.
2. *Counting-up-to* is where the starting point and the end point are given, and the task is to find out the difference by keeping track of how many counts are made. This often involves finger counting, and this should be encouraged. For example, '8 and what is 11?' 'Begin with 8 and then count to 11; so that is 9 . . . 10 . . . 11. I counted three numbers, so the answer is 3.' In this type of problem, one of the numbers (or one part of the amount) is given, as well as the total amount. The task is to find out the other number (or the other part of the amount) to equal the total. As the amounts (the two parts) are given and are added, they are referred to as *addends*. In this particular problem, only one addend is given, so this is referred to as a missing addend problem.

Counting back

Counting back is the reverse of counting on. Students often experience more difficulty with this aspect of counting than with counting on. It is of similar difficulty for young students as it is for adults when they attempt to say the alphabet backwards. Counting back requires being able to identify 'what comes before' any given number. It is important for the development of the subtraction process.

Two counting back strategies are useful for simple subtraction:

1. *Counting-back-from* is where two numbers are given and the student counts back the smaller amount beginning with the larger number—e.g. 11 take away 3. 'Begin with 11, and then count back 10, 9, 8. So the result is 8.'
2. *Counting-back-to* is where the starting point and the end point are given, and the task is to find out the difference by keeping track of how many counts are made. This often involves finger counting, and this should be encouraged. For example, 'I started with 11 and there are now 8, so how many were removed?' 'Begin with 11 and then count back to 8, so that is 10 . . . 9 . . . 8. I counted three numbers, so the answer is 3.' Like the counting-on problems, where the unknown quantity is within the problem, this type of subtractive task is known as a missing subtrahend problem.

Skip counting

Skip counting is counting in multiples of any number apart from 1. Skip counting is promoted because students listen to the patterns and observe the rhythm of counting in various multiples. For example, counting in tens is a simple pattern, since each count after 10 ends in '-ty'. For fives, the 5 and '-ty' endings are heard in an AB pattern. After students have skip counted orally, record the numbers in the skip for students to explore. They will see how the end digits in the number start to take on a repeating pattern. For example, when counting in fives, they will note that each number ends in either a 5 or a 0. When they skip count in fives starting at, for example, 2, they will see that the repeating pattern is 2, 7, 2, 7. This helps number sense and understanding of divisibility rules. Skip counting forwards and backwards represents an important learning experience for students; it is the precursor to multiplication and division. For example, to find out what 6 groups of 5 totals, students can skip count in fives—5, 10, 15, 20, 25, 30. Similarly, to find out how many groups of 2 there are in 10, students would begin at 2 and skip count to 10, keeping track of how many counts they make.

Opportunities for practising counting

Students should be given plenty of opportunities to practise counting. Where possible, links to the world beyond school should be encouraged. Ask students whether they watch the counter at the petrol bowser when they get petrol with their parents, and what patterns they notice. Ask them whether they watch kilometre markers on the sides of roads when they are travelling on long trips; whether they take the stairs two at a time, and whether they count in twos while doing it. Ask them to think of devices that count in the real world. What patterns do they notice on a calendar?

Students should also be given opportunities to practise counting (particularly skip counting) using technology such as calculators or apps. Many calculators have an inbuilt constant key, so that when '2 + 6' is entered and the 5 symbol is then pressed, the answer will appear. Pushing the 5 symbol will keep adding 6 to the last figure. This allows students to see skip counting in action, and to experience larger numbers than would be possible if they were to undertake the skip counting themselves.

Number benchmarks

The capacity to be flexible with numbers means knowing how numbers work, how they are related to other numbers, and how they can be rearranged and regrouped. Such knowledge supports mental computation and is a major component of number sense. Being able to relate numbers to other numbers is about building number benchmarks. Common number benchmarks are 5 and 10 (and, later, multiples and powers of 10). When numbers become benchmarks, they provide a reference point to use for calculation or conceptualisation purposes. For example, to calculate 9 + 4, a common strategy is to add 1 to make 10, and then add 3. Because 9 is one less than 10, the strategy of adding 1 to make 10 and reducing the 4 to 3 is applying flexible thinking around 10. Knowing the combinations that make 10 (1 + 9, 1 + 1 + 8, 2 + 1 + 7, etc.) is also using 10 as a benchmark, particularly when adding two-digit numbers. For example, when adding 38 and 74, the first strategy might be to recognise that 3 and 7 is a combination that makes 10, even though, in this example, these two numbers are 30 and 70 and the total is 100. Then children can add 8 + 2 + 2 for a total of 112. Five is a benchmark, when seen as half of 10 (and, later, 50 as half of 100), but supporting children to see the number combinations that make 5 (1 + 4, 2 + 3) helps to promote 5 as a benchmark in its own right that can be used as a reference point in other situations.

To build familiarity with 5 and 10 as benchmarks, and hence to promote number sense, teachers need to provide students with a range of materials and activities that allow students to experience the size of numbers and to make connections between numbers. Teaching strategies need to be enacted that will enable students to develop these benchmarks to facilitate more effective counting and operation strategies. Encouraging children to count on their fingers, and hence to develop finger patterns, takes account of the strategies and knowledge students bring to school (Fuson 1988).

Subitising—the capacity to recognise the number of objects within a collection without counting—can be facilitated through the use of counters, where a collection of counters is 'flashed' before children and they then need to state how many counters are in the collection. Different arrangements of counters totalling 5 will build 5 as a benchmark. Educators recognise the value of counters to support the development of early number knowledge (Steffe & Cobb 2012). Reforms in the Netherlands also encourage the use of abacuses and strings of beads (Gravemeijer 1994).

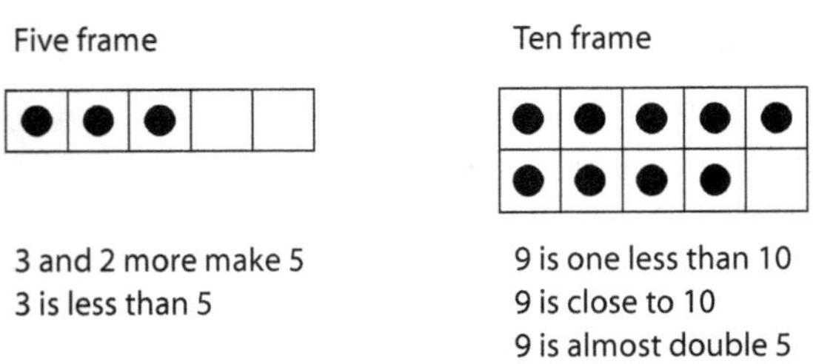

Reinforcement of our base 10 numeration system can occur through promoting 5 and 10 as benchmarks, and this is supported through the use of *five frames* and *ten frames*. A five frame presents five counters in a line, with one counter placed in each cell of the five frame. Students use the five frame to show the combinations to 5 using two different coloured counters (5 + 0, 4 + 1, 3 + 2, 2 + 3, 1 + 4, 0 + 5). A ten frame is a double five frame, and is used in a similar way to show combinations to 10. The ten frame has the added feature of being able to show that 10 is 2 groups of 5, or 5 groups of 2 (if displayed vertically). It also shows clearly that 5 is half of 10. The frames allow students to develop spatial patterns for numbers.

More and less

When exploring numbers and relationships between numbers, the words 'more' and 'less' may be used often (e.g. 'Is 6 more or less than

7?'). Increasingly, research indicates that the use of these terms in early number work can create difficulties for some students because of their unfamiliarity with the term 'less'. Rarely are children asked whether they want 'less' of anything; rather, they are asked, 'Do you want more ice-cream?' or 'Do you want more toast?' Teachers need to spend some extra time when undertaking early number work that has a language component to ensure that all students have equal access to the intended learning.

Types of numbers

There are three different types of numbers that students will encounter in primary mathematics: cardinal, ordinal and nominal. It is not necessary for students to learn these names, but they need to be exposed to the three different types in their study of number:

- *Cardinal numbers* refer to the amount. 'Cardinality' refers to finding 'how many'. Through early counting activities, students are immersed in cardinal numbers.
- *Ordinal numbers* refer to the consideration of 'where or which one'. 'Ordinality' refers to a position, or the order of an object in a series—for example, the first car in the race, the third house in the street or the fifth rung on a ladder. Often, the position relates to a cardinal number, and it can relate to equal quantities separating items (such as rungs on a ladder) or unequal quantities between positions (such as a race). This can be an issue when students are using number lines and/or rulers, and the misuse of number lines and rulers shows this misconception. For example, some students will use their ruler to add numbers (e.g. 5 and 7). Due to poor number sense, the strategy used will be a counting-on strategy, which will be misapplied. The student will locate the number 5 on the ruler/number line, and count on from that point, starting with the number 5. By the time the student counts on seven numbers, he/she is at number 11. The student is counting the line markers in the ruler and not taking into account that each distance between the line marker is the actual value of the number. The student is misapplying a rote counting strategy to a concrete representation that they do not fully understand. Students often experience difficulty with this aspect of number study, so experiences need to be organised

to facilitate appropriate understandings to develop. Students often come to school with early ordinal numbers (first, second and third). A further difficulty with ordinal language is that many of the words used for ordinal numbers are also used for fractional concepts (e.g. third, fifth, sixth), so the distinction needs to be taught explicitly to students in order to avoid confusion.

- *Nominal numbers* refer to number names. Some objects—such as racing cars, postcodes and telephone numbers—have numbers as their names. These types of numbers are useful for identifying important information but do not have the same properties as cardinal and ordinal numbers. For example, a baseball player may wear a shirt with the number 18 on it, but this does not mean that she is 18 years of age, or that she is the eighteenth person in the team. Instead, this number is for identification purposes.

Students need to have a range of experiences with these types of numbers. Early work with number is typically done through cardinal numbers, so students operate on such numbers. Students need to realise that they cannot undertake operations with ordinal or nominal numbers. For example, it is not possible to add two postcodes, or two places in a race, to arrive at an answer.

Learning about numerals

Through a range of experiences with counting, students come to see the symbolic form of a number. This is referred to as the *numeral*. As they count beyond nine, students learn the patterns for combining numerals to make representations of larger numbers. Through relating the number name for the items counted and the symbolic representation, students learn intuitively about place value. There is some debate about how place value is best taught—whether as a formal process or when learned intuitively. What is perhaps more important is that teachers know of different approaches so that when one approach is not successful, they will have other ideas of how to approach the concept.

Writing numerals

As students come to develop a sense of number and cardinality, they also need guidance for writing numerals. There are ten numerals that make up our base 10 numeration system. Depending on location, there are some particular ways of writing the numerals. In most countries with an Anglo-Saxon background, a similar format tends to be used (see the following diagram). However, with increasing globalisation, recognition of European numerals also needs to be given. For example, in most European countries, the numeral 7 is written with a bar through the upright, and 1 is written with a lead-in 'tail'.

1 2 3 4 5 6 7 8 9 0

Explicit instruction must be given to ensure that students learn to write the numerals correctly, as displayed above, and, since symbol formation is a part of handwriting, class handwriting time should include a focus on correct numeral formation. As a result, students will value the importance of correct writing of numerals and see this as a key component of good handwriting. This will ensure that mathematics time focuses on conceptual development rather than symbol development.

Variety in exercises for promoting good symbol development is important to ensure students' interest and concentration are maintained. Suggested experiences include:

- tracing over numerals already written on paper
- drawing on a partner's back and then asking the partner to identify the numeral
- placing the numeral on an overhead projector and projecting its image on a wall for students to trace
- cutting numerals out of sandpaper for students to trace with their fingers
- having students trace numerals on an overhead transparency with a water-soluble pen
- making class numerals
- making big numerals on the floor using rope or other thick cord.

Students should be encouraged to talk through the writing process. For example, to make a 7, 'Start at the top, then you go across the top and then you go on a diagonal across to the bottom on the other side.' As students gain confidence in counting and writing numerals, they need to have experiences of counting on from the numbers given, and writing them, as well as inserting numerals into a sequence where numerals are missing. This is often more difficult than counting on, so support needs to be offered to students when they undertake this aspect of writing their work. A digital resource for early number recognition and writing of symbols is the ABC app Eggy Numbers (https://readingeggs.com.au/apps/eggy-numbers/), which offers early learning opportunities for the numerals 1–10.

Numeration and place value

Numeration

Developing numeration is developing students' understanding of the Hindu-Arabic number system, which is the primary system of numbers in our society. In some curricula, children also learn how to count in base 2 (binary), the counting system that underpins computing. The Hindu-Arabic system is based on five principles. A student would be expected to appreciate all five to understand numeration:

- *Base*—The system is a base 10 system, and uses exactly ten digits (0, 1, 2, 3, 4, 5, 6, 7, 8, 9). These ten digits enable us to create any number.
- *Position*—The position of a digit in a number dictates its value, and every place in the system is related to other places by powers of 10. To understand the meaning of a number, students use the position of the digit. For example, in the number 435, the symbol 4 means that there are four hundreds, since it is in the hundreds place, or position. Similarly, the 3 is in the tens position and the 5 is in the ones position.
- *Multiplicative*—The value of each digit is multiplied by the value of its place (in 435 the value of the 4 is 4×100, the value of the 3 is 3×10, and value of the 5 is 5×1).
- *Additive*—Digits in a number are added to give the total amount (435 is $400 + 30 + 5$).
- *Odometer*—The Hindu-Arabic system is a counting system, in that every place counts. Thinking of how an odometer on a vehicle works helps with this notion. On each reel on an odometer are the digits 0 to 9. On each count, the reels turn—the ones reel turns at every count, the tens reel on every tenth count, the hundreds reel on every

hundredth count, and so on. This principle assists in counting and mental computation.

Developing place value

The Hindu-Arabic system is a place value system. The development of place value understanding begins through exploration and recording of two-digit numbers. To promote number sense beyond 10, students need to be immersed in activities to experience the size of numbers and to develop the principles of numeration listed above.

Bundling icy-pole sticks into groups of ten with a rubber band and holding ten bundles in the hand is an excellent way to feel the size of 100. When students say, 'I'd never be allowed to eat this many icy-poles', they are demonstrating a strong sense of number in relation to 100. Bundles of ten sticks also provide a physical referent when counting in tens. For numbers greater than 100, ten bundles of ten sticks can be fastened with a rubber band to further promote a sense of the size of numbers. The icy-pole sticks are a 'grouping' material, in that they can be grouped and ungrouped with the use of a rubber band. Students can use this material to explore numbers between 10 and 100, demonstrating how they can be represented as tens and ones. For example, 35 is made up of three groups of ten sticks and five loose sticks. By unbundling one of the groups of ten, students can see that 35 is also made up of two groups of ten sticks and 15 loose sticks. Unbundling another group of ten shows that 35 is the same as 10 and 25.

In the photograph, a range of materials has been used to represent 35. All materials are proportional, since the base 10 longs are equivalent to ten ones, a stick of Unifix blocks is the same as ten single blocks and a bundle of sticks is the same as ten single sticks. Representing place value in these forms encourages students to see the relationship between the representation and the materials.

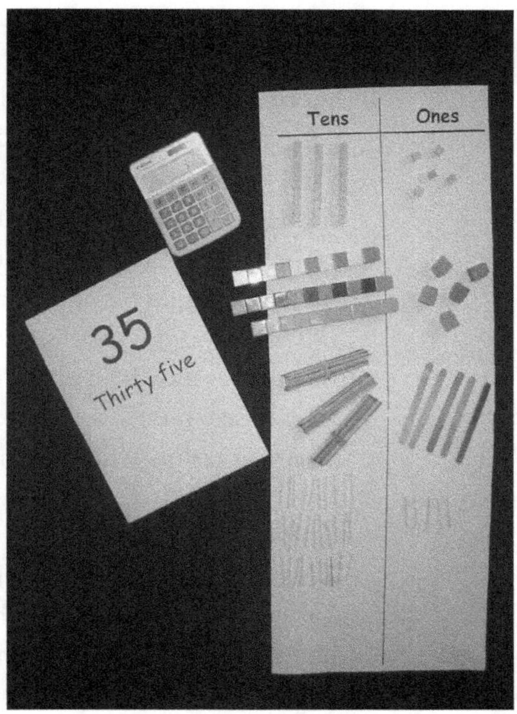

Ways to represent 35 with various materials

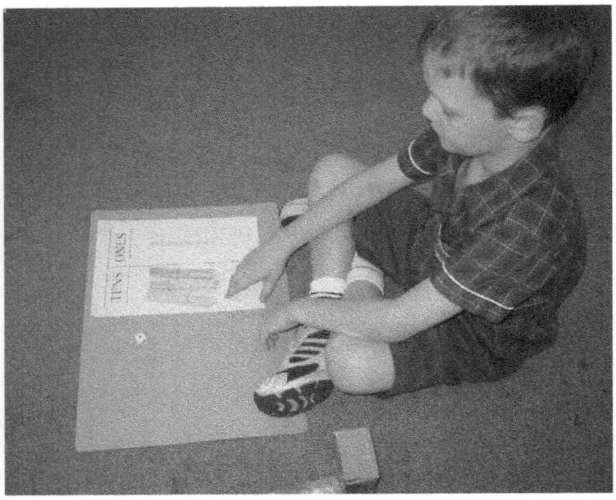

Using MAB to represent two-digit numbers

Another common material for developing numeration and place value is through base 10 blocks. These blocks are also referred to as MAB or Dienes blocks (Zoltan Dienes created the blocks in the 1960s). The letters MAB stand for 'multibase arithmetic blocks', because they were created in a series of bases and not just base 10. However, MAB other than in base 10 are uncommon in schools, and when teachers refer to MAB, they are usually referring to the base 10 MAB blocks. A set of base 10 MAB consists of ones, tens, hundreds and thousands blocks that can be used to represent numbers. For example, 453 would be represented with four hundreds blocks, five tens blocks and three ones blocks. MAB cannot be pulled apart or reassembled, and so should be used later in the teaching sequence.

A number of quality apps can be used for developing place value understanding. One we recommend is Place Value Chart (https://apps.apple.com/au/app/place-value-chart/id568750442). This app caters for early experiences with place value as students can use their fingers to place counters on a place value chart. Once ten dots have been placed in one column, they are automatically grouped and replaced with one counter in the column to the left. If one counter is moved to a column to the right, it splits into ten counters. If it is moved two columns, it splits into 100 counters. As students build an understanding of place value, the app extends to include decimal places. It can also be set to work in other bases. Base 2 will become important when children begin to learn binary notation when coding.

Symbolising numbers

When numbers are represented with concrete materials, the placement of objects may not reflect the way the digits are recorded. For example, the number 42 can be represented with base 10 blocks by arranging four tens and two ones blocks on the desk in any order. To relate the value of the blocks to the placement of digits in symbolic form, a place value chart is useful to organise the material. Relate the digits in the symbolic recording to the concrete representation.

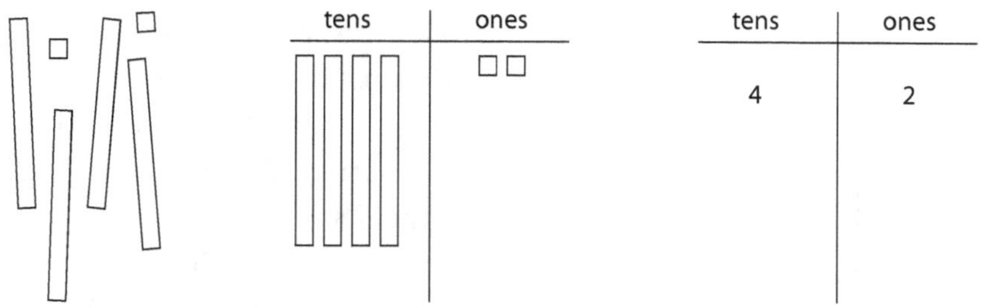

Grouping and regrouping

To emphasise relationships between numbers and to build place value understanding, students need a lot of experience with grouping and regrouping numbers. Using bundling sticks and place value charts, students can explore how numbers can be grouped and regrouped, and thus develop flexibility in thinking about numbers. With concrete materials, the number 34, for example, can be seen as consisting of three tens and four ones; with regrouping, it can also be seen as two tens and fourteen ones, and also as 34 ones. Flexibility with numbers lays the foundation for mental computation, and for understanding the written procedures for addition and subtraction.

Being able to regroup numbers into various component parts is evidence of good number sense. The process of swapping or trading ten items in a collection for one of the next item (e.g. MAB ten ones for one ten) to develop place value knowledge has been used for many decades with differing levels of success. Since the mid-1980s, researchers (e.g. Cobb 1991; Treffers 1991) have questioned the effectiveness of using base 10 blocks to teach place value.

The empty number line

An alternative approach to developing children's place value knowledge is Realistic Mathematics Education (RME), from the Netherlands. In this approach, the focus is on using a number line—not expressly for developing place value, but rather for developing mental computation around addition and subtraction, and using children's counting and number knowledge to support working flexibly with numbers. Through this approach, students' place value knowledge has been seen to develop without explicit teaching (Gravemeijer 1994).

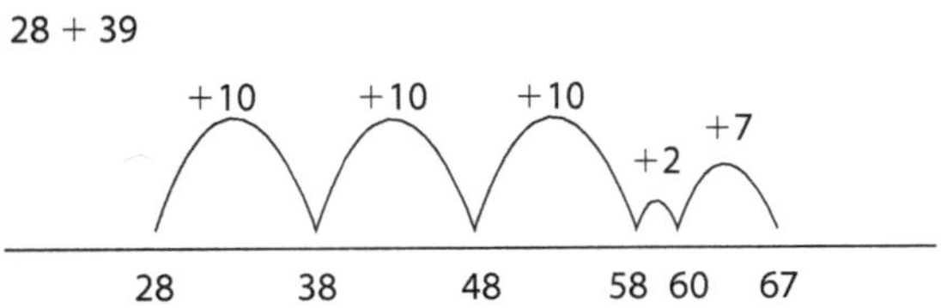

Example of jump strategy

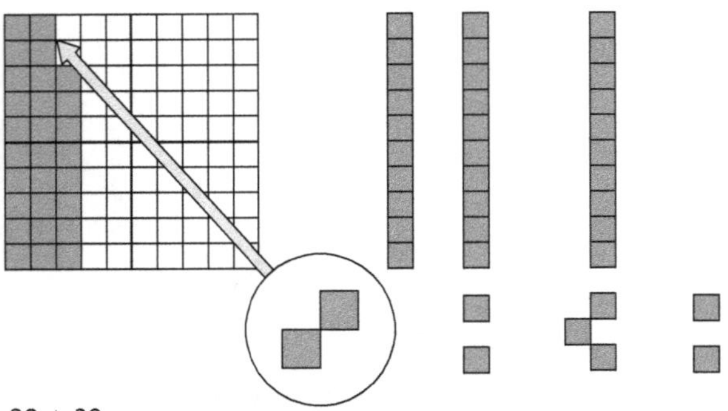

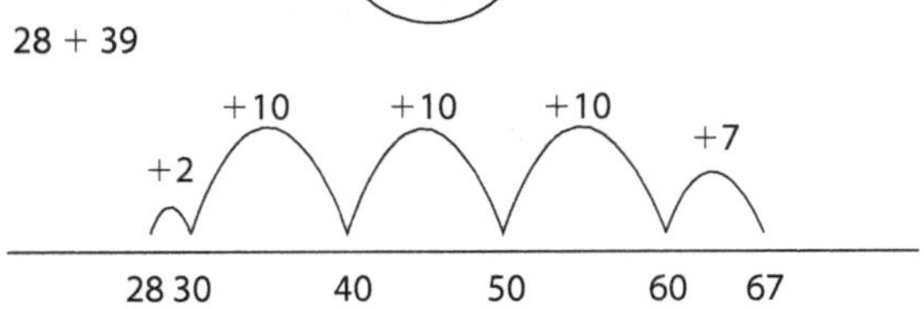

Example of split strategy

Two main strategies for working with the empty number line are the jump and the split strategies. The *jump strategy* is where a two-digit addition problem is given and the empty number line is used to track the situation. For example, in 28 + 39, the 28 is written on the empty number line and 'jumping' in tens occurs three times to arrive at 58 (which means that 30 has been added to 28). To add the remaining 9, there is a jump from 58 to 60, followed by a jump of 7, to arrive at the solution of 67. A *split strategy* is where the number is broken into numbers that fit with the other number so as to add to 10. Using a split strategy for 28 + 39 means that the 39 is split into 2 + 30 + 7, and this is the way it is represented on the number line. The split strategy encourages students to work flexibly with numbers by splitting them in ways that make them more usable, depending on the context of use. Combinations of jump strategies and split strategies can be used.

Another strategy is that of *compensation*, where the number may be rounded up and the rounding up (or down) factor then adjusted. For example, in the case here, the student may recognise that 39 is 40 – 1, and so add (using jump, split or combination strategies) to the 28 (28 + 40), and then take the 1 away so that it is 68 – 1. Through these strategies, students come to see the effectiveness of the tens and ones. They use 10 as a number benchmark, and they come to experience numeration informally, so that the introduction of formal place value study builds on their intuitive understandings. The Netherlands reforms have drawn heavily on the intuitive methods of mental computation to show that, in the world beyond school, this is the predominant mode of thinking when adding (or working with) numbers.

Base 10

An important aspect of understanding place value is knowing the relationship between each place in the Hindu-Arabic system. Each place is related to the place next to it by a power of 10. Explorations with a calculator can assist comprehension. Enter a number on a calculator, then progressively multiply this number by 10. Watch as the digits move progressively across each place, with zeros appearing to fill the gaps where the digits previously were located:

$24 \times 10 = 240$
$24 \times 100 = 2400$
$24 \times 1000 = 24000$
$24 \times 10000 = 240000$

This knowledge is much more powerful than the blindly applied strategy of 'adding zeros'. By progressively dividing by 10, the number becomes less, and the digits move back to their original position (or even become decimal numbers). Base 10 understanding assists in mental computation and understanding of the process of multiplication and division. (See also Chapter 10 on written algorithms.)

The Number Framework

In New Zealand, a Number Framework (Ministry of Education, New Zealand 2008) was developed as a result of extensive research and professional development between 2002 and 2007. The Number Framework provides a summary of knowledge and strategies required by children to operate successfully in the number strand. The framework emphasises the importance and interconnectedness of knowledge and strategies. New number knowledge is a result of using advanced strategies, but strong knowledge is essential for broadening a student's strategy repertoire. The framework presents nine global stages, further categorised as relating to the counting stage or the part-whole stage. The counting stages are:

- emergent
- one-to-one counting
- counting from one on materials
- counting from one by imaging
- advanced counting.

The part-whole stages are:

- early additive part-whole
- advanced additive part-whole
- advanced multiplicative part-whole
- advanced proportional part-whole.

The global stages subcategorised as counting and part-whole reflect children's ever-increasing knowledge and understanding of number. The part-whole stages are signified by a student's capacity to 'recognise numbers as abstract units that can be treated simultaneously as wholes or can be partitioned and recombined' (Ministry of Education, New Zealand 2008, p. 4). Number sense and thinking flexibly about numbers, as outlined in the preceding sections of this chapter, align children's developmental movement between the counting and part-whole stages in the New Zealand framework. The part-whole stages reflect children's increasing knowledge and strategy development around number, moving from whole numbers to rational numbers. Teaching to promote the development of rational number knowledge and understanding is detailed in Chapter 11. It should be noted, however, that, while the counting stages of the Number Framework have been validated by research, the part-whole stages have not (Young-Loveridge & Wright 2002). Nevertheless, the stages provide a valuable way of thinking about children's number knowledge development and the growth of strategic thinking.

Big numbers

As students' understanding of place value develops, conceptualisation of the size of numbers being dealt with must be promoted. As each new place in the Hindu-Arabic system is encountered, students need to investigate the following:

- How big is 1000?
- How big is 10 000?
- How big is 100 000?
- How big is 1 000 000?

The following Teaching Idea lists some questions that can be posed to students to encourage them to think about big numbers. An excellent way to visualise one million is to create a cubic metre from metre rulers. Place four one-metre rulers on the floor, and then ask for student volunteers to hold four more one-metre rulers vertically at each corner of the square metre on the floor. Then use four more one-metre rulers to finish the skeletal construction of the cubic metre. Place one MAB one cube (1 cm) in the cubic metre, and ask students to state how many cubic centimetres can be contained within the cubic metre. Gradually, students will realise that, with dimensions of 100 cm × 100 cm × 100 cm, the cubic metre holds one million cubic centimetres. A similar strategy can be completed using pieces of wooden dowel, each one metre in length, that can be connected with PVC connectors and Styrofoam blocks to represent the thousands MAB. A very useful resource when introducing large numbers is the book *Counting on Frank* by Rod Clement.

TEACHING IDEA

Big numbers

Estimate:

- how many hairs you have on your head
- the number of words in a 200-page book
- the number of people who would fit on the deck of a large cruise ship
- what you could buy for $1 million
- how many days/hours/minutes you have been alive
- the number of grains in a cup of sand
- how much water you would use in a shower over one year
- how many centimetres of fencing would be needed to enclose a paddock
- how many kilometres a rider would ride when training for a bike race such as the Tour de France
- how many grains of rice are in a kilogram bag.

These types of questions are sometimes referred to as *Fermi questions* (named after famous scientist Enrico Fermi). The following website contains more than 30 questions suitable for an upper-primary class: www.edgalaxy.com/journal/2012/5/29/an-excellent-collection-of-fermi-problems-for-your-class.html.

Reading big numbers

The Hindu-Arabic system enables counting to continue beyond any given number. Reading big numbers is assisted by knowing that each three places (beginning at the ones) is a period, and that each period has its own label.

billions	millions	thousands	ones	thousandths	millionths	billionths
TEN BILLION BILLION	HUNDRED MILLION TEN MILLION MILLION	HUNDRED THOUSAND TEN THOUSAND THOUSAND	HUNDRED TEN ONE	tenth hundredth thousandth	ten thousandth hundred thousandth millionth	ten millionth hundred millionth

The periods are signalled by a comma or a space. The periods of numbers are indicated in the table. The period names assist in naming very large numbers and also decimal numbers. Reading numbers occurs by considering the digits within each period as hundreds and then referring to the period in which that group of digits occurs. The following are two examples of how large numbers are read:

- 459 346 278: four hundred and fifty-nine *million*, three hundred and forty-six *thousand*, two hundred and seventy-eight
- 6 589 320 400: six *billion*, five hundred and eighty-nine *million*, three hundred and twenty *thousand*, four hundred.

Once students see this structure, they enjoy the challenge of reading very big numbers.

Addition and subtraction

Through number and counting activities, students develop skills in simple addition and subtraction. Yet understanding of addition and subtraction is more than the skill of being able to add two or more groups together, or being able to take away an amount from a different amount. Informal counting activities and the development of more complex counting strategies are precursors for operating on numbers. A rich conceptual understanding of addition and subtraction is necessary in order for students to be successful problem-solvers in situations involving addition and subtraction.

The language of addition and subtraction

For many students, addition is narrowly conceptualised as 'joining', and subtraction as 'taking away'. When presented with problems, words such as add, join, meet, altogether, arrive, take away, subtract, remove, eat, run away and leave will signal whether the required operation for the solution is addition or subtraction. The strategy of looking for cue words (joining words and/or take-away words) is precarious, since it will not assist in all cases. Consider the following word problems:

- I went to the bank and withdrew $50. This left $100 in my account. What was my original balance?
- I added 50 grams of currants to my cake mixture, which made a total of 750 grams of dried fruit. How much dried fruit was in my mixture before I added the currants?

In these examples, the words 'withdraw' and 'add' are used, which are subtraction and addition words, respectively. Yet the operations required to solve each problem are addition and subtraction, respectively.

Students need to experience a variety of addition and subtraction problems that do not follow a standard format. A range of research has been undertaken in this field, resulting in classification of addition and subtraction problems under such labels as compare problems, change problems, combine problems, part-part-whole problems and unknown start problems (see Chapter 3). While the labels are possibly confusing, the salient point of such research is that, in the real world, addition and subtraction problems are not always packaged in simple 'join' or 'take-away' language.

Types of word problem

A simplified classification system that works well for both addition and subtraction problems involves labelling problems as joining, comparison, inaction or take-away problems. The labels provide some indication of how the problem is modelled with concrete material. Joining problems have a joining action, inaction problems are where no action takes place, comparison problems are where there are two groups to be compared, and take-away problems have a take-away action. The following table provides examples of these problems for addition and subtraction.

	Addition	**Subtraction**
Joining	I have 5 mini-cars. My mother gave me 4 more. How many do I have now?	I had 7 mini cars, and I gave some to my brother. Now I have 3. How many did I give to my brother?
Inaction	I have 6 soccer cards in my pocket and Lisa has 3. How many soccer cards do we have between us?	I have 9 soccer cards in my collection. I have my favourite 3 in my pocket. How many are at home?
Comparison	I have 6 lollipops and Ben has 2 more than me. How many lollipops does Ben have?	I have 4 lollipops; Sam has 2 less than me. How many does Sam have?
Take-away	I ate 6 cherries; now I have 7 left. How many did I have to start with?	I had 14 cherries and I ate 7. How many do I have left?

Examples of four types of addition and subtraction problem

When working with addition and subtraction word problems, concrete materials may assist with representation of the elements within the problem, or acting out the situation may assist in finding meaning. The important thing to remember is that the problem-solving process, whereby students are engaged in discussion, justification, verification and/or reasoning, is the most valuable aspect. Immersion in a variety of problems with different language structures promotes familiarity with addition and subtraction problems and leads to greater mathematical understanding. (See Chapter 3 for further details about word problems, language, equity and mathematical access.)

Multiplication and division

Just as exposure to a range of addition and subtraction word problems promotes conceptual understanding and problem-solving skills, this is also the case for multiplication and division.

Multiplication problems

There are five basic types of multiplication problem: grouping, rate, scalar, cross-product and array problems. Some examples are presented below.

- *Grouping*—There are three bags that have 5 sweets in each bag. How many sweets altogether?
- *Rate*—Sara bought 4 lollipops that cost 5 cents each. How much did it cost her?
- *Scalar*—Jan has 3 times as many baseball cards as Peta. Peta has 8 cards. How many cards does Jan have?
- *Cross-product*—Brian has 3 shirts and 2 pairs of shorts. How many different outfits can be created?
- *Array*—In a theatre, there are 6 chairs in each row and 12 rows in total. How many chairs are there in the theatre?

It is important that children be exposed to a range of multiplication problems so that they can experience the language of multiplication situations. Concrete materials support understanding of the problem and awareness of the multiplicative aspect of different types of problems. For the grouping problem above, students put five counters (or blocks) in three specific piles. They use skip counting in fives, or recall the basic fact, to attain the solution. For rate problems, students need to put a coin next to each of the four lollipops. They then count up the value of the coins, or recall the basic fact, to reach the solution. For the scalar problem, two groups of material are required, with the owner of each designated clearly. This is actually a ratio situation in which the two quantities are compared multiplicatively. For every card that Peta has, Jan has three cards. These cards should be displayed so that the learner can see that Jan has 24 cards when Peta has eight. It is important that the arrangement of the material matches the problem. For the array problem, the materials need to be arranged in grid formation.

Cross-product problems are the most difficult multiplication situations. The best way to support students' understanding of cross-product solutions is to act out the problem using real or simulated materials. For the problem above, provide students with cut-out pictures of red and black shorts, and white, green and yellow shirts. Students then arrange the shorts and shirts together on the table to ensure that they have accounted for all possible combinations. They will come to realise that there are $2 \times 3 = 6$ possible combinations, and that, in general, you multiply the number of choices to find the solution. Organising this information in an array can also assist in representing and conceptualising the cross-product situation. The information in the array can be organised either way—as shorts × shirts or shirts × shorts, as seen in the diagram.

	Shirts		
Shorts	**White**	**Green**	**Yellow**
Red	R,W	R,G	R,Y
Black	B,W	B,G	B,Y

	Shorts	
Shirts	**Red**	**Black**
White	W,R	W,B
Green	G,R	G,B
Yellow	Y,R	Y,B

Division problems

There are two types of division problem: partition and quotition. The classification is made according to the action that occurs when acting out, or modelling, the problem to assist with interpretation. Some examples follow:

- *Partition*—There are 12 lollipops and 3 children. How many lollipops will they each get if they are shared equally?
- *Quotition*—There are 12 lollipops and each child receives 3. Between how many children can the lollipops be shared equally?

The difference between these two types of problems is the action of sharing. In partition division, the number of groups is known, but how many will be in each group is unknown, so a dealing or one-at-a-time sharing action is invoked. In quotition division, the amount in each group is known, but the number of groups is unknown, so that amount is taken each time to determine how many groups will be made of the specified amount.

Meaning and concrete representations

In multiplication, an initial meaning is often grouping, and making groups can assist in representing multiplication situations. As students gain experience with a variety of multiplication problem types, organising the group into arrays also provides a structure for showing the commutative nature of multiplication (e.g. that 8 × 5 results in the same total as 5 × 8), and, depending on students' facility with basic multiplication facts, can assist in determining the solution. For division, the basic meaning is sharing. Eventually, as students' understanding of multiplication and division grows, they will dispense with materials for representing situations and come to rely on their knowledge of basic facts or computational methods to solve the problems they encounter.

Review questions

9.1 Describe why 5 and 10 are benchmarks. What other number benchmarks might be useful for supporting number understanding? Give three examples of when 5 or 10 could be used as a benchmark.

9.2 Write a definition of number sense. Compile a list of questions that could determine whether (a) a young child, and (b) an adult have good number sense.

9.3 Compile a sequence of activities for skip counting that starts with students exploring skip-counting patterns in large numbers. Justify your approach.

9.4 What is subitising? What activities promote this skill? How does awareness of subitising inform your approach to teaching addition?

9.5 Compile a list of activities, including those available on websites, that promote conceptualisation of large numbers (10 000, 100 000, 1 000 000).

Further reading

Clements, D.H. & Sarama, J. (2014). *Learning and teaching early math: The learning trajectories approach.* New York, NY: Routledge.

Curtis, R., Okamoto, Y. & Weckbacher, L.M. (2009). Preschoolers' use of count information to judge relative quantity. *Early Childhood Research Quarterly, 24*(3), 325–36.

Kastberg, S.E. & Walker, V. (2008). Insights into our understandings of large numbers. *Teaching Children Mathematics, 14*(9), 530–6.

Martin, B. (2000). What is counting? How do we teach it? *Prime Number, 15*(1), 5–7.

Phillipson, S., Gervasoni, A. & Sullivan, P. (2017). *Engaging families as children's first mathematics educators.* Singapore: Springer.

Schleifer, P. & Landerl, K. (2011). Subitizing and counting in typical and atypical development. *Developmental Science, 14*(2), 280–91.

Taylor-Cox, J. (2016). *Math intervention P–2: Building number power with formative assessments, differentiation, and games, grades pre K–2.* New York, NY: Routledge.

Wang, J., Lin, E., Tanase, M. & Sas, M. (2008). Revisiting the influence of numerical language characteristics on mathematics achievement: Comparison among China, Romania, and US. *International Electronic Journal of Mathematics Education, 3*(1), 24–46.

References

Allen, J.M. & Wright, S.E. (2014). Integrating theory and practice in the pre-service teacher education practicum. *Teachers and Teaching, 20*(2), 136–51.

Anghileri, J. (2006). *Teaching number sense* (2nd ed.). London: Continuum.

Clarke, B., Clarke, D. & Cheeseman, J. (2006). The mathematical knowledge and understanding young children bring to school. *Mathematics Education Research Journal, 18*(1), 78–103.

Clarke, B., Gervasoni, A. & Sullivan, P. (2000). The early numeracy research project: Understanding, assessing and developing young children's mathematical strategies. Paper presented at the AARE Annual Conference, Sydney, 2000.

Cobb, P. (1991). Reconstructing elementary school mathematics. *Focus on Learning Problems on Mathematics, 13*(2), 3–32.

Fuson, K. (1988). *Children's counting and concepts of number.* New York, NY: Springer-Verlag.

Gelman, R., Gallistel, C.R. & Gelman, R. (2009). *The child's understanding of number.* Cambridge, MA: Harvard University Press.

Gervasoni, A. & Perry, B. (2013). Children's mathematical knowledge prior to starting school. In V. Steinle, L. Ball and C. Bardini (eds), *Mathematics education: Yesterday, today and tomorrow. Proceedings of the 36th Annual Conference of the Mathematics Education Research Group of Australasia.* Melbourne: MERGA.

Gravemeijer, K. (1994). *Developing realistic mathematics education.* Utrecht: CD-B Press.

Mason, J. (2004). Book review: *Principles and practices in arithmetic teaching: Innovative approaches to the primary classroom* by J. Anghileri. *British Educational Research Journal, 30*(6), 855–7.

Ministry of Education, New Zealand. (2008). *The number framework.* Wellington: Ministry of Education.

Perry, B. & Dockett, S. (2013). Reflecting on young children's mathematics learning. In English, L. and Mulligan, J. (eds). *Reconceptualizing early mathematics learning* (pp. 149–61). Dordrecht: Springer.

Steffe, L.P. & Cobb, P. (2012). *Construction of arithmetical meanings and strategies.* Springer Science & Business Media.

Treffers, A. (1991). Meeting innumeracy at primary school. *Educational Studies, 22*(4), 333–52.

von Glasersfeld, E. (1982). Subitizing: The role of figural patterns in the development of numerical concepts. *Archives de Psychologie, 50*, 191–218.

Wright, R.J., Stanger, G., Stafford, A.K. & Martland, J. (2006). *Teaching number: Advancing children's skills and strategies.* SAGE Publications.

Wright, R.J., Martland, J. & Stafford, A. (2000). *Early numeracy: Assessment for teaching and intervention.* London: Paul Chapman Publishing.

Young-Loveridge, J. & Wright, V. (2002). Validation of the NZ Number Framework. In B. Barton, K. Irwin, M. Pfannkuch and M. Thomas (eds), *Mathematics education in the South Pacific* (pp. 722–9). Sydney: MERGA.

CHAPTER 10

BASIC FACTS, MENTAL COMPUTATION AND ALGORITHMS

In the past, school mathematics programs were dominated by pen-and-paper practise of the four operations of addition, subtraction, multiplication and division. Mental mathematics also featured, with chanting of times tables and 'ten quickies' signalling the commencement of mathematics lessons. In the 1980s, students' understanding of written calculation methods came under scrutiny, with research indicating that flawless written computation performance did not necessarily equate to understanding of the procedures involved (e.g. Carpenter & Moser 1982). Calls were made for mathematics instruction that would ensure that, in promoting computational skill, rich conceptual knowledge was also developed (Skemp 2006).

In a technological age, the necessity of teaching standard procedures (often regarded as algorithms) for written computation is often questioned when calculators are ubiquitous. However, some parents question teachers about the mathematics program if their children cannot perform written computations flawlessly. While research continues to provide insight into the development of computation and to draw implications for teaching (e.g., Blote et al. 2000; Torbeyns et al. 2018), there is no question that building number sense and enabling students to develop understanding of the operations they perform is vital (Hickendorff et al. 2019; Lemonidis 2015).

Working flexibly with number

One example of working flexibly with number is New Zealand's Number Framework (Ministry of Education, New Zealand 2008), introduced in the previous chapter. This framework outlines nine global stages for learning number that describe flexible thinking strategies characteristic of each stage. The Number Framework stresses that children should not be exposed to standard written algorithms until they can use part-whole mental strategies. Here, the importance of mental strategies is emphasised.

It becomes possible to work flexibly with numbers in mental computation situations when children have good number sense. Students with good number sense are able to see numbers in a range of combinations and groupings. For example, rather than see the addition of 6 and 7 as a fact to be remembered, they see 6 and 4 and 3, which enables them to add the 4 to the 6 to make 10 and then add the remaining 3 to the 10. In another context, where 7 is added to 8, the 7 is seen as 2 and 5. In order for students to work flexibly with numbers in this way, they need to have strong number sense and the confidence to work with numbers so that they can make use of this knowledge.

Depending on the context, greater or lesser emphasis is placed on developing number sense. In Dutch research and curriculum development, the first five years of schooling focus on developing number sense (over place value) and on promoting students' intuitive strategies for mental computation. This work has the potential to have a profound impact on how number work is conceptualised.

In most Western classrooms, teaching operations such as 65 + 28 has been undertaken through a place value model. That is, the ones are regrouped for tens and the answer is duly noted. The Dutch model encourages numbers to be manipulated more flexibly, depending on the context. In this case, the 28 will need to be broken into components that can readily be added to the 65. It is possible that this may be 20

+ 5 + 3, so that 20 is added to 65, and then the 5 (to round to 70) and then the remaining 3 is added. In another context, if the 28 were added to 47, then it would make sense to break the 28 into 20 + 3 + 5. In these cases, 28 is not seen as 2 tens and 8 ones, but rather is broken into components that will make it easy to add to the other number. Students are encouraged to think flexibly with numbers (recall the jump and split strategies discussed in Chapter 9). A strategy used to support this flexibility with numbers is that of the empty number line (van den Heuvel-Panhuizen 2008). The number line can assist in tracking and thinking about numbers and operations. Students will use the number line differently depending on their number sense. A useful app for both empty number line work and number line work with whole numbers, fractions, decimals and, later, negative numbers, is the Number Line App (https://apps.apple.com/au/app/number-line-by-the-math-learning-center/id751816884).

Fluency

In the Australian Curriculum: Mathematics, fluency is listed as one of the four proficiencies to be developed through the teaching and learning of mathematics in the classroom. The other three proficiencies are understanding, problem-solving, and reasoning. The inclusion of proficiencies within the curriculum serves, in the teaching of any mathematics topic, to take mathematics instruction away from an emphasis on algorithms, rules and procedures. It is important to consider what is meant by fluency, particularly in developing mental and written computation.

The term 'fluency' is often misunderstood. It is not about automatic recall of number facts. Having the ability to automatically recall facts fluently does not mean students will have strong conceptual understanding (Hurst & Hurrell 2016). It is not about speed and accuracy in performing written computation. Instead, fluency is about skills in 'choosing appropriate procedures, [and] regrouping out procedures flexibly, accurately, efficiently and appropriately' (ACARA 2019). Students demonstrate fluency when they 'calculate answers efficiently, when they recognise robust ways of answering questions, when they choose appropriate methods and approximations' (ACARA 2019). Fluency in computation is being able to use a range of strategies to perform calculations, and being confident to do so. Fluency is nested within a strong sense of number (Boaler 2015).

Basic facts as the foundation for computation

'Basic facts' is the term used to refer to simple addition, subtraction, multiplication and division calculations that are expected to be automatic. For example, 6 + 8 is a basic fact of addition; 4 × 5 is a basic fact of multiplication. Multiplication facts are often referred to as 'the tables'. Automatic recall of basic facts assists computation of larger numbers, which is why this chapter has a strong focus on teaching for developing basic fact fluency. Fluent number fact knowledge spans far beyond conventional automatic mastery practices involving rote learning (Baroody 2006).

The approach suggested here is to provide students with thinking strategies for basic fact calculation that, in turn, assist in mental computation of larger numbers (two digits and beyond). It is about building fluency through the development of number sense. Through a strong sense of number, students are being provided with a 'license to think' about number in different ways (Boaler 2015; Buchholz 2004). Achieving fluency in number facts through a strong foundation of number sense 'grows out of discovering the numerous patterns and relationship that interconnect the basic combination' (Baroody 2006, p. 24). Activities that focus on additive and multiplicative thinking, encourage rich number discussions and engage with number sense in a non-fearful environment have had a profound effect on students' automaticity (Boaler 2015; Godfrey & Stone 2013). Practices that promote discussion of patterns, relationships and connections outweigh the blind memorisation of number facts.

A *strategies approach* to teaching the basic facts of addition and multiplication is elaborated in this chapter. This is described as instruction that aims to promote strong visual images for basic facts through the use of concrete materials. The commutativity principle (i.e. 3 + 5 = 5 + 3) is emphasised to reduce the number of basic facts to be learned. The basic facts of subtraction and division are addressed by using the inverse principle (think addition and think multiplication, respectively) because subtraction is the inverse of addition, or 'undoes' it; division is the inverse, or 'undoes' multiplication. This capitalises on the part-part-whole composition of addition and the factor-factor-product of multiplication. For example, suppose that, in addition, two (or more) parts are given and the whole is to be determined (3 + 4 = ?). In subtraction, the whole and at least one part are given, and the other part is to be determined

(7 – 3 = ?). If the basic fact of 3 + 4 is automatically recalled as equalling 7, then to determine 7 – 3, thinking addition of '3 plus what makes 7' makes learning basic facts of subtraction relatively simple and draws on number sense.

Basic facts of addition

The basic facts of addition involve the addition of two addends up to 10. The basic facts of subtraction are the addition facts in reverse. There are 100 basic facts of addition—or 121, if facts with zero are also included. In this section, the facts have been categorised according to a thinking strategy that can assist in understanding and recall of the facts. Similar to the categorisation of the 121 basic addition facts by Thornton and colleagues (1983), thinking strategies for seven categories of facts are presented here as follows:

1. count-on facts
2. doubles
3. tens facts
4. adding 10 facts
5. bridging 10 (9, 8) facts
6. doubles + 1, + 2 facts
7. last facts.

The 121 facts, sorted according to category, are displayed in the table, which serves to highlight overlaps of facts between categories. For example, 6 + 4 is in the tens facts category and also the doubles + 2 category. Similarly, 8 + 3 is a count-on fact as well as a bridging 10 (8) fact. In some cases, particular facts fall within two or more categories, and this will be apparent as each fact category is presented. The teacher's role in this situation is to encourage discussion of the new fact category and the ways in which two different strategies can be used to recall the same fact. Students should be encouraged to express their opinions on particular strategies that they might employ in certain cases, and why.

The table of facts also lists the *spin-around* (the initial fact in reverse order) for each fact in each category. This is to reflect the efficiency of learning facts by strategy and visualisation. Facts are not seen as isolated and individual, but rather as flexible, by simultaneously considering the related spin-arounds at the time of learning the initial facts. This covertly promotes students' understanding of the commutative principle.

The fact categories are not necessarily hierarchical, but particular groups of facts must be in place before some other fact categories are learned, since they provide the foundation for deriving other facts. The sequence presented is a suggestion only. The introduction of groups of facts according to strategy will depend upon the needs of the students and the teacher's own professional judgement.

Basic addition facts sorted by strategy

Count-on facts

Count-on 1	2 + 1	3 + 1	4 + 1	5 + 1	6 + 1	7 + 1	8 + 1	9 + 1			
Spin-around	*1 + 2*	*1 + 3*	*1 + 4*	*1 + 5*	*1 + 6*	*1 + 7*	*1 + 8*	*1 + 9*			
Count-on 2	3 + 2	4 + 2	5 + 2	6 + 2	7 + 2	8 + 2	9 + 2				
Spin-around	*2 + 3*	*2 + 4*	*2 + 5*	*2 + 6*	*2 + 7*	*2 + 8*	*2 + 9*				
Count-on 3	4 + 3	5 + 3	6 + 3	7 + 3	8 + 3	9 + 3					
Spin-around	*3 + 4*	*3 + 5*	*3 + 6*	*3 + 7*	*3 + 8*	*3 + 9*					
Count-on 0	0 + 1	0 + 2	0 + 3	0 + 4	0 + 5	0 + 6	0 + 7	0 + 8	0 + 9	0 + 10	
Spin-around	*1 + 0*	*2 + 0*	*3 + 0*	*4 + 0*	*5 + 0*	*6 + 0*	*7 + 0*	*8 + 0*	*9 + 0*	*10 + 0*	
Doubles	1 + 1	2 + 2	3 + 3	4 + 4	5 + 5	6 + 6	7 + 7	8 + 8	9 + 9	10 + 10	
Tens facts	10 + 0	9 + 1	8 + 2	7 + 3	6 + 4	5 + 5	4 + 6	3 + 7	2 + 8	1 + 9	0 + 10
Adding 10	10 + 0	10 + 1	10 + 2	10 + 3	10 + 4	10 + 5	10 + 6	10 + 7	10 + 8	10 + 9	10 + 10
Spin-around	*0 + 10*	*1 + 10*	*2 + 10*	*3 + 10*	*4 + 10*	*5 + 10*	*6 + 10*	*7 + 10*	*8 + 10*	*9 + 10*	
Bridging 10 (9)	9 + 2	9 + 3	9 + 4	9 + 5	9 + 6	9 + 7	9 + 8				
Spin-around	*2 + 9*	*3 + 9*	*4 + 9*	*5 + 9*	*6 + 9*	*7 + 9*	*8 + 9*				
Bridging 10 (8)	8 + 3	8 + 4	8 + 5	8 + 6	8 + 7						
Spin-around	*3 + 8*	*4 + 8*	*5 + 8*	*6 + 8*	*7 + 8*						
Doubles + 1	2 + 3	3 + 4	4 + 5	5 + 6	6 + 7	7 + 8	8 + 9				
Spin-around	*3 + 2*	*4 + 3*	*5 + 4*	*6 + 5*	*7 + 6*	*8 + 7*	*9 + 8*				
Doubles + 2	2 + 4	3 + 5	4 + 6	5 + 7	6 + 8	7 + 9					
Spin-around	*4 + 2*	*5 + 3*	*6 + 4*	*7 + 5*	*8 + 6*	*9 + 7*					
Last facts	7 + 4										
Spin-around	*4 + 7*										

Count-on facts

The count-on strategy is where, when given two numbers, one of the numbers is the starting point for counting on the amount of the second number. The count-on strategy is used frequently, but is often the most difficult strategy to overcome once learned. Many students use the count-on strategy to mentally add two numbers that are far too large to ensure accuracy or efficiency. When used appropriately, the count-on strategy is very useful for mental computation, but students need to know when and how to use this strategy, and when it is more efficient to use other strategies.

Counting-on 1, 2 and 3

The count-on strategy is only suitable for counting on small numbers such as 1, 2, 3 and zero. In the early years, special instruction needs to be given for counting on zero, and this should be taught after the count-on strategy for numbers 1, 2 and 3 has been consolidated. When teaching basic fact groups, selection of the examples used needs to be considered carefully so that the usefulness of the strategy will be demonstrated clearly. To teach the count-on facts, select examples in which one of the two numbers in the pair is considerably greater than 1, 2 or 3. The following facts match this description: 4 + 1, 5 + 1, 6 + 1, 7 + 1, 8 + 1, 5 + 2, 6 + 2, 7 + 2, 5 + 3 and 6 + 3. These are the best facts for initial experiences with the strategy. Visual, auditory and tactile experiences can also be used to introduce students to the count-on strategy:

- Using marbles in a tin is a useful teaching approach for developing the count-on strategy. Show students a group of five marbles, then place them in an empty tin. Let the students see and hear the group of five marbles rolling around in the bottom of the tin. Add one more marble and have the students count on from five to six as they hear the marble drop into the tin. Practise with other examples. Having the marbles in the tin discourages students from counting the first collection of objects, as they cannot see them. The noise created when a marble is added assists in exemplifying the counting-on action.
- Generate a group of counters on an IWB. Flash the screen to students and ask them to state how many counters are on the screen. This action capitalises on students' natural ability to *subitise*—that is, to recognise the size of a collection of objects without counting. Check by counting this collection if necessary. Cover the initial collection on the IWB, place two further counters on the screen and flash the screen again so that only the two additional counters are visible. Ask students to state how many counters there are altogether. This action

discourages the urge to recount the first collection and to count on from this number.

- Flash sets of counters in various count-on fact representations on the IWB and ask students to state the solution. Ask the students to explain their strategy. Encourage students to use their subitising skills to determine the amount of counters in each group, and then to select the largest group and count on from there.
- Provide students with the fact in symbolic form and ask them to show that fact with their counters. Ask students to cover the biggest amount with their hand and to count on the remaining amount.
- As students are presented with particular facts, encourage them to press their fingers on their skin, or to slap their thigh, as they count on each number to help keep track of the amount they are counting on.

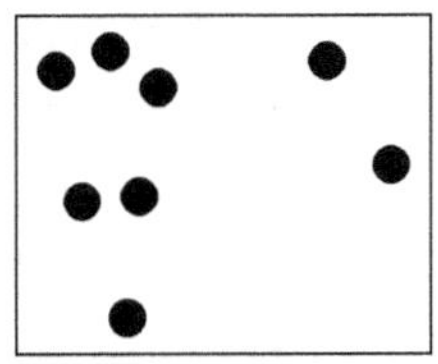

What can you see?

The commutative law (spin-arounds or turn-arounds)

The count-on strategy is where the larger of the two numbers is the starting point for counting on the other number. As children explore this strategy through use of concrete materials, they will naturally engage with one of the laws of arithmetic: *commutativity*—the concept that order does not matter (e.g. 7 + 2 results in the same total as 2 + 7). The commutative law must be made explicit to students so they can see that, as they learn each basic fact, they are actually learning *two* facts: each fact and its *spin-around.*

Student working with spin-arounds—here the student has 4 and 2, or 2 and 4, depending on which way the Unifix blocks are turned

To explicitly model commutativity, and therefore the spin-around or turn-around notion for the count-on facts, use a piece of paper or a card that is divided into two parts. Place counters on each half of the card to model a count-on fact. Ask students to state the fact being modelled. Then physically turn the card around and ask the students to state the fact being modelled. Similarly, as shown in the photo, students can use Unifix to represent the task and twist this in either direction to demonstrate the commutativity principle.

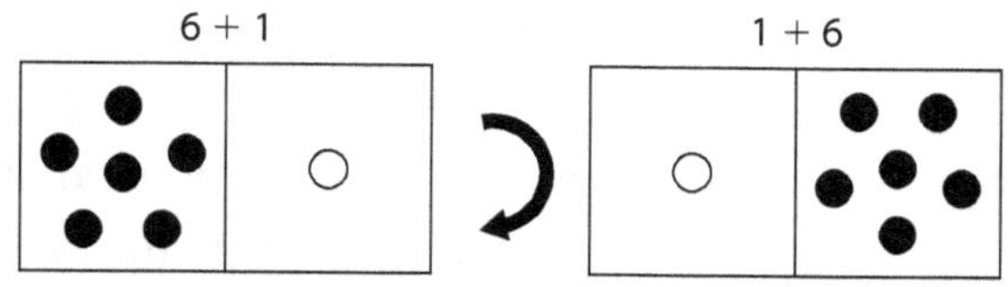

Cards or dominoes reinforce spin-arounds

Reinforce the count-on strategy: select the larger of the two numbers and count on the other amount. This strategy can be further reinforced by providing students with practice sheets of count-on facts. Encourage students to circle the larger of the two numbers as the first step, before they count on, and then record the answer.

Counting on zero

Special attention needs to be given to counting on with zero. The count-on strategy is just that—counting on; that is, an action is expected and required. However, when counting on zero, no action is immediately evident. This may be a potential source of confusion to some students. Through use of concrete materials and careful use of language, students will come to see that counting on with zero results in the original amount that they started with.

Doubles facts

The doubles are usually the most easily learned facts for students. Often, students know their doubles facts up to 5 + 5 before they commence formal schooling. When introducing the doubles thinking strategy, encourage students to brainstorm *things* that *come in* a particular amount that can be doubled. For example, the number of wheels on a bicycle is two—bicycle wheels *come in* twos. So two bicycles is four wheels (double two is four). The following is a list of suggestions that can be used to help children visualise a particular number and then double it:

- 1—person, stop sign, sun
- 2—drumsticks, eyes, ears, bicycle wheels
- 3—clover leaf, cricket stumps, tricycle wheels
- 4—car tyres, legs on a dog, cat, horse, etc.
- 5—fingers
- 6—legs on an insect
- 7—calendar (2 weeks)
- 8—legs on an octopus/spider
- 9—Channel 9 TV symbol.

When teaching doubles, focus on the language and a visual image—'5 fingers on one hand; 10 fingers on two hands; 5 and 5 is 10; double 5 is 10'. When students falter in thinking of the solution to 8 + 8, have them think of a spider: 'How many legs does a spider have? Now double it.' For 7 + 7, think of a calendar: 'How many days is 2 weeks?' For 6 + 6: 'How many legs do two insects have?' The images students use to assist in learning the doubles should come from them. They need to select the item that most appeals to them and then double it; however, too many images may be a source of confusion.

Doubles plus one and doubles plus two

Once the doubles are mastered, they can be used as a platform for facts that are almost doubles. Facts that fall into this category are the doubles plus one facts (2 + 3, 3 + 4, 4 + 5, 5 + 6, 6 + 7, 7 + 8, 8 + 9) and the doubles plus two facts (2 + 4, 3 + 5, 4 + 6, 5 + 7, 6 + 8, 7 + 9). As a category, these facts should be introduced once the doubles facts have been consolidated. Recognising doubles plus one and doubles plus two facts requires considerable cognitive effort, so this category should be delayed until students are familiar with easier categories of facts.

As students learn more and more facts, it may become apparent that particular facts have previously been learned through another strategy. This negates the need to relearn the same fact with a different and possibly more difficult strategy. Through examination of the list of doubles plus one and doubles plus two facts, it can be seen that some facts fit within other fact categories. The following facts have been covered through the count-on strategy: 2 + 3, 2 + 4, 3 + 4, 3 + 5. The 4 + 6 fact is a simple tens fact, and is discussed below. The facts 7 + 8, 8 + 9 and 7 + 9 belong to the bridging 10 facts category, and may be simpler for some students to learn through that strategy than as an almost double fact. The almost double facts have been presented here because of their link to the doubles facts.

To assist students to visualise facts that are almost doubles, use counters on 2 cm grids. Use different coloured counters to show the two numbers in the fact. Display counters as a mirror image, with an extra counter clearly visible for doubles plus one facts and two extra counters for doubles plus two facts. When the counters are displayed, describe the fact as being a particular double plus one or plus two. From the examples given below, it can be seen that 5 + 6 gives the same total as double 5 + 1 more. Similarly, 6 + 8 gives the same total as double 6 + 2 more.

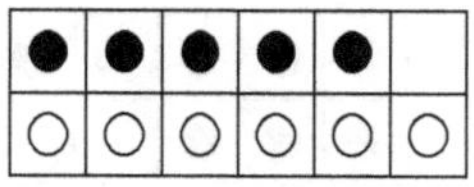

5 + 6 is the same as double 5 + 1

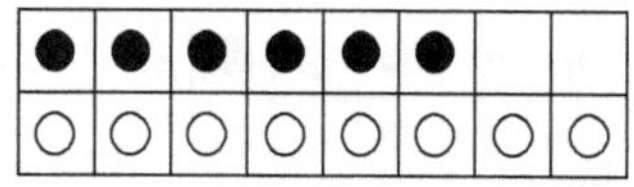

6 + 8 is the same as double 6 + 2

For some students, these facts may be more appropriately recognised as doubles *less* 1 or 2. That is, 5 + 6 may be seen more easily as double 6 less 1, 6 + 8 may be double 8 less 2, and so on. Promote this type of discussion around the visual images provoked by the concrete representations, and encourage students to use the strategy that most appeals to them. It is often better if students select and use one strategy until particular facts become consolidated; otherwise, confusion may occur.

Tens facts

The tens facts are those facts where the number pair combines to give a total of 10. Recognising combinations to 10 is a frequently used strategy in mental computation. Tens facts are often readily acquired, but it is important to ensure that all tens facts are equally familiar to students so that they have access to a valuable strategy for mental computation.

Orientation to the tens facts is best achieved through use of a tens frame. A tens frame is a 2 × 5 grid with grid cells large enough to accommodate a coloured counter. Before using the tens frame to represent tens fact combinations, students need to be able to instantly recognise numbers from 0 to 10, as represented with counters on the grid. This is promoted by positioning counters in a particular, consistent arrangement and by drawing attention to the number of counters, and the number of counters missing, as follows (for the numbers 10, 9, 8, 5, 6 and 7):

- 10—no counters missing
- 9—one missing
- 8—two missing
- 5—top line filled
- 6—five and one more
- 7—five and two more.

The numbers 0 to 4 are readily recognised by students. The arrangement of counters on the ten frame for the numbers 0 to 10 are presented in the diagram.

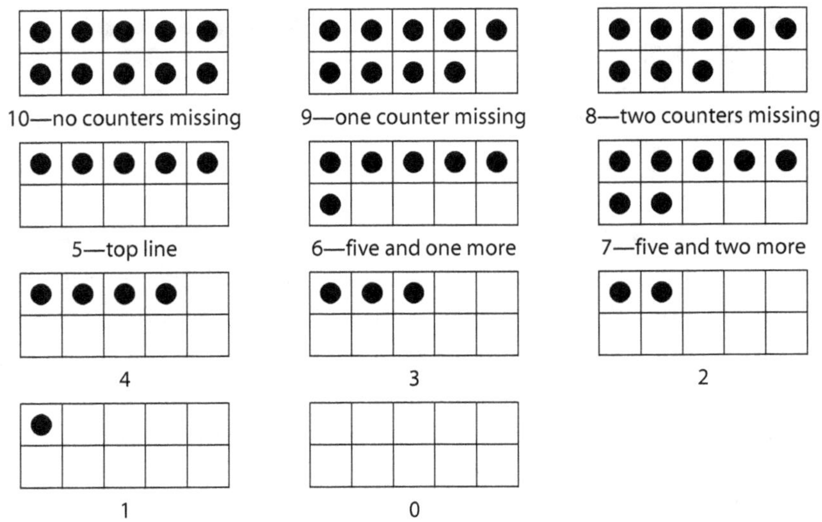

A digital version, Number Frames, is available here: https://apps.apple.com/au/app/number-frames-by-the-math-learning-center/id873198123. In addition to the tens frame, this resource includes options for a five frame, a 20 frame and a range of different numbered frames for later fraction work.

Once students can readily recognise numbers 0 to 10 on the tens frame without counting, link tens fact pairs by emphasising the amount needed to fill the tens frame for the following five facts: 10 and 0, 9 and 1, 8 and 2, 5 and 5, 7 and 3, 6 and 4. Once these five number combinations have been consolidated, the rest of the tens facts are the spin-arounds.

The sequence for presentation of the tens facts and accompanying language is as follows:

$10 + \square = 10$ What goes with 10 to give 10?
$9 + \square = 10$ What goes with 9 to give 10?
$8 + \square = 10$ What goes with 8 to give 10?
$5 + \square = 10$ What goes with 5 to give 10?
$6 + \square = 10$ What goes with 6 to give 10?
$7 + \square = 10$ What goes with 7 to give 10?
$0 + \square = 10$ Zero and what gives 10?
$1 + \square = 10$ One and what gives 10?
$2 + \square = 10$ Two and what gives 10?
$3 + \square = 10$ Three and what gives 10?
$4 + \square = 10$ Four and what gives 10?

Adding tens facts

Understanding what happens when a ten is added to a number relates to place value and numeration knowledge. To explore this notion with students, counting activities with a calculator or spreadsheet are useful. If a calculator or a spreadsheet has an inbuilt constant function, it can be programmed to count by entering a starting number, pressing the addition key and then entering the number by which you wish to count. Try the functions listed in the Teaching Idea box.

TEACHING IDEA

Adding tens facts

To make your calculator count in tens:

- Enter any number (e.g. 6).
- Press the add key.
- Enter 10.
- Press the equals button repeatedly.
- Observe the changes occurring.

Alternatively:

- If that doesn't work, try entering the starting number, then press the add key twice before entering the counting number.
- Then press equals repeatedly.

In the example of counting in tens from 6, the idea is to make students focus on what is happening every time the equals button is pressed. Students will see that when adding ten, the number in the ones place does not change, but the digits to the left of the entered number increase by one every time. The pattern for 6 + 10 repeatedly is 16, 26, 36, 46, 56, 66, and so on. Try other starting numbers and see which digits change and which stay the same. Link this to a general principle of adding ten to a number—the digit in the ones place does not change, but a ten is added to the digit in the tens place. From this generalisation, students can quickly determine the solution to facts in which ten has been added. This is an extremely useful strategy for mental computation of numbers with more than two digits.

Bridging ten facts

The bridging ten facts are when one number of the fact pair is close to 10. Bridging ten is very useful when one digit is 9 (9 + 4, 9 + 5, 9 + 6, 9 + 7, 9 + 8) or 8 (8 + 4, 8 + 5, 8 + 6, 8 + 7), but for numbers smaller than 9 or 8 the strategy loses its usefulness and efficiency. To show students how to use the bridging ten strategy and to provide them with a strong visual image of this strategy in action, use a double tens frame. A double tens frame consists of two tens frames displayed one above the other. Place in the top frame nine (or eight) counters in one colour. (Note that 'filling' of ten frames is consistent with that suggested for developing tens facts.) On the bottom frame, display the second number in the fact pair with counters in a different colour. The top frame represents the number in relation to 10. Clearly, when the number is 9, one more counter is required to fill the ten frame. For 8, two more counters are required. Take the required number of counters from the bottom frame to fill the top frame. The thinking behind this strategy is to think of the first number as 10 and alter the second number accordingly. For example, 9 + 4 is 10 + 3; 8 + 5 is 10 + 3. The arrangement of counters on the tens frames for the facts 9 + 4 and 8 + 5 are presented in the diagram.

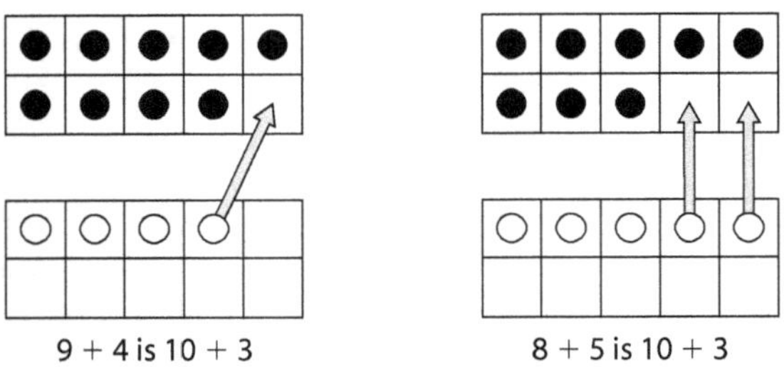

Double tens frame to show the thinking behind the bridging ten strategy

Last facts

By highlighting facts on an addition grid as each fact category is introduced, all of the 121 facts are covered by the preceding strategies—except one. That fact is 4 + 7, or 7 + 4, the only one that does not fit simply into any category. In this light, this fact is special, and it should be indicated to students as such. With this fact, strategies for solution are a discussion point. Some students may see it as an almost ten fact, some as a bridging ten fact, and some as double four add three; other students may simply see it as a special fact that gives 11.

Basic facts of subtraction

Once all addition facts have been consolidated, subtraction facts should follow easily. Each fact should be considered as a set of four facts—two addition and two subtraction facts—generated through rearrangement of the three numbers in the fact. For each fact encountered, students learn two facts using the commutative law—for example, 4 + 6 + 10; therefore, 6 + 4 + 10. Two subtraction facts are generated from the addition facts: 10 – 6 + 4, 10 – 4 + 6. Within each fact, the three numbers share a special relationship. These sets of four facts are sometimes referred to as *fact families*.

The most efficient strategy for subtraction facts is to *think addition*. The think addition strategy is best applied once the addition facts have been fully consolidated—that is, when students can rapidly determine the solution to the addition fact. When they are presented with a subtraction fact, attention is drawn to the related, and previously learned, addition fact and the number combinations within that fact. For example, when presented with 10 – 6, to think addition is to think of the number that, when combined with 6, makes 10.

Addition and subtraction: mental computation

Having ready recall of basic facts assists in mental computation. The strategies outlined in the previous section for promoting recall of basic addition facts were presented as ways in which learning basic facts can be facilitated through provision of visual images and thinking strategies. The thinking strategies are also useful for mental computation. Consider how you would mentally compute the examples in the diagram.

Common strategies for mentally computing solutions to these examples will include the strategies outlined for learning the basic facts. In the first example, it is possible to see that the bridging ten strategy can be used when adding 90 and 50; in the second example, the adding ten strategy can be used; for the third example, knowledge of tens facts is useful; and in the fourth example, doubles are presented. For the last example, another strategy may be useful: the *count-back* strategy for subtraction. Like the count-on strategy, this strategy is useful for counting back 1, 2 or 3, but becomes inefficient and prone to error when used to count back larger amounts.

How would you work these out?

The suggested methods for mental computation are not the only ways in which solutions may be determined, and mental methods are diverse and often quite idiosyncratic. The important thing is that learners are given support and encouragement to determine their own solution methods, and that they have the confidence to do so. Armed with automatic recall of basic addition and subtraction facts, as well as thinking strategies for computation, learners are in a strong position to invent their own procedures for computation that will be inherently meaningful to them.

Basic facts of multiplication

Automatic recall of basic multiplication facts is an important objective of primary-school mathematics, and is vital for mental computation and estimation. There is no question about the value of students learning these basic facts, but the approach taken must be considered. Why students need to learn their tables beyond the ten times must also be questioned. For efficient mental computation, students need automatic recall of all multiplication facts up to ten times. From this, they can derive solutions to eleven and twelve times tables and beyond.

Foundations for basic multiplication facts

Counting patterns lay the foundation for multiplication facts. Skip counting in various amounts encourages students to look for patterns in counting, and this builds number sense. Early skip counting activities should provide students with some form of concrete reference, so that the size of numbers being counted can be conceptualised. The following are some ideas for early counting in tens, fives, and twos where the students are standing in a line and count progressively in a specified amount:

- For counting in tens, the first student in line brings her/his hands forward, stretching out the fingers while shouting 'ten'. The second child in line does the same, but shouts 'twenty'. This continues down the line until all students have raised their hands.

- For counting in fives, the first student brings one hand forward and shouts 'five', then the other hand forward and shouts 'ten'; the second student brings one hand forward and shouts 'fifteen', then 'twenty', and so on down the line.
- For counting in twos, the students stand in line and, at the count of 'two', 'four', 'six', etc., each student blinks in turn (i.e. using two eyes), with a nod of the head.

Follow-up activities for these counting sequences can include making posters using cut-outs of students' hands, and cutting out magazine pictures of eyes that are displayed in a linear fashion with the counting sequence displayed underneath.

Displays of counting sequences enable patterns to be determined. When counting in fives (from 0), the numbers in the sequence end in 5 or 0; when counting in tens (from 0), each counted number ends in 0; and when counting in twos (from 0), the sequence is 2, 4, 6, 8, 0 before the sequence repeats again, starting with 2. From previous exploration of odd and even numbers, students will recognise that the pattern of counting in twos (from 0) is also a way of identifying even numbers. Understanding of number relationships and patterns in number builds number sense.

Counting patterns and the one-hundred board

Displaying counters on a one-hundred board is useful for skip counting and patterning. On a one-hundred board, the numbers 1 to 100 are displayed in a 10 × 10 array, with the numbers 1 to 10 displayed along the first row and the numbers 91 to 100 displayed along the bottom row. The one-hundred board is actually a number line represented as an array rather than in a linear representation.

Select a number between 3 and 9 and place a transparent counter or block on each multiple of the starting number to create a visual representation of the counting sequence. Focus students' attention on the digits in the first row in the sequence, noting the pattern of digits and when the pattern recurs. For counting in nines, the one-hundred board emphasises how counting in nines is one less than counting in tens each time, since the placement of counters on the board shows a diagonal pattern. Key ideas associated with counting in nines emerge through analysing this arrangement of the last digit of numbers in the sequence and

Students working on the one-hundred board

noting that they follow a pattern of counting backwards in ones, and that the digits in each number in the nines sequence total 9 (18 is 1 + 8, which equals 9; 27 is 2 + 7, which equals 9; 36 is 3 + 6, which equals 9, and so on). Later, students can examine why this pattern is generated. Some counting patterns are quite complex, and the one-hundred board display may add to their complexity. For some students, the one-hundred board may not be a useful representation. And, clearly, not all patterns should be explored on the one-hundred board at once; this would lead to information overload. The one-hundred board is only one way to explore patterns. An interactive, digital version of the one-hundred board, along with a teacher's guide, is available here: www.venturaes.com/iosapps/hom2.html.

Sequencing basic facts

Certain groups of multiplication facts are easier to learn than others. The following is a suggested sequence for instruction whereby simpler facts are consolidated in order to reduce the number of individual facts to be learned. For example, students find the twos, fives and tens facts relatively easy to recall. When these facts are learned, students can readily access 44 of the 121 multiplication facts. In the following section, ideas for promoting understanding and recall of the basic facts of multiplication are described in the following order:

- twos, fives and tens facts
- ones and zeros facts
- pattern facts
- square numbers
- threes and sixes facts
- the last fact.

Twos, fives and tens facts

The easiest multiplication facts are the twos, fives and tens. Experience with skip counting in these amounts lays the foundation for learning the two, five and ten times tables. The first ten numbers in each counting sequence need to be signalled to students as being of particular interest for learning the multiplication facts. Although these are the easiest facts, each set of facts should be treated separately until that set is consolidated.

Twos facts (doubles)

The twos facts should be the first facts that students learn in the formal study of multiplication facts. From the study of addition facts, students should already know their two times tables, since these facts are actually the addition doubles. Linking the twos facts to addition doubles connects multiplication to repeated addition. The symbolic representation of the addition doubles fact and counters in appropriate arrays should be presented with the multiplication fact in symbolic form so that connections can be seen.

$6 + 6 = 12$

$6 \times 2 = 12$

A group of 6, times 2

As each double is related to multiplication, its associated spin-around fact is also presented. The answer to 6 × 2 results in the same total as 2 × 6.

Fives facts

Early explorations of counting in fives (from 0) reveal that all numbers in the sequence end in 5 or 0, and this is a check when learning the multiply by five facts. An analogue clock can also be used as a reference point, in that the numbers 1 to 12 on the clock are related to five-minute intervals. When multiplying by five, the number of fives to be calculated can be located on the clock and counting in fives can occur from the one to that point to give the answer. For example, for 7 × 5, locate the 7 on the clock and skip count in fives to reach 7 (35).

Tens facts

Counting in tens from zero has a simple rhythm (10, 20, 30 . . .) and is learned easily by students. Making bundles of ten icy-pole sticks secured with rubber bands provides a physical referent for counting in tens, and also builds number sense through conceptualising the size of numbers. As each group of ten is added, the number of groups and the total amount need to be recognised consciously. If there are three groups of ten, there are 30 icy-pole sticks. A quick check of tens facts will reveal which students in your class need further support for their tens facts.

Ones and zeros facts

Multiplying by one and multiplying by zero often cause difficulty for students, who may confuse the solution with that which occurs when adding one or zero. Special attention needs to be given to these two groups of multiplication facts so that students will have something to which they can refer when they are feeling confused as they recall the one and zero times tables.

Ones facts

As with previous facts, the interpretation of the symbolic representation with the concrete representation provides a mental clue for checking answers. A group of counters is displayed and discussed as one group of a particular number of counters. The matching written form is given and analysed. If there are seven counters, this is described as 'a group of seven taken one time'. The answer is 7. From this, the spin-around fact is derived so that 7 × 1 results in the same total as 1 × 7. The generalisation is that, upon multiplying by one, the solution is the starting number. This can be confirmed with a calculator. A real-world situation that can be invoked when multiplying a number by one is to think of queues. Students line up at the tuckshop and know their place in the line. If there are eight others ahead of them in the line, then they are modelling 9 × 1, which equals 9.

Zeros facts

Multiplying by zero is difficult in that finding a real situation in which to contextualise why multiplication by zero results in zero is challenging.

Initially, students can model multiplying by zero using a story in which either the number of objects in a group or the number of groups is slowly reduced to zero. Thus, by presenting multiplication equations in a particular sequence, students can predict that the next number in the sequence is zero:

$5 \times 3 = 15$
$5 \times 2 = 10$
$5 \times 1 = 5$
$5 \times 0 = 0$

Use of a calculator can support students' knowledge that a number multiplied by zero is zero, that zero multiplied by a number is zero, and that zero multiplied by zero is zero. Patterning may also help.

Division by zero

One of the tricky facts to learn is division by zero. When students enter the following in a calculator, they may be confronted by an unexpected result:

$0 \div 12 = 0$
$12 \div 0 =$ error

Patterning can assist students to make meaning of the second situation. By observing the following sequence, it can be seen that, as you divide by increasingly smaller amounts, the solution gets bigger and bigger:

$12 \div 12 = 1$
$12 \div 6 = 2$
$12 \div 4 = 3$
$12 \div 2 = 6$
$12 \div 1 = 12$
$12 \div 0.5 = 24$
$12 \div 0.1 = 120$
$12 \div 0.01 = 1200$
$12 \div 0.001 = 12\,000$

When dividing by a very small number—a number that gets closer and closer to zero—the solution becomes very large. We can only imagine the size of the solution if we actually did divide by zero. The solution is undefined.

Pattern facts: fours, nines, square numbers

Fours facts

The fours facts require facility with the twos facts because the strategy is to think of doubling the twos.

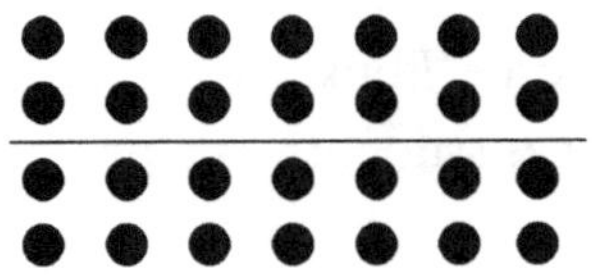

Visually, the twos facts are shown and then the fours facts shown as a mirror image of the twos facts. The diagram shows 7 × 4, where this fact is considered as 7 × 2 (which equals 14) and is then doubled (to give 28).

Nines facts

Learning the nines facts is a process of using two thinking strategies. The first is to relate the fact to known tens facts to give an estimate of the correct answer. For example, to determine 9 × 6, find 10 × 6 (60). Because 9 is less than 10, the solution will be 50-something. To determine what the correct solution will be is to know the patterns made when multiplying by nine, since the answer to each fact gives a combination of digits that add to nine:

$$9 \times 2 = 18 \qquad 1 + 8 = 9$$
$$9 \times 3 = 27 \qquad 2 + 7 = 9$$
$$9 \times 4 = 36 \qquad 3 + 6 = 9$$
$$9 \times 5 = 45 \qquad 4 + 5 = 9$$
$$9 \times 6 = 54 \qquad 5 + 4 = 9$$
$$9 \times 7 = 63 \qquad 6 + 3 = 9$$
$$9 \times 8 = 72 \qquad 7 + 2 = 9$$
$$9 \times 9 = 81 \qquad 8 + 1 = 9$$

As the solution to 9 × 6 is 50-something, the answer must be 54, because 5 + 4 = 9 and this fits the pattern of multiplying with nines. Prior experience with skip counting and exploration of the representation of counters on the one-hundred board will provide further confirmation of the accuracy of the solution.

Square numbers

Many students do not realise that square numbers are given their name because that is the shape that the numbers make when they are represented using blocks or counters. It is important for students to know the square numbers as a group of special numbers, because square number patterns are encountered frequently in algebra and other number study. Early exploration with square numbers should provide students with the opportunity to discover the first ten square numbers themselves through the use of concrete materials.

1 4 9

Square numbers

When linking square numbers to the multiplication facts, the image of blocks or counters in square arrays should be a visual referent point that students can use to assist in solution recall.

x	**0**	**1**	**2**	**3**	**4**	**5**	**6**	**7**	**8**	**9**	**10**
0	0	0	0	0	0	0	0	0	0	0	0
1	0	1	2	3	4	5	6	7	8	9	10
2	0	2	4	6	8	10	12	14	16	18	20
3	0	3	6	9	12	15				27	30
4	0	4	8	12	16	20	24	28	32	36	40
5	0	5	10	15	20	25	30	35	40	45	50
6	0	6	12		24	30	36			54	60
7	0	7	14		28	35		49		63	70
8	0	8	16		32	40			64	72	80
9	0	9	18	27	36	45	54	63	72	81	90
10	0	10	20	30	40	50	60	70	80	90	100

Remaining facts

Remaining facts

After learning the twos, fives, tens, ones, zeros, fours, nines and square number facts, almost all of the 121 multiplication facts have been covered, as seen on the following grid. The remaining facts are indicated by the grey cells. There are 12 empty cells in the grid, but only six facts remain to be learned, since the other six are spin-arounds. There are three facts from the threes (3 × 6; 3 × 7; 3 × 8), two facts from the sixes (6 × 7; 6 × 8) and one from the sevens (7 × 8).

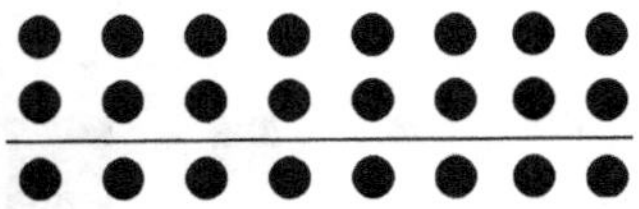

3 × 8 = (2 × 8) + (1 × 8)

Threes facts

One way to help students learn the remaining threes facts is to encourage the use of a build-up strategy, whereby the fact is linked to a known twos fact. For example, the 3 × 8 fact can be considered in terms of 2 × 8, or doubling 8, and then adding another set of 8 to that double. The visual representation can also provide a mental picture to assist in the thinking process.

Sixes facts

A similar strategy can be used for learning the remaining sixes facts, whereby the related fives fact is used as the starting point. For example, to find 6 × 8, think 5 × 8 (40) and then add another set of 8 (48).

The last fact

The only fact not covered by any of these strategies is 7 × 8 (and its spin-around, 8 × 7). For this fact, students have a variety of thinking strategies at their disposal, acquired through learning all previous facts. This fact can be solved using any number of strategies, but, because it is the only fact in this category, students should feel free to use whatever method they prefer and to discuss their choice.

Basic facts of division

As with deferring subtraction facts until addition facts are known, the division facts should be deferred until the multiplication facts are consolidated. Learning division facts is assisted by automatic recall of multiplication facts and number sense, in terms of understanding the relationship between the three numbers in each multiplication fact.

Each multiplication fact consists of two factors, which are multiplied together to give a third number: the *product*. The structure of multiplication facts is *factor* × *factor* = *product*. Rearrangement of the three numbers in each multiplication fact gives two corresponding division facts—for example:

$4 \times 7 = 28 \qquad 28 \div 4 = 7$
$7 \times 4 = 28 \qquad 28 \div 7 = 4$

The strategy for division facts is to *think multiplication*. When provided with a division problem, students should think of what number, when multiplied by the given answer, equals the starting amount. The structure of division facts is *product* divided by a *factor* to give the other *factor* (product ÷ factor = factor).

Mental computation and estimation

Automatic recall of the 121 basic multiplication facts assists mental computation and estimation. Consider the following exercises:

7 × 50	5 × 25
5 × 19	4 × 26
10 × 40	6 × 49
3 × 99	7 × 104

With ready recall of basic facts, and with number sense that incorporates place value, mentally determining the solution to the above exercises should be a relatively straightforward exercise. Addition strategies are also seen to play a role in attaining a solution. For example, to calculate 5 × 19, a simple path may be to round 19 to 20, multiply 20 by 5 and then take away the one group of 5 ones that were added when 19 was rounded to 20.

Providing students with opportunities and encouragement to practise mental computation with multiplication and division helps them come to rely on their own mental ability to calculate. With confidence in mental computation ability, students are in a position to choose the most appropriate computational method in any given situation—pen-and-paper, mental or calculator.

Estimation

Estimation assists in all computation situations involving written, mental or calculator methods. Having an estimate of the answer provides a checking mechanism against which the calculated answer can be compared and the reasonableness of the result determined. Although very useful for computation, estimation is not a natural process, and students need to be taught specific strategies to estimate.

Frequently, the processes of estimation and mental computation are considered to be the same; however, this is not the case. Estimating the cost of two music CDs at $27 each means coming up with an approximate answer, possibly by rounding $27 to $30 and then multiplying by two to give $60. Mental computation is to find the exact answer. The estimated answer is around $60; the exact answer is $54.

Front-end estimation

One of the skills of estimation is called *front-end estimation*, where the first digits in each number are the point of focus. For example, to estimate 56 × 74, the front digits of 50 and 70 are considered and the multiplication is carried out. When estimating the sum of three numbers such as 43, 49 and 46, the front four in each number are added together (or one of the fours is multiplied by three) to give an estimate of the solution. Front-end estimation is useful as a beginning point for estimating answers, but further estimation often needs to be conducted for a better estimate to be made.

Compatible numbers

Another strategy for estimation is finding compatible numbers. For example, to estimate 25 × 28 × 4, a beginning point would be to multiply 25 × 4, because this yields 100. These two numbers are compatible. The ability to see compatible numbers is a product of number sense and having a wide-ranging repertoire of knowledge about number relationships.

Rounding

Rounding numbers can also assist in seeing compatible numbers. In our base 10 system, rounding involves rounding to the nearest ten, hundred, thousand, and so on. However, for assisting in mental computation, 'rounding' numbers to five can assist in creating compatible numbers for mental computation purposes. In the following examples, the numbers in each exercise have been rounded to be compatible. Once the numbers are compatible, estimation becomes a much simpler task:

38 + 67 + 49 + 56	becomes	35 + 65 + 50 + 50
64 × 8	becomes	60 × 8
60 ÷ 8	becomes	64 ÷ 8
2637 ÷ 4	becomes	2800 ÷ 4

Good estimates and better estimates

Students often find rounding and estimating to be a tedious process, one that is usually required of them before they are asked to complete written calculation procedures. Because students often find written calculation procedures to be tedious in themselves, they may perceive rounding and estimating to make the task even longer and more tedious.

To encourage students to develop and use estimation skills, they need to be convinced that such skills are valuable. This frequently means finding alternative reasons for estimating and rounding that are not extra tasks when practising written procedures. Estimation is a process of finding good estimates and then making better estimates. Students need to be engaged in discussion about when it is appropriate to overcompensate or undercompensate as they make estimations.

The daily mentals block

A structured, daily focus on mental computation will promote students' facility with basic facts and build number sense. The first ten minutes of every mathematics lesson should be devoted to mental mathematics activities and exercises. In the past, mental mathematics consisted of speed and practice exercises, with students chanting the basic facts of multiplication and then being tested on their recall of the facts through timed tests. Often, punitive action fell on those students who did not know their 'tables'. However, a new approach to mentals and the ten 'quickies' can benefit students when implemented with care and consideration of students' self-concept.

Sets of ten 'quickies' can be used to settle the class at the commencement of the lesson, and, if little attention is drawn to those who do not get ten out of ten correct, students will continue to put in effort and energy to enhance their mentals score. The ten quickies are a great way to warm up the brain. Each set of quickies should be previously learned maths facts.

After a few sets of ten quickies have been used as a warm-up, introduce the targeted strategy of focus. This should be the class focus for the week. As instant recall of basic facts is the foundation for mental computation, it is important that you support students to acquire these facts. Provide opportunities for students to explore the new strategy, discuss it and share similar strategies. Students should then be provided with varied practice activities so that they can apply the strategy and reinforce the basic facts that relate to it. An assessment task can provide students with feedback on progress. Once a fact category has been targeted, it can be included in future sets of ten quickies used at the start of the lesson. The aim is for automatic recall of basic addition and subtraction facts to support mental computation, estimation and number sense activities. Once mastery of basic addition and multiplication facts has been attained, build other mental computation activities into the daily mentals block (e.g. fractions, percentages, decimals, two-digit addition and subtraction, measurement conversions, shape names and properties).

Promoting mental computation in the middle years

Mental computation provides fluency in number and flexibility in using and applying number knowledge in various situations. In the middle years, a structured mental computation program is an essential component of the yearly mathematics curriculum. In a classroom in which mental computation is an integral part of a mathematics lesson, where students can share their thinking strategies and learn new strategies and approaches without ridicule, students' number knowledge and confidence in learning will flourish. In this chapter, a strategies approach for teaching the basic facts has been outlined. Through such an approach, students develop strategies for thinking about the basic facts that support recall of the facts when that recall fails. The aim of a basic facts program is automatic recall, but the pathway to this is not through rote drill and practice. The middle-years teacher must diagnose students' difficulties with basic fact recall and target those groups of facts that are not part of students' automatic repertoire. After basic fact mastery, two-digit addition and subtraction exercises should follow; after that, target multiplication and division using single-digit numbers multiplied or divided by two-digit numbers. Every new number topic in the middle-years curriculum is an opportunity to promote mental computation. When doing fractions, promote addition and subtraction where mental images assist students to solve the exercises:

$2 - \frac{5}{6}$
$\frac{3}{4} + \frac{6}{8}$
$\frac{5}{8} + \frac{5}{8}$

Ask students to visualise what is happening when you are trying to determine how many halves there are in 3. When they say that they can see three pizzas divided into two pieces each, resulting in six pieces, show them that they have just calculated 3 divided by ½. Ask students to think of a number line and determine how many 0.2s in one whole. Then ask them to mentally calculate 3 divided by 0.2 by asking them how many 0.2s are in 3. By posing fraction and decimal division questions in language that supports meaning, students' ability for mental computation of these problem types will be enhanced. For multiplication, think of groups. For example, 6 × ½ can be regarded as six groups of ½. This means that the solution is 3. When studying percentages, provide students with opportunities to mentally calculate percentages of given amounts, but carefully sequence the exercises. Begin with asking students to find 50 per cent of given amounts, linking 50 per cent with ½. Then extend this to finding 25 per cent, where students find half and then half again. But also discuss other strategies. For example, students may find it easier to find one-quarter of an amount when the amount is readily recognised as being a multiple of 4 (e.g. 25 per cent of 800). The halving and halving strategy may work better for other numbers (e.g. 25 per cent of 968). Include mental computation exercises when studying measurement topics. Provide students with opportunities to convert between units of measure. For example, when studying length, reinforce the place value nature of SI metric units, by providing students with practice in changing centimetres to metres (multiply by 100), metres to centimetres (divide by 100), centimetres to millimetres (divide by 10), and so on. As outlined above, daily mental computation practice is an important element of a mathematics lesson, and this is especially so in the middle years.

Written computation and standard algorithms

An *algorithm* is a prescribed sequence of steps that, when executed correctly, results in the desired outcome. Real-world examples of algorithms include using an ATM, building a model plane, putting together a trampoline and even, possibly, making a cup of coffee. When the steps are executed in these algorithms in the prescribed manner, the results are favourable. If particular steps are not executed as prescribed, the result may be less than satisfactory. In mathematics, algorithms are procedures for performing calculations to achieve the correct solution to a problem or exercise. In the primary-school mathematics curriculum, students meet algorithms as they are taught standard pen-and-paper calculation procedures for addition, subtraction, multiplication and division.

The debate about the value of teaching standard written algorithms in a digital age will be ongoing. The Australian Curriculum: Mathematics encourages students to use a range of flexible mental and written strategies, and the term 'algorithm' is not used. There is considerable evidence suggesting that students should be at liberty to devise their own methods for written computation (e.g. Clements & Sarama 2014; Csikos 2016; Torbeyns & Verschaffel 2016). The reality is that many students may not be provided with such an opportunity. Many of the calculations undertaken outside the school environment use very different strategies from those used in the classroom. Facility, confidence and competence in these methods would be greatly enhanced if there were a balanced approach to teaching calculations.

The standard algorithms for the four operations are extremely efficient computational procedures that, for many students, have stood the test of time. They are legitimate operations and, once learned, enable computation when the numbers are too large for mental computation and there is no calculator close at hand. Teaching standard algorithms provides children with an efficient means for computation.

Instruction in the algorithms must be planned carefully to enable students to develop an understanding of the steps. It must always be remembered that accuracy in performance of written computation is not a measure of understanding (Lemonidis 2016). The approaches to teaching the algorithms described below are just some of many. We strongly recommend that teaching written algorithms not overshadow an emphasis on mental methods, and that the value of using technology for calculations if poor written computation impedes problem-solving is not downplayed.

The value of concrete materials

Concrete materials are typically used to support students' understanding of the steps in the standard algorithm. The use of appropriate concrete materials to embody the steps in the standard algorithms has been the focus of much research. Typical concrete materials for teaching algorithms include sticks that can be bundled into tens with an elastic band, and base 10 blocks. Early models are proportional—that is, size correlates with value—so that the tens block is equivalent in size to ten ones blocks or the ten sticks bundled together represent ten single sticks. These materials can exemplify the processes of algorithms and assist children to understand the algorithms, but specific research into the high cognitive demand and the confusion they cause for children (Bolton-Lewis & Halford 1992) should not be dismissed. The how, why and for how long of concrete (and later digital) materials for teaching the algorithms need to be considered carefully.

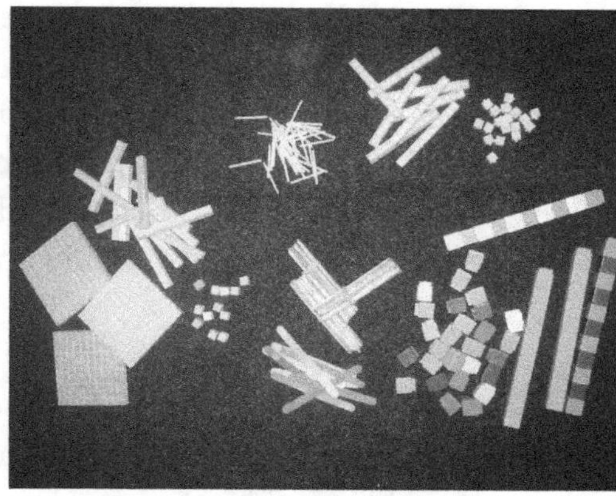

Some of the materials available for teaching algorithms

Principles for using concrete materials

1. Use concrete materials to support students' understanding of the steps of the algorithm.
2. Carefully select examples for demonstrating the steps in the algorithm—use simple examples that contain small numbers.
3. Once students demonstrate understanding of the steps in the algorithm, as exemplified through the concrete materials, the concrete materials should no longer be used.

4. Use concrete materials to support understanding of the steps only; concrete materials should not be used for the purpose of obtaining solutions.
5. If students require the use of concrete materials to obtain answers, they are not ready to practise written computation methods.

The addition algorithm

Prerequisite knowledge

The prerequisite knowledge for developing meaning of the standard algorithm is as follows:

1. place value and numeration understanding—that is, understanding of how numbers can be grouped into tens (e.g. 10 and 10 and 10 is 30) and regrouped into tens and ones (25 is 2 tens and 5 ones, or 1 ten and 15 ones; see Chapter 9)
2. understanding of the concept and language of addition (see Chapter 9)
3. basic addition fact recall
4. estimation, rounding and mental computation skills.

Sequencing

There is a sequence for teaching the standard addition algorithm, since knowledge of adding with particular numbers lays the foundation for adding with other numbers. Learning the written algorithms for adding numbers such as 25 and 32 where there is no trading (or regrouping) between columns lays the foundation for adding numbers such as 26 and 38.

For students who have been encouraged to develop mental methods, teaching the addition algorithm for numbers where there is no regrouping may appear to be a useless skill, since calculation of the solution would be a simple mental exercise. However, when using concrete materials to exemplify the written procedure, examples involving small two-digit numbers are required in order to reduce unwieldy manipulation of materials. Once addition with two-digit numbers has been mastered, the procedure can be applied to adding larger numbers, and this is the rationale for teaching the written procedure.

Linking concrete to written

Materials required for linking the concrete to the written algorithm include place value charts (large enough to accommodate base 10 blocks), base 10 blocks (MAB—see Chapter 9 for a description of this material), pen and paper, and a calculator. Since the concrete materials are exemplifying the steps in the written procedure, they must mirror the written procedure as closely as possible. Linking the concrete to the written for the addition algorithm proceeds as follows:

1. Select a simple example that is easily represented with concrete materials.

   Add 36 and 28

2. Write the computation in vertical form, emphasising the importance of aligning the digits in the correct places.

$$\begin{array}{r} 36 \\ +\ 28 \\ \hline \end{array}$$

3. Calculate the answer on the calculator and leave this display visible for ready reference.
4. Display the two numbers to be added (the addends) on the place value chart with concrete materials, arranging the materials to mirror the written representation.

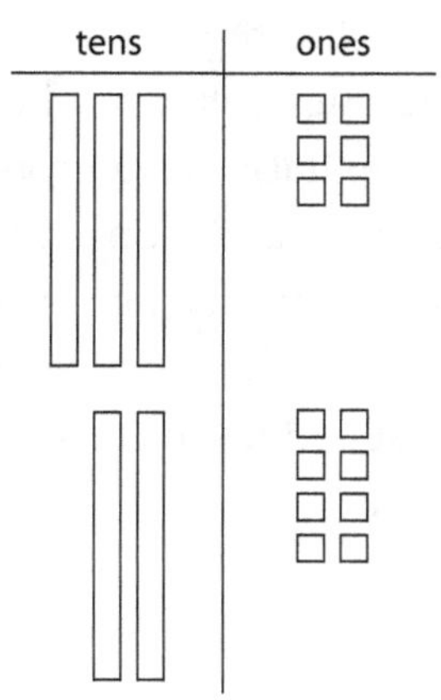

$$\begin{array}{r} 36 \\ +\ 28 \\ \hline \end{array}$$

5. Focus on the ones place in both the concrete and symbolic representation. Verbalise the two numbers and ask students to state the solution not by counting, but as a basic fact.

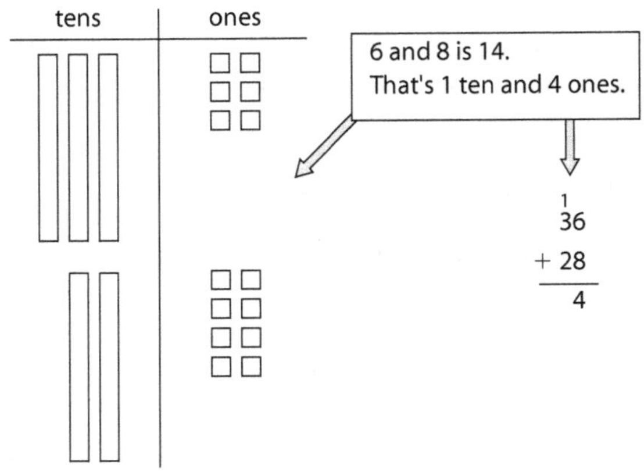

6. If the solution is greater than 10, regroup the ones into tens. Draw attention to how this step is recorded on paper.
7. Repeat steps 5 and 6 as attention moves to the digits in the tens place:
   (a) Verbalise the two numbers, and recall the solution as a basic fact.
   (b) If the solution is greater than 100, regroup the tens into hundreds and tens.
   (c) Draw attention to how this is recorded on paper.
8. Check that the written solution is the same as the calculator display. Repeat this sequence with other examples.

As students manipulate the blocks at the same time as recording in symbols, they are building meaning for the steps in the algorithm. For example, why is it easier to start in the ones column when we so often start in the tens column for mental computation? Regrouping provides this reason. Once students have mastered addition of two numbers, extend this to addition of more than two numbers, but this time, use of concrete materials should not be necessary. The same processes apply, as shown in the sequence outlined in this section, which is a further reason for the sequence of steps in the algorithm.

The subtraction algorithm

Prerequisite knowledge

The prerequisite knowledge for understanding the meaning of the standard subtraction algorithm is similar to that for addition:

1. place value and numeration understanding—that is, understanding of how numbers can be grouped into tens (e.g. 10 and 10 and 10 is 30) and regrouped into tens and ones (25 is 2 tens and 5 ones, or 1 ten and 15 ones; see Chapter 9)
2. understanding of the concept and language of subtraction (see Chapter 9)
3. basic subtraction fact recall
4. estimation, rounding and mental computation skills.

Linking concrete to written

In the modelling of the addition algorithm, both numbers to be added are displayed on the place value chart. Modelling the subtraction algorithm is different in that only the minuend (the starting amount) is shown. The subtrahend (the amount taken from the starting amount) is not shown. Materials required for linking the concrete to the written algorithm are the same as for addition: a place value chart (large enough to accommodate base 10 blocks), base 10 blocks, pen and paper, and a calculator. Since the concrete materials are exemplifying the steps in the written procedure, they must mirror the written procedure as closely as possible. Linking concrete to written for the subtraction algorithm proceeds as follows:

1. Select a simple example that is easily represented with concrete materials.

   43 take away 28

2. Write the computation in vertical form, emphasising the importance of aligning the digits in the correct places.

$$\begin{array}{r} 43 \\ -\ 28 \\ \hline \end{array}$$

3. Calculate the answer on the calculator and leave this display visible for ready reference.

4. Display only the number from which an amount is going to be subtracted on the place value chart.

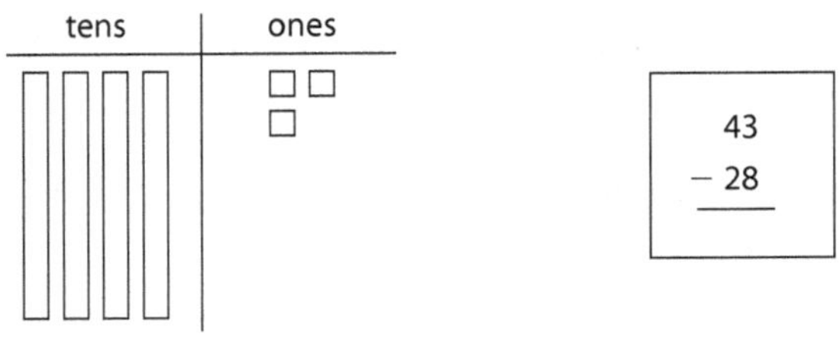

5. Focus on the ones place in both the concrete and symbolic representation. Are there enough ones to take the required amount of ones from? If yes, use basic fact recall and record the remaining amount in the ones place. If no, go to the tens place and regroup one ten in that column as ten ones. Place the ten ones in the ones column and subtract the required amount from the ones.
6. Draw attention to how this step is recorded on paper.

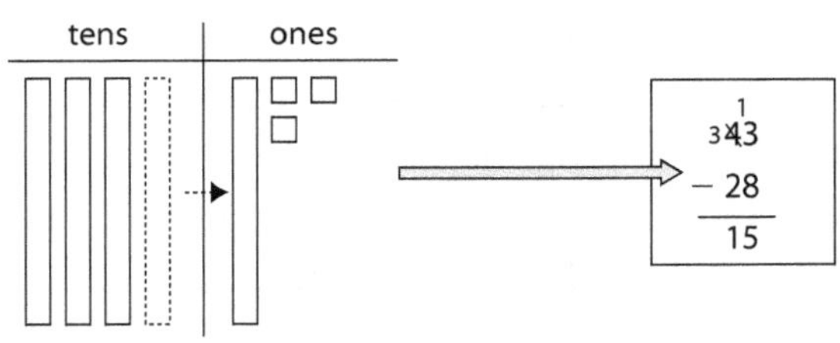

7. Repeat steps 5 and 6 as attention moves to the digits in the tens place.
8. Check that the written solution is the same as the calculator display. Repeat this sequence with other examples.

This process of subtraction is referred to as the *decomposition* method. There is an alternative algorithm for subtraction that many people still remember from their time at school: the *equal additions* method. This method is also sometimes referred to as the 'borrow and pay back' method. It was taught without recourse to materials for meaning, and is not readily modelled with base 10 blocks. In this method, when a number is too small to subtract an amount from, a 1 is 'borrowed' and written next to the small number in the column under focus, with the 1 paid back to the next number below and to the immediate left. The 1 that was borrowed became a value of 10 and was added to the small number. This then enabled the subtraction to occur. The 1 'paid back' at the bottom was added (as 1) to the number to be subtracted in the next operation. For example, in the exercise 65 – 17, start in the far-right column, and note that you can't take 5 from 7. Borrow a one and write it as a small notation to the left of the 5; 'pay back' this 1 by writing a 1 as a small notation next to the 1 that is immediately to the left of the 7. Now read the 5 as 15 and take 7, leaving 8 in that column. Next, consider the 1 and the notated 1 in the next column as 2, and take this from 6, which leaves 4. The result is 48. Even though this approach, if followed precisely, ensures a correct result every time, its steps are difficult to model meaningfully. If students know this method and can execute it flawlessly, then there is no need to teach the decomposition method of subtraction as well, since this would cause confusion and be a redundant piece of knowledge (unless there is student interest).

Generalising the addition and subtraction algorithms

The steps in the addition and subtraction algorithms follow a pattern whereby, for efficiency, the digits in the far-right column are dealt with first, before the next column to the left of the first digit, and so on, until the digits in all the columns have been dealt with. An understanding of the legitimacy of the steps within the algorithms promotes conceptual understanding of the mathematical principles underlying the algorithms. In the addition algorithm, the ones digits are added to the ones digits, the tens digits are added to the tens digits, and so on.

Similarly, in subtraction, the ones digits are dealt with together, as are the tens digits, then the hundreds digits, and so on. It is illegitimate to subtract a ten from a one in the subtraction algorithm; the reason for aligning the digits in their appropriate columns is so that this type of error will not occur. A further principle is that of regrouping across the columns. When there are more than ten ones in the ones column upon adding, groups of ten are 'carried' over to the tens place. This pattern recurs across all places as we proceed from right to left. The principles of the algorithms with whole numbers provide the foundation for understanding algorithms with other types of numbers, such as decimals and fractions; measurements such as time and distance; and algebra, as outlined in the following examples:

1. *whole numbers*—ones added to ones, tens added to tens, hundreds added to hundreds. Regrouping of ones to tens, tens to hundreds, hundreds to thousands.

$$\begin{array}{r} 469 \\ +\ 948 \\ \hline \end{array}$$

2. *decimals*—hundredths added to hundredths, tenths added to tenths, ones added to ones, tens added to tens. Regrouping of hundredths to tenths, tenths to ones, ones to tens.

$$\begin{array}{r} 25.7 \\ +\ 18.45 \\ \hline \end{array}$$

3. *fractions*—parts added to parts, wholes added to wholes. Regrouping of parts into wholes.

$$\begin{array}{r} 6\ \tfrac{1}{4} \\ +\ 5\ \tfrac{3}{8} \\ \hline \end{array}$$

4. *lengths*—centimetres added to centimetres, metres added to metres. Regrouping of centimetres into metres.

$$\begin{array}{r} 59\text{ m } 84\text{ cm} \\ +\ 26\text{ m } 39\text{ cm} \\ \hline \end{array}$$

5. *time*—minutes added to minutes, hours added to hours. Regrouping of (60) minutes into one hour.

$$\begin{array}{r} 6\text{ hrs } 29\text{ mins} \\ +\ 4\text{ hrs } 35\text{ mins} \\ \hline \end{array}$$

6. *algebra*—collection of like terms. Like terms are added to like terms.

$$\begin{array}{r} 2 + 3y + 7 + 2y \\ 2 + 3y \\ +\ 7 + 2y \end{array}$$

The multiplication algorithm

The multiplication algorithm is very complex, but extremely efficient. Before students tackle the multiplication algorithm, the following prerequisite knowledge must be in place:

1. place value and numeration knowledge, particularly in relation to understanding the effects upon numbers through multiplying by powers of 10 (see Chapter 9)
2. expanded notation
3. understanding of the concept and language of multiplication
4. basic multiplication facts
5. the distributive property
6. estimation, rounding and mental computation skills.

Place value

The effect of multiplying numbers by powers of 10 provides the foundation for the multiplication algorithm. Entering a single-digit number in a calculator and observing what happens to that number as it is progressively multiplied by ten promotes understanding of movement across the places, rather than a generalisation that zeros are added.

What happens to each number when we multiply by a power of 10?

Th	H	T	O.	t
			6.	
		6	0.	
	6	0	0.	

The number jumps one place to the left and the zero is a place holder.
As we multiply by powers of 10, we move across places to the left.

Using basic fact knowledge and exploring the patterns of the zeros upon multiplication of numbers in multiples of ten promotes number sense in relation to the effects of the operations upon number. Progressive examples enable students to make generalisations about multiplication:

$$6 \times 4 = 24$$
$$6 \times 40 = 240$$
$$6 \times 400 = 2400$$

$6 \times 4000 = ?$
$60 \times 4 = ?$
$60 \times 40 = ?$

As larger numbers are multiplied, the size of the solution becomes anticipated, and this assists in checking the reasonableness of solutions upon using written methods.

Expanded notation

Expanded notation refers to expressing numbers in their component parts—that is, elaborating the number of ones, tens, hundreds, and so on, that are contained in a number. The number 345 in expanded notation is expressed as comprising three hundreds, four tens and five ones (300 + 40 + 5), or $3 \times 100 + 4 \times 10 + 5 \times 1$. The way numbers are represented in symbolic form assumes knowledge of the base 10 numeration system, and that the place in which a digit in a number is located indicates its value. Writing numbers in expanded notation assists in developing understanding of the distributive property.

The distributive property

In symbolic form, the distributive property is represented as follows: $(a \times b) + (a \times c) = a \times (b + c)$. The meaning of the distributive property can be exemplified using arrays. The following example shows the solution to 3×13 using the distributive property, where 13 is presented in expanded notation as $10 + 3$.

$3 \times 13 = 3 \times (10 + 3)$
$(3 \times 10) + (3 \times 3)$
$30 + 9$
39

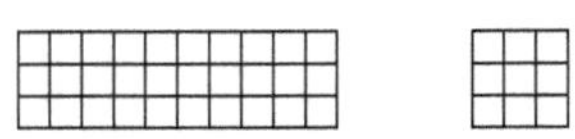

3 × 10 3 × 3

Developing the multiplication algorithm

The multiplication algorithm is efficient and concise, and is extremely difficult to teach meaningfully without materials. Sometimes, in an effort to ensure that students understand each step in the process, the algorithm is developed through convoluted and indirect means. If students have facility with all prerequisite knowledge for the multiplication algorithm, presenting them with various models and extended recording methods may result in greater confusion than a more direct approach.

A combination of symbolic, concrete and calculator representations to exemplify particular steps in the standard algorithm in a sequential manner is outlined in the following sections.

Simple multiplication

1. Select a simple example that is easily represented with concrete materials.

32 multiplied by 3

2. Write the computation in vertical form. Use the calculator (or mental computation) to determine the answer and have this answer displayed for ready reference.

$$\begin{array}{r} 32 \\ \times\ 3 \\ \hline \end{array}$$

3. Display the number that is to be multiplied on a place value chart.

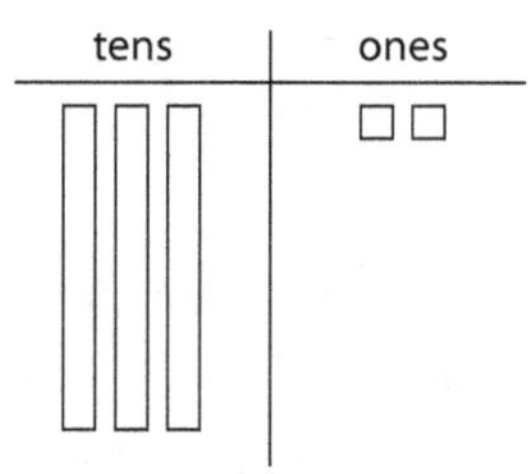

4. Focus on the ones place in both the concrete and symbolic representation. Refer to the written form to check how many times the ones are being multiplied.
5. Multiply the ones by recalling the basic fact.
6. Use base 10 blocks to represent that number.
7. Draw students' attention to the place for recording the number of ones.

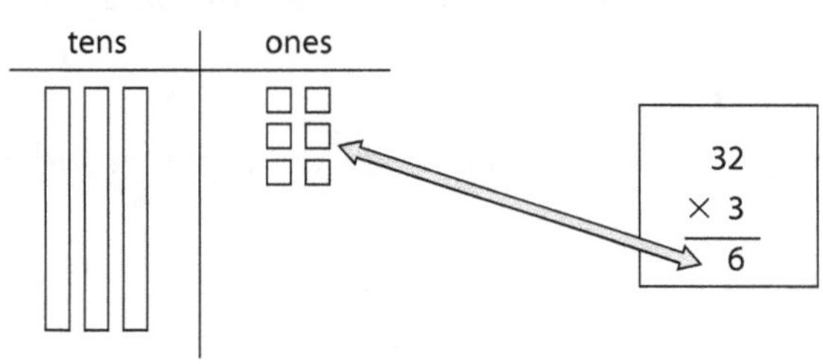

8. Check the ones digit on the calculator display and the number recorded in the ones place.
9. Focus on the number in the tens place in both the concrete and written forms. Refer to the written form to check how many times the tens are multiplied.
10. Multiply the number in the tens place by the multiplier, by recalling the basic fact.
11. Use base 10 blocks to represent that number and draw students' attention to the place for recording the number of tens.

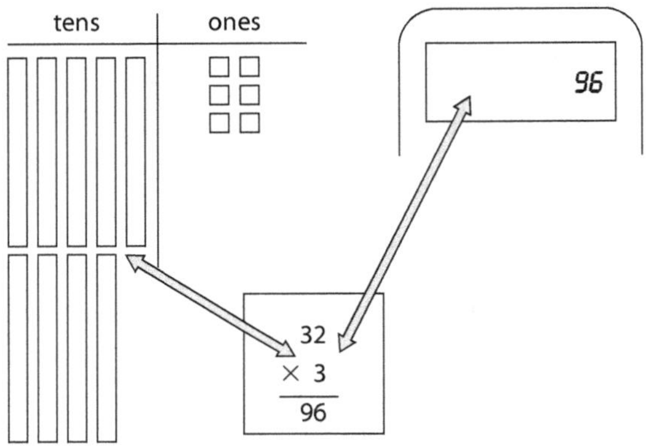

To consolidate the process, repeat with other examples until students can execute the algorithm without the use of the base 10 blocks. Once the pattern has been established, provide examples with larger numbers, such as 124 × 2; 232 × 3; 1124 × 2. When the process of multiplying with single digits without regrouping is established, the process of multiplying with single digits involving regrouping should be relatively simple, but attention needs to be drawn to the recording of *regrouped* digits. Continued reference to the calculator display will provide a further check for the legitimacy of the steps in the algorithm. Steps for teaching the multiplication algorithm for numbers involving regrouping are described below.

Multiplication with regrouping

1. Select a simple example that is easily represented with concrete materials.

24 multiplied by 3

2. Write the computation in vertical form.

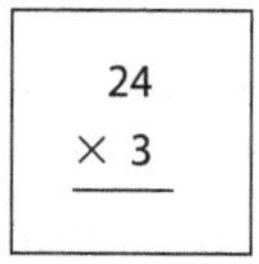

3. Use the calculator (or mental computation) to determine the answer and have this answer displayed for ready reference.
4. Display the number that is to be multiplied on a place value chart.

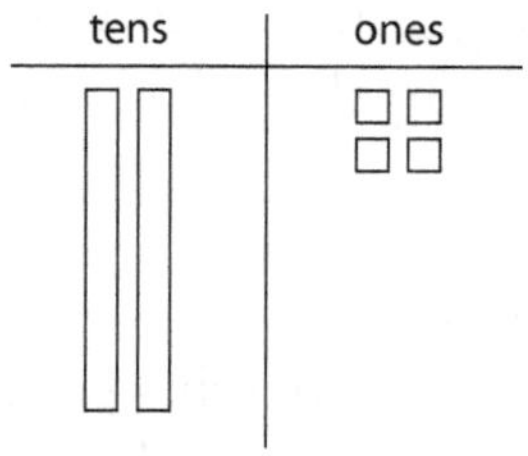

5. Focus on the ones place in both the concrete and symbolic representations. Refer to the written form to check how many times the ones are being multiplied.
6. Multiply the ones, by recalling the basic fact.
7. Use base 10 blocks to represent that number.
8. Draw students' attention to the place for *regrouping* the number of tens and recording the numbers of ones.

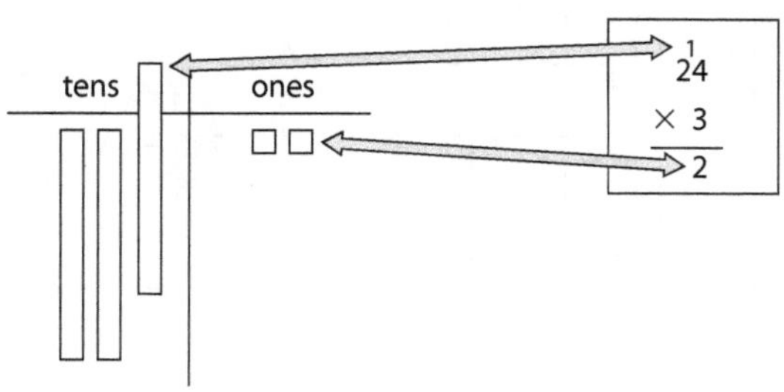

9. Check the ones digit on the calculator display and the number recorded in the ones place. Following the established pattern for multiplying without regrouping, ask students to look at the number in the tens place and determine the solution through basic fact recall, emphasising that they must disregard the regrouped figure. Then add the regrouped figure to the basic fact solution and record. Compare the answer with the calculator display.

When multiplying with regrouping, the established pattern from multiplying without regrouping is drawn upon. For many students, confusion occurs when deciding what to do with the regrouped digit. During instruction, the teacher must be extremely diligent in ensuring that students fully understand the role of the regrouped digit, and students should not be provided with practice examples until the teacher is confident of their understanding of that step in the algorithm. As students become more familiar with the process of regrouping in the tens place, provide further examples in which the solution requires an answer in the hundreds place (e.g. the solution to multiplying 34 × 6). Refer to the calculator display to show students how to record the answer.

Multiplying by two-digit numbers

Multiplying by two digits is a case of using prior knowledge of multiplying by single digits and possible new knowledge of multiplying by tens. From experience with number patterns, students should know the effect of multiplying a number by ten. Simple patterning activities, such as those described for place value, should lay the foundation for generalising the effect of multiplying by powers of 10.

The multiplication algorithm occurs in two parts, with the ones multiplied first and then the tens. These two totals are then added. Arrays can assist in providing a visual image of two-digit multiplication.

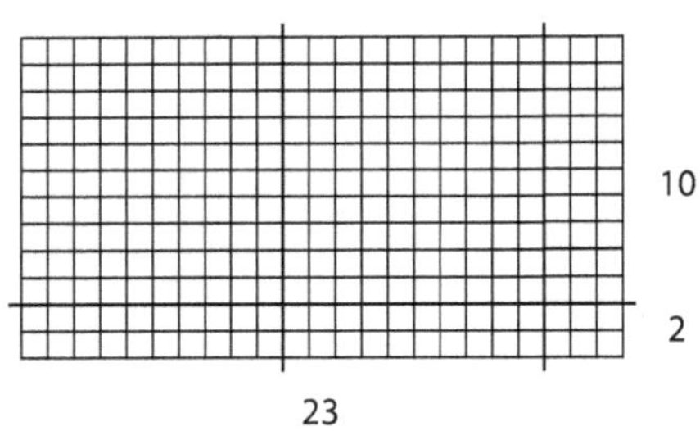

$$23 \times 12 = (23 \times 10) + (23 \times 2) \\ = 230 + 46 \\ = 276$$

When making links between the algorithm and the array, deliberately draw on students' prior knowledge of the sequence of the algorithm for single-digit multiplication. In two-digit multiplication, the distributive law comes into play. The sequence is to consider the multiplier in extended form (in tens and ones), and then to multiply each digit as in single-digit multiplication.

1. Select a simple example that is easily represented as an array.

23 multiplied by 12

2. Write the computation in vertical form.

$$\begin{array}{r} 23 \\ \times\ 12 \\ \hline \end{array}$$

3. Present the computation in array form, exemplifying the distributive property.

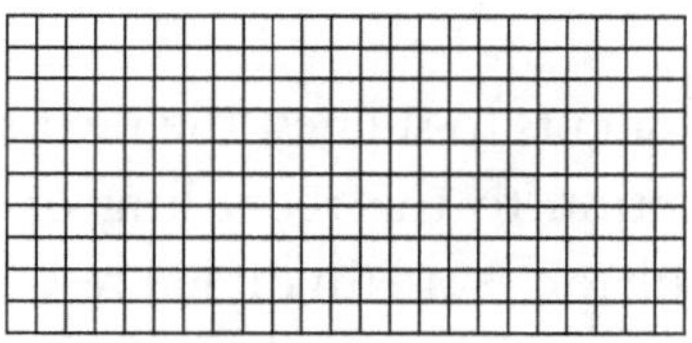

23 × 10

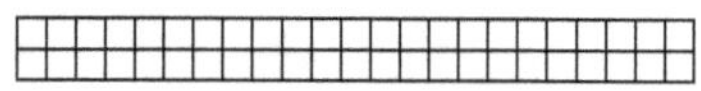

23 × 2

4. Use the calculator to determine the answer and have this answer displayed for ready reference.

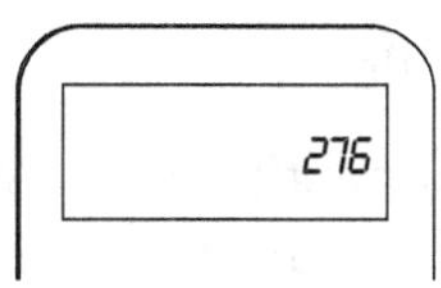

5. Focus on the digit in the ones place of the multiplier, and on the array in which the product of that number and the multiplicand (the number being multiplied) are presented. Discuss in relation to single-digit multiplication. Complete the multiplication as if for single-digit multiplication and draw students' attention to the recording of the product for that part of the algorithm.

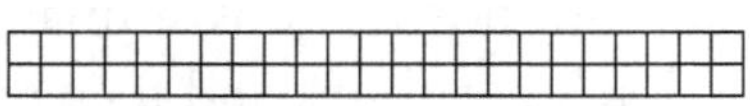

23 × 2

$$\begin{array}{r} 23 \\ \times\ 2 \\ \hline 46 \end{array}$$

6. Focus on the digit in the tens place of the multiplier, and on the array in which the product of that number and the multiplicand is presented. Discuss in relation to multiplication with tens and the fact that the solution will result in a zero in the ones place. Record the zero under the ones place, and complete the multiplication as if for single-digit multiplication. Draw students' attention to the recording of the product for that part of the algorithm.

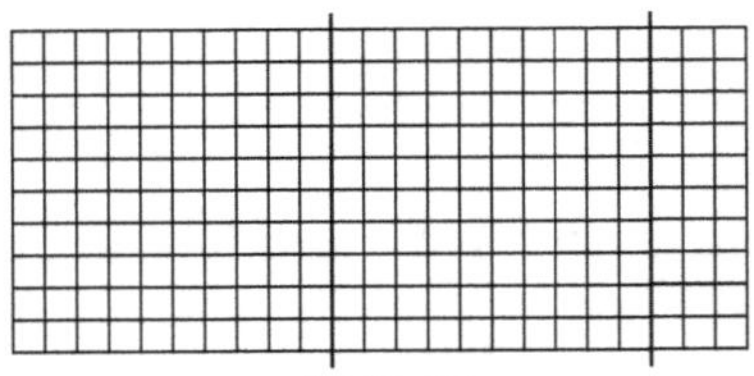

23 × 10

$$\begin{array}{r} 23 \\ \times\ 12 \\ \hline 46 \\ 230 \end{array}$$

7. Consider the array in total and invite students to suggest the next step in the algorithm. Draw students' attention to the recording of the solution. Check the solution on the calculator. A useful digital resource to support the development of the array model is available from the National Council of Teachers of Mathematics (NCTM) Illuminations website (www.nctm.org/Classroom-Resources/Illuminations/Interactives/Understanding-Multiplication-Using-Dynamic-Sketches-of-an-Area-Model/). This model is also appropriate when teaching multiplication with decimals.

When learning two-digit multiplication, students need to be provided with plenty of time to become familiar with the sequence of steps. The process of the algorithm may need to be revisited several times before students feel confident to execute it unassisted.

The division algorithm

Prerequisite knowledge

The division algorithm is relatively simple compared with the multiplication algorithm. Prerequisite knowledge is similar to that required for the addition and subtraction algorithms, and includes the following:

1. place value and numeration knowledge, particularly in relation to understanding the effects upon numbers through dividing by powers of 10

2. understanding of the concept and language of division
3. basic division facts
4. estimation, rounding and mental computation skills.

Place value

The effect of dividing numbers by powers of 10 provides the foundation for the division algorithm. Explorations similar to those for the multiplication algorithm are useful for assisting students to develop understanding of the effects of dividing numbers by powers of 10. Entering a single-digit number in a calculator and observing what happens to that number as it is progressively divided by ten promotes understanding of movement across the places, rather than a generalisation that zeros are deleted (or that the decimal point moves).

Th	H	T	O.	t
	6	0	0.	
		6	0.	
			6.	
			0.	6

The number jumps one place to the right. As we divide by powers of 10, we move across places to the right.

Using basic fact knowledge and exploring the patterns of the zeros upon division of numbers in multiples of ten promotes number sense in terms of the effects of the operations upon number. Progressive examples enable students to make generalisations about division:

$24 \div 6 = 4$
$240 \div 6 = 40$
$2400 \div 6 = 400$
$24\,000 \div 6 = ?$
$24 \div 4 = 6$
$240 \div 40 = ?$

As larger numbers are divided, the size of the solution becomes anticipated, and this assists in checking the reasonableness of solutions upon using written methods.

Linking concrete to written

Unlike the addition, subtraction and multiplication algorithms, the division algorithm begins with the digit in the biggest place. The emphasis for the division algorithm is in the use of *sharing* language:

1. Select a simple example that is easily represented with concrete materials.

496 divided by 4

2. Write the computation in standard form.

$4\overline{)496}$

3. Use a calculator to determine the answer and have this answer displayed for ready reference.

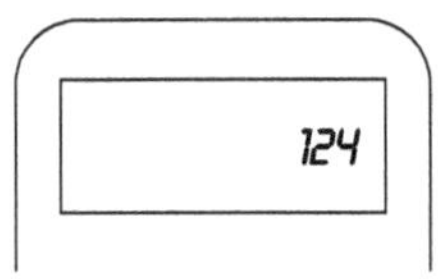

4. Represent the number to be divided with base 10 blocks.

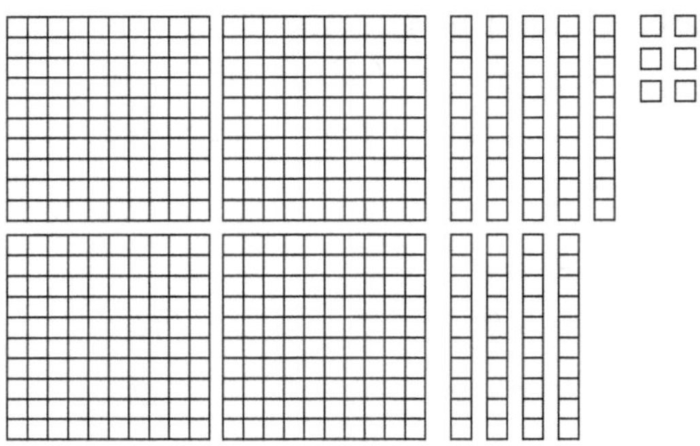

5. Start with the biggest group (400) and calculate the share by recalling the basic fact. Record this solution, drawing attention to the position of the solution and how it indicates the numbers of shares made from that amount.
6. Focus on the next biggest group to be shared, in both the written and concrete forms. Use basic fact recall to determine the number of shares. Draw attention to the position for recording the solution. Emphasise meaning of the steps in the algorithm by drawing attention to the size of the numbers being shared and the placement of the recorded answer at each step. At each step in the algorithm, check

the calculator display. Provide further examples until the process is familiar and students feel confident to compute division exercises using the algorithm without reference to the materials. Encourage students to estimate the solution prior to undertaking any computation and to check their resulting answer with their estimated answer. Calculators should always be available for checking results. Once students are familiar with the written algorithm, their learning can be reinforced via the NCTM Quotient Café digital manipulative found here: www.nctm.org/Classroom-Resources/Illuminations/Interactives/The-Quotient-Cafe/.

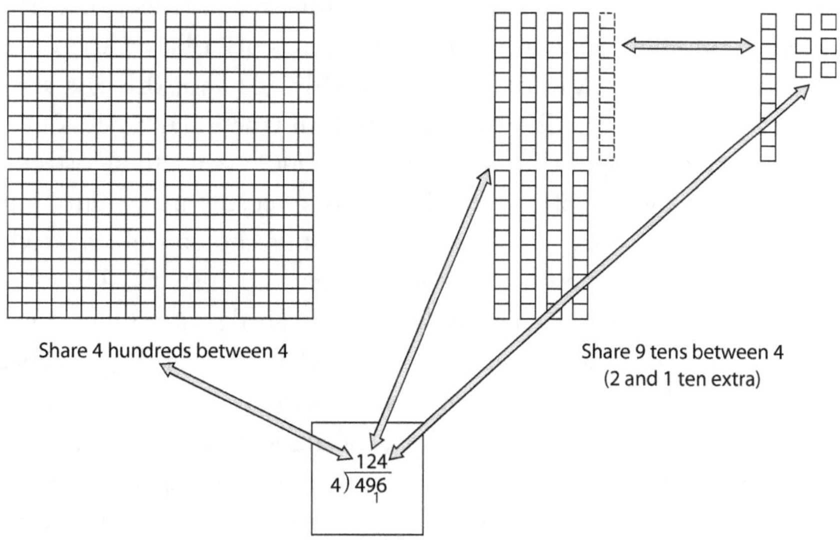

Long division

The long division algorithm has been a topic of considerable debate, and is often cited by students as the turning point at which they began to dislike mathematics. The long division algorithm is just that—long. It is an algorithm that takes a long time to execute. The long division algorithm can be taught meaningfully, and can promote students' understanding of mathematics and build number sense, as well as be an interesting program of study (Lampert 1992). For many students, however, the long division algorithm may be a piece of knowledge that they do not need. It is not highly likely that students will need to complete long division in non-classroom contexts (i.e. where an exact answer is required), and if they do, it is likely that they will have access to devices that they can use for such calculations. Therefore, teaching the long division algorithm is no longer a given, and the decision about whether to teach it should be left to the teacher's professional opinion, in accordance with an assessment of the needs of the students.

Teaching written algorithms in the middle years

Teaching children standard written algorithms in the middle years can be a tedious process. The question that must be asked is whether the time and effort required to re-teach standard algorithms are justified, or whether promoting number sense and estimation represents a better use of time in the middle years. With calculators readily at hand, this is a serious consideration for the middle-years teacher. If instructional time is devoted to teaching written algorithms in the upper-middle years, then instruction must be extremely well planned and focused, to ensure maximum impact in minimum time. The examples selected must be chosen well, with MAB used to exemplify the process but not for getting answers. That is, the MAB should be used to legitimise the steps in the algorithm, but students should then practise the algorithm without MAB so that the process can be internalised. However, if students exhibit errors in written algorithms, using concrete materials to demonstrate the algorithm's meaning may not be a simple solution. Research has shown that, despite explicit instruction with appropriate materials, students' errors in written algorithms are extremely difficult to eradicate (Dole 2003). Other than remedial and intervention instruction, learning new algorithms can be a rewarding problem-solving investigation in the middle years, where students gain further appreciation of the structure and beauty of mathematics. Lampert (1992) has described the interest her Year 5 students showed in learning long division through an investigative approach. The message, then, is to approach the standard written algorithms with caution in the middle years, and to question the value involved for students in relation to this decision.

Summary

Before teaching algorithms, a substantial amount of prerequisite knowledge must be in place. Recall of the basic facts is necessary to reduce cognitive overload when learning the steps in the algorithms. Estimation and rounding skills are important because they provide a reference point for determining the reasonableness of the calculated solution.

Concrete (and, later, digital) materials are extremely useful for promoting an understanding of the steps in the algorithms, but they can also be a source of confusion. The use of these materials must be considered carefully in instruction for algorithms, and they should never be used by students to obtain answers. Instead, these materials are used only to exemplify the process.

There is current debate around the teaching of standard algorithms. Encouraging students to invent their own algorithms is a trend in mathematics education that is gathering some momentum. Teaching standard algorithms has traditionally occupied a large proportion of the mathematics program. With digital technology, spending this amount of instructional time on algorithms becomes questionable. However, the standard algorithms are efficient and embody a rich history of development. The task for educators is to consider the impact on students if standard algorithms continue to be, or cease to be, taught explicitly.

Review questions

10.1 Why is number sense so important for learning basic multiplication facts? In your response, provide a definition of number sense and provide examples of how number sense might be used to assist in developing multiplication facts.

10.2 What is the purpose of concrete materials in teaching standard written algorithms?

10.3 Describe the distributive law. Give five examples of computation that are simpler using the distributive law. How does knowledge of this law assist understanding of the multiplication algorithm?

10.4 Provide an argument for teaching the standard addition algorithm in primary school.

10.5 Outline the prerequisite knowledge for learning the standard multiplication algorithm.

10.6 Imagine you are teaching in a multi-age setting in which there is considerable diversity in achievement levels. Select one of the operations and outline the sequence through which you would teach students.

Further reading

Anghileri, J. (2001). *Principles and practices in arithmetic teaching: Innovative approaches to the primary classroom.* Buckingham: Open University Press.

Clarke, B., Nelson, N. & Stanley, L. (2016). Mathematics fluency—more than weekly timed tests. In K. Cummings and V. Petscher (eds), *The fluency construct.* New York, NY: Springer.

Ebby, C.B. (2005). The powers and pitfalls of algorithmic knowledge: a case study. *The Journal of Mathematical Behavior, 24*(1), 73–87.

Huinker, D. (2002). Calculators as learning tools for young children's exploration of numbers. *Teaching Children Mathematics, 8*(6), 316–21.

Randolph, T. & Sherman, H. (2001). Alternative algorithms: Increasing options, reducing errors. *Teaching Children Mathematics,* 7(8), 480–4.

Star, J.R. & Madnani, J. (2004). Which way is best? Students' conceptions of optimal strategies for solving equations. In *Proceedings of the twenty-sixth annual meeting of the North American chapter of the International Group for the Psychology of Mathematics Education* (pp. 483–9).

References

Australian Curriculum and Reporting Authority (ACARA) (2019). *Australian Curriculum: Mathematics v8.3.* http://www.australiancurriculum.edu.au/f10-cirriculum/mathematics/.

Baroody, A. (2006). Why children have difficulties mastering the basic number combinations and how to help them. *Teaching Children Mathematics, 13*(1), 22–31.

Blote, A., Kleih, A. & Beishuizen, M. (2000). Mental computation and conceptual understanding. *Learning and Instruction, 10,* 221–47.

Boaler, J. (2015). *Fluency without fear: Research evidence on the best ways to learn math facts.* Retrieved from www.youcubed.org/fluency-without-fear/.

Bolton-Lewis, G. & Halford, G. (1992). The processing loads of young children's and teachers' representations of place value and implications for teaching. *Mathematics Education Research Journal, 4*(1), 1–23.

Buchholz, L. (2004). Learning strategies for addition and subtraction facts: The road to fluency and the license to think. *Teaching Children Mathematics, 10*(7), 362–7.

Carpenter, T.P. & Moser, J.M. (1982). The development of addition and subtraction problem solving skills. In T.P. Carpenter, J.M. Moser and T.A. Romberg (eds), *Addition and subtraction: A cognitive perspective.* Hillsdale, NJ: Lawrence Erlbaum.

Clements, D.H. & Sarama, J. (2014). *Learning and teaching early math: The learning trajectories approach.* New York, NY: Routledge.

Csikos, C. (2016). Strategies and performance in elementary students' three-digit mental addition. *Educational Studies in Mathematics, 91*(1), 123–39.

Dole, S. (2003). Applying psychological learning theory to helping students overcome learned difficulties in mathematics: An alternative approach to intervention. *School Psychology International, 24*(1), 95–114.

Godfrey, C. & Stone, J. (2013). Mastering fact fluency: Are they game? *Teaching Children Mathematics, 20*(2), 96–101.

Hickendorff, M., Torbeyns, J. & Verschaffel, L. (2019). Multidigit addition, subtraction, multiplication and division strategies. In A. Fritz, V. Haase and P. Rasanen (eds), *International handbook of mathematical learning difficulties* (pp. 543–60). Switzerland: Springer.

Hurst, C. & Hurrell, D. (2016). Multiplicative thinking. *Australian Primary Mathematics Classroom, 21*(1), 34–8.

Lampert, M. (1992). Teaching and learning long division for understanding in school. In G. Leinhardt, R. Putnam and R. Hattrup (eds), *Analysis of arithmetic for mathematics teaching* (pp. 221–82). Hillsdale, NJ: Lawrence Erlbaum.

Lemonidis, C. (2016). *Mental computation and estimation: Implications for mathematics education research, teaching and learning*. New York, NY: Routledge.

Ministry of Education, New Zealand. (2008). *The number framework*. Wellington: Ministry of Education.

Skemp, R.R. (2006). Relational understanding and instrumental understanding. *Mathematics Teaching in the Middle School, 12*(2), 88–95.

Thornton, C.A., Jones, G.A. & Toohey, M.A. (1983). A multisensory approach to thinking strategies for remedial instruction in basic addition facts. *Journal for Research in Mathematics Education, 14*(3), 198–203.

Torbeyns, J., Peters, G., De Smedt, B., Chesquiere, P. & Verschaffel, L. (2018). Subtraction by addition strategy use in children of varying mathematical achievement level: A choice/no choice strategy. *Journal of Numerical Cognition, 4*(1), 215–34.

Torbeyns, J. & Verschaffel, L. (2016). Mental computation or standard algorithm? Children's strategy choices on multi-digit subtractions. *European Journal of Psychology of Education, 31*(2), 99–116.

van den Heuvel-Panhuizen, M. (2008). Learning from 'Didactikids': An impetus for revisiting the empty number line. *Mathematics Education Research Journal, 20*(3), 6–31.

CHAPTER 11

RATIONAL NUMBER

Topics that fall under the heading of rational number include fractions, decimals, ratio and proportion, rate and per cent. These topics are linked mathematically, but conceptually they are subtly different. In this chapter, key ideas associated with the topics of fractions, decimals, ratio, rate, proportion and per cent and their interlinked nature are presented. Approaches for enhancing students' knowledge of these topics are described.

Common and decimal fractions

Fractional numbers can be represented in fraction form (e.g. ¼) and in decimal form (e.g. 0.25), and the terms 'common fraction' and 'decimal fraction', respectively, are used to distinguish the two symbolic representations. The word *fraction* is frequently applied to numbers in both fraction form and decimal form, yet there are subtle conceptual differences between common fractions and decimal fractions. Common fraction understanding is based on the part-whole concept. Decimal fraction understanding stems from a combination of an understanding of common fractions, and whole number and place value knowledge. For simplification, in this chapter common fractions are referred to as 'fractions', and decimal fractions as 'decimals'.

Whole number and rational number connections

Whole number understanding provides the foundation for understanding of rational numbers. Particular rational number topics provide a foundation as well as a link to other rational number topics. Decimal understanding is connected to both fraction and whole number knowledge. Ratio and proportion understanding links to fractions, as well as to multiplicative thinking developed through the study of whole numbers. Rate links to ratio. Per cent links to decimals and fractions, and to ratio and proportion. The interconnected nature of rational number topics to each other and to whole number is depicted in the accompanying flowchart. New Zealand's Number Framework (Ministry of Education, New Zealand 2008) includes nine global stages of number knowledge and strategy understanding that encompass the development of rational number knowledge.

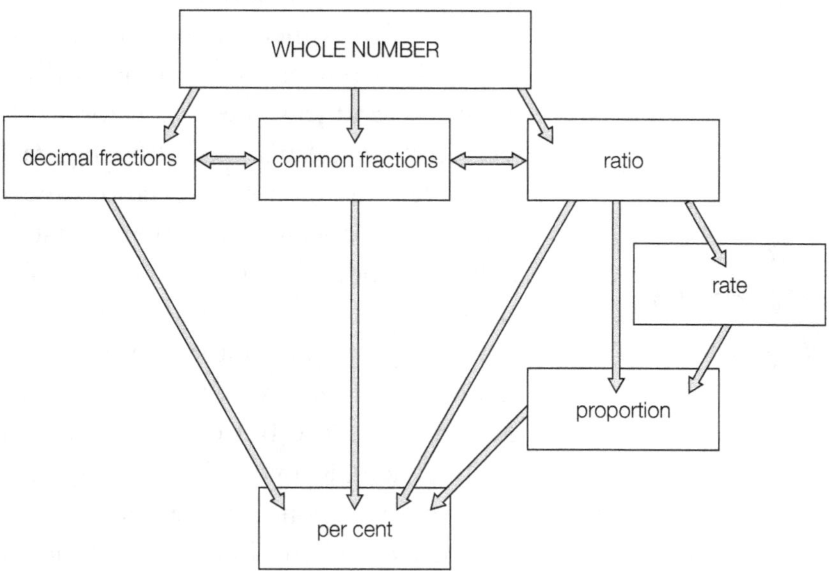

The category of rational numbers includes all numbers that can be expressed in the form a/b, where a and b are whole numbers and b is not zero. From the study of whole numbers, the conceptualisation of numbers falling on a number line that can stretch infinitely to accommodate increasingly larger numbers is promoted. Through the study of fractions and decimals, a further dimension to the number line occurs through locating fractions and decimals between whole numbers on the number line. The linking of fractions and decimals to per cent is further consolidated through locating percentages on a number line. Ratio, rate and proportion link to fractions, decimals and percentages, but understanding of these topics is not necessarily enhanced through the number line concept.

Fractions

The basis of fraction understanding is that fractions are a part of a whole—that is, when one whole thing is split into equal parts, each part is a fraction in relation to the original whole. Fractions can also be created from a set of objects numbering more than one. The *whole* in this sense is more than one, but the collective set of objects must be regarded as *one whole* in order to operate with the objects in a fraction sense.

Student representing one quarter

One of the other crucial understandings of fractions relates to fractions as division situations, and this is essential for working with algebraic calculations. Other big fraction ideas are that fractions are ratios, and that fractions are operators (fractions as ratio are addressed under 'Ratio and proportion' later in this chapter). Fractions as operators refers to the way the numerator (the top number) in a fraction is that which multiplies, and the denominator (the bottom number) is that which divides. Knowing fractions as operators should be a natural by-product of a rich conceptual understanding of fractions; however, it is not elaborated on here. In this section, ideas for developing students' understanding of the following three notions are the focus:

- fractions as part of a whole
- fractions as part of a set
- fractions as division.

Intuitive fraction ideas

Many students come to school with intuitive fraction ideas. The most commonly used fraction is one-half, and children are exposed to this from an early age through natural language and experience. When cutting toast or sandwiches, children are often asked if they want their bread cut in half or if they want it left whole. They may be cajoled into eating half their vegetables in order to get ice-cream. When they are given a bag of sweets, they are told to give half to their sibling. The notion of one-half is usually well established, but mathematically, it may not be as precise as required. A young student, provided with a diagram of a rectangle cut into three equal parts, two of which were shaded, was asked to state how much of the rectangle was shaded. The response was 'one-half'. Is it because two parts of the rectangle are shaded and one part is unshaded that the student sees one-half? Or is it because the student names every 'fractional' picture as one-half? Capitalising on students' intuitive mathematics knowledge is important, but ascertaining the mathematical correctness of that knowledge is equally important.

Beginning fraction activities

Early activities with fractions must be informal and linked to the world of the students. Any situation in which fractions can be identified and the language of fractions used lays the foundation for more formal fraction investigations. When cutting up fruit, the teacher's commentary on the situation would draw students' attention to the fact that the fruit is whole, but is going to be shared—it is going to be cut up into equal pieces. When it is cut up, there are eight pieces, and the teacher emphasises that these eight pieces are part of the whole apple, and that they are called eighths. These pieces can be counted (one-eighth, two-eighths, three-eighths . . . eight-eighths—one whole), and counting in eighths shows how the eight pieces can be recombined to make the whole. Through such situations, students can be exposed to a number of fractional amounts with varying denominators. They can count pieces to complete the whole, and become familiar with fraction names beyond unit fractions (fractions with a numerator of 1).

Fraction models

To build understanding of fractions, students need to create fractions of various denominators using a range of two- and three-dimensional materials and, later, digital manipulatives. Examples of various two- and three-dimensional models for the study of fractions include the following:

1. *Region models*—These models are usually regular two- or three-dimensional geometric shapes where all fraction pieces are congruent (i.e. the same size and shape).

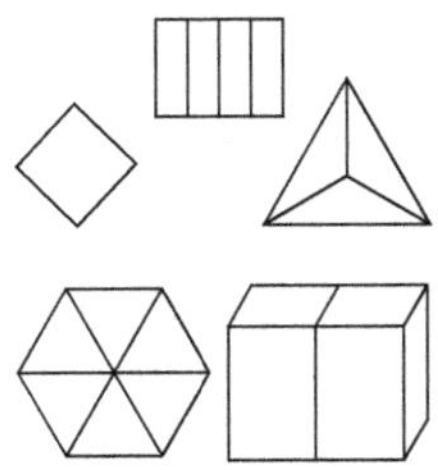

2. *Length models*—Examples include string, paper strips, rope and liquorice straps. Fractions in this representation lead to locating fractions on the number line and counting in fractional amounts.

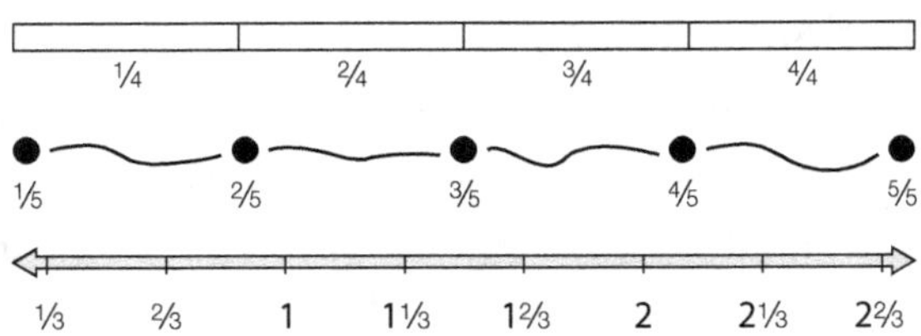

3. *Area models*—These models are similar to region models, but the whole is divided into equal parts that have the same area, though not necessarily the same shape.

4. *Set models*—These include any collection of objects such as counters and blocks. Experience with set models is vital to extend the part-whole concept, and is addressed specifically in the section 'Fractions as part of a set'.

Real-world fraction models

Finding appropriate real-world models for developing mathematical concepts is always an aim in good teaching. However, care must be taken when referring to fractions using real-world objects, because these objects rarely come divided into equal pieces. Is a pizza usually divided into any amount of pieces other than eight (or four, for smaller pizzas), and are these pieces ever even? When a birthday cake is divided in the real world, does everyone always get the same amount, or is a smaller amount cut for Aunt Jan because she is on a diet, and a larger piece for Uncle Jim because he loves cake? The typical way in which the real-world item is divided needs to be considered when assessing its value as a model for promoting fraction understanding. Once students are familiar with real-world fraction models, they can reinforce and practise their fraction understanding using digital manipulatives. One worthwhile app for supporting fraction understanding is Fractions App (www.mathlearningcenter.org/resources/apps/fractions), which supports early conceptual development as well as later fraction concepts of equivalence and decimal fractions.

Fractions as part of a whole

The part-whole concept

Exploration of a variety of fractions with different denominators can be achieved simply with scrap paper folded and cut or torn to create fractions of various denominators. Such activities provide a platform for discussing whether various representations can show the same fractional amounts. As students create different fractions, they are moving through a sequence of five steps:

1. identifying the whole (this is one whole)

2. partitioning into equal parts (5)

3. naming the parts (fifths)
4. selecting the required number of parts (3)

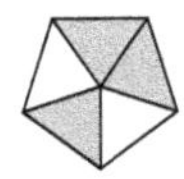

5. naming the fraction (three-fifths).

Each step in this sequence highlights an important aspect necessary for a rich conceptual understanding of fractions. Fractions are parts of a whole (step 1). A fraction is created when the whole is divided into equal parts (step 2). Each part within the whole is named according to the number of parts into which the whole was divided (step 3). The required number of parts is then selected (step 4). The selected parts are named as a fractional amount of the whole (step 5). If three parts out of five are selected, the fraction name is three-fifths. The three indicates the number of parts selected, and the five the number of parts into which the whole was divided (see the accompanying Teaching Idea).

TEACHING IDEA

Finding part of a whole

- Provide students with sheets of scrap paper (each about half of an A4 sheet of paper).
- Students wave the piece of paper in the air, stating, 'This is one whole.'
- Students create various fractions on different pieces of scrap paper (by folding, tearing, cutting, ripping, shading).
- The different representations of each fraction are checked and discussed in terms of mathematical appropriateness.

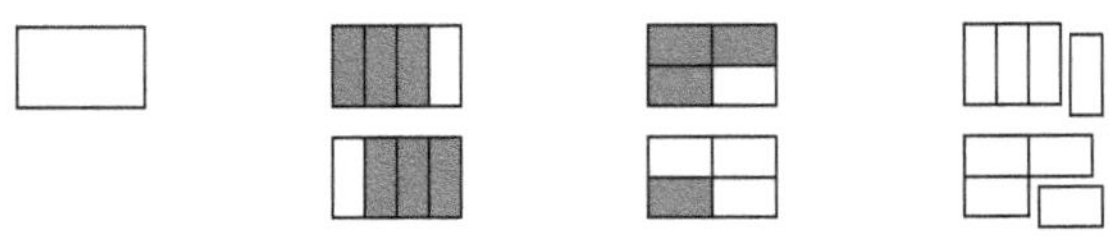

Part-whole scrap-paper fractions:

- Provide students with sheets of scrap paper (each about half of an A4 sheet).
- Students wave the piece of paper in the air, repeating the fraction name given to the piece of paper by the teacher.
- Students re-create the whole from more scrap paper in terms of the identified fraction part (encourage students to work in pairs).
- Different representations of each whole, in terms of its initial parts, are checked and discussed in relation to their mathematical appropriateness.

Finding a whole from a part

Finding a part of a whole from paper can be extended by reversing the activity. The task is for students to re-create the whole in relation to the fraction name given to the piece of paper (see the previous Teaching Idea). In order to find the whole in relation to the specified part, students need to consider the number of parts required to make the whole and the number of parts embodied in the piece of paper they hold in their hands. This activity is an invaluable means through which individual students' conceptual understanding of fractions can be informally determined by watching as students interact with each other, ask questions of each other and the teacher, and work with the material.

Fraction family investigations

Manipulation of written fraction symbols has been the dominant approach to formal fraction study, whereby students learned and practised procedures for fraction computation. Instruction typically involved manipulation of symbols to find common denominators, generate equivalent fractions, reduce fractions to their lowest terms, convert improper fractions to mixed numbers and vice versa. These skills were applied to fraction addition, subtraction, multiplication and division. To many students, the study of fractions is a series of confusing, meaningless exercises, as attested to by the range and types of errors students make in fraction calculations (see Dole 2003, for examples of common fraction calculation errors).

Investigations of a family of fractions can assist in providing meaning for written fraction procedures. Four different coloured paper circles can be used to create a family of fractions, such as halves, quarters, eighths and one whole (see the following Teaching Idea), and manipulated to explore a number of ideas.

TEACHING IDEA

Fraction families

Halves, quarters, eighths

- Cut out four equally sized circles from different coloured pieces of paper.
- Take one circle and fold and cut it into two equal parts. Write the name of each fraction piece in words on one side and in symbols on the other side.
- Repeat for the second circle, but divide and cut it into quarters.
- Repeat for the third circle, but divide and cut it into eighths.
- Keep the last circle whole. Write its fraction name in words on one side (one whole) and its fraction symbol on the other side (1/1).
- Use the pieces to generate equivalent fractions, to convert mixed numbers to improper fractions and vice versa, to perform simple addition and subtraction calculations, and to find common denominators.

Using fraction family materials, explore equivalence, the part-whole concept, improper fractions and mixed numbers, compare fractions, and engage in simple addition and subtraction tasks. Students manipulate the materials in response to the teacher's questions. Discussion of the process and students' thinking strategies should be integral to these explorations.

Equivalence

- How many eighths in one-quarter?
- How many eighths in three-quarters?
- One-half is equal to how many eighths? How many quarters?

Part-whole concept

- How many quarters make one whole?
- How many eighths in one whole?
- What is one whole less seven-eighths?

Improper fractions and mixed numbers

- How many wholes are there if I have five-quarters?
- What is six-quarters the same as?
- What is eleven-eighths the same as?

Comparison

- Which is bigger, ½ or ¼?
- Which is bigger, ⅚ or ½?

Simple addition and subtraction

- How many more pieces do I need to make one whole if I have three-quarters?
- What is three-quarters and three-quarters?
- What different combinations of fraction pieces can you use to make one whole?

Through such explorations, the whole circle remains clearly visible as a reference point. The language of fractions is emphasised through discussion of fraction names, symbolic representation of fraction solutions and symbolic representation of fraction calculations. Encourage students to close their eyes and visualise manipulation of the fraction pieces as various questions are posed to promote mental computation. Different fraction families should only be introduced and explored after students can mentally operate with the 1, ½, ¼, ⅛ fraction family.

Fractions as part of a set

For some students, finding one-third of 6, two-fifths of 20, and so on is an extremely difficult task. The difficulty frequently stems from poor fraction understanding and limited work with finding fractions of a whole where the whole is a collection of objects numbering more than 1. Explorations and investigations using sets of concrete material enhance and extend the part-whole concept to build *part of a set* understanding.

The teaching sequence for developing the part of a set concept is the same as for part of a whole. Using counters or blocks, sets of material in specified amounts are identified as *one whole* (see the following Teaching Idea). Fractions of that set are found through the process of dividing the set into equal parts, naming each part in relation to the whole, selecting the required number of parts and naming the resulting fractions. Finding part of a set has an added cognitive dimension to finding part of a whole, in that the resulting solution is not a fraction part *per se*, but is a number of individual blocks that can be held in the hand and counted. For example, when ten counters is the whole, two-fifths of ten is four individual blocks that must be seen as four parts of ten. The counters within the set must be disconnected and reconnected to the whole interchangeably.

TEACHING IDEA

Developing part of a set concept

Whole ⟶ part

- Use counters or blocks to make sets of a specified number.
- Identify each set as one whole.
- Find a specified fraction of that set.

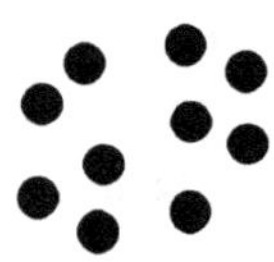

Ten counters is one whole

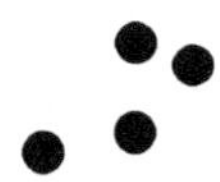

Four counters is two-fifths

Part ⟶ whole

- Use counters or blocks to make sets of a specified number.
- Identify each set as a specific fraction.
- Find a whole in relation to the specified fraction.

Six counters is two-thirds

Nine counters is one whole

To further extend the notion of part of a set, reverse the order of the investigation by finding the whole from a beginning fractional amount. Encourage mental calculation by asking students to visualise the counters in their mind as they explore solutions to given 'part of a set' exercises.

Fractions as division

Consider the problem '3 divided by 5'. Put it into a 'real' context, and see what happens: share three pizzas between five people. Create three pizzas from paper to act out this situation and discuss how these three 'pizzas' could be shared. Divide each pizza into five pieces and distribute the pieces among three people. The result is that three shared between five gives three-fifths of a pizza each. This adds a further dimension to the conceptualisation of fractions whereby each symbolic fraction can be regarded as part of a whole, but also as a division. By acting out the situation with the pizzas, attaining the solution shows that the operation is division and that it is partition division (sharing).

To develop the fractions as division concept, attention needs to be drawn to the fact that a fraction written in symbolic form means the top number divided by the bottom number. This is easier to see when related to improper fractions. The improper fraction $\frac{4}{3}$ means 4 divided by 3. The solution is simply $1\frac{1}{3}$. The 'rule' for changing an improper fraction into a mixed number takes on meaning through considering the fraction as a different symbolic representation of a division exercise. In a pizza situation, $\frac{4}{3}$ is considered as four pizzas shared between three people, where each person gets one full pizza and one-third of the fourth pizza each.

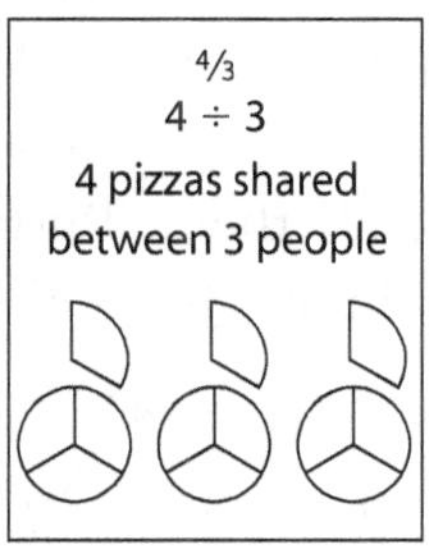

Fraction computation

Fraction addition and subtraction

Informal addition and subtraction of fractions can be explored through the use of fraction circles, as outlined previously. Use questioning to guide and encourage mental images. Also work with fraction pieces so that students see that there are, for example, four quarters in a whole, that five quarters mean that there is one whole and one quarter left over, that one-half is the same as two quarters and that one-fifth is smaller than one-third.

For simple addition and subtraction, students should be able to *see* the additions and subtractions in their heads, without having to resort to calculation. Pose questions such as the following, and ask students about the mental images they see in their mind as they determine the solution:

$1 - \frac{1}{3} =$
$\frac{3}{5} + \frac{1}{5} =$
$\frac{3}{4} + \frac{3}{4} =$
$\frac{3}{7} + \frac{4}{7} = 1$ whole
$\frac{7}{8} - \frac{3}{4} =$

Draw attention to the fraction name and what that means. If adding fifths, how many fifths are required to make one whole? When subtracting quarters from eighths, think of the quarters in an equivalent form with a denominator of eighths.

Fraction multiplication and division

Introducing multiplication and division of fractions needs to be done in a meaningful way that capitalises on prior knowledge. Consider finding $6 \times \frac{1}{2}$. This statement makes more sense when linked to the language used for whole number multiplication: '6 groups of ½'. One representation of this situation is to skip count in halves along a number line six times to reach a solution of 3. Another representation is to think of halves as half-circles and imagine six of them. Putting two halves together to make one whole, there are three wholes altogether.

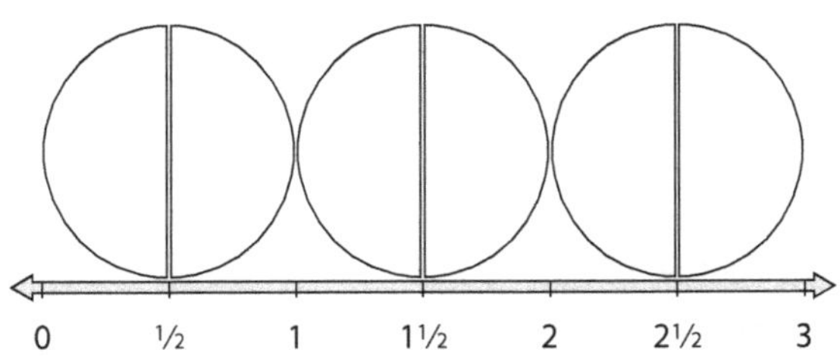

For fraction division (e.g. $6 \div \frac{1}{2}$), referring to prior language of whole number division assists in solution attainment. Rather than state that this is 6 divided by ½, state that the problem is, 'How many halves in six'? Ask students to describe or draw the mental picture that comes into their minds as they are asked this question. Do they imagine a number line, where they skip count in halves to 6 to find there are 12 halves? At this stage, simple multiplication and division exercises should be confined to those that can be done mentally.

Decimals

Three notions underpin decimal understanding:

1. Decimals are numbers that can be located on the number line (as can common fractions).
2. Decimals are fractions with denominators that are powers of 10 (tenths, hundredths, etc.).
3. Place value indicates the value of decimal numbers in symbolic form.

Decimals as fractions

Prior experience with fractions provides the basis for conceptualising decimals as fractions. The sequence is as follows:

- exploration of halves, fifths and tenths as a family of fractions
- symbolic representation of tenths linked to visual image and number line
- symbolic representation of hundredths and tenths linked to visual image and number line.

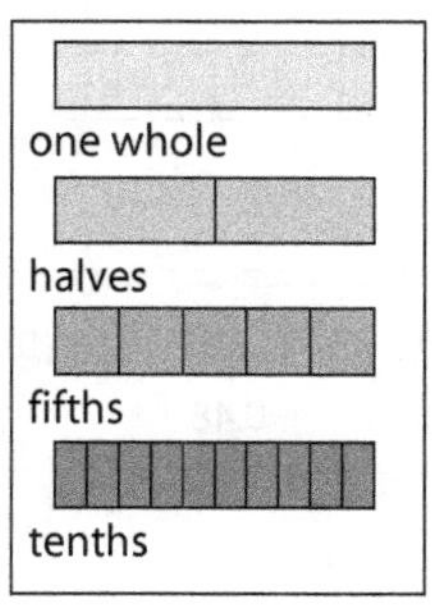

A set of fractions made from paper enables the relationship between halves, fifths and tenths to be explored. When such fractions are created using a linear model, the foundation is laid for visualising decimals on a number line. Encourage manipulation of the fraction pieces until students can determine answers to the following types of questions mentally:

- How many tenths in one whole?
- What is three-tenths and three-fifths?
- How many tenths is the same as one-half?
- What is seven-tenths and nine-tenths?
- What is another name for fifteen-tenths?

Shading pictures of tenths provides a link between the fraction picture, the symbolic fraction form and the symbolic decimal form.

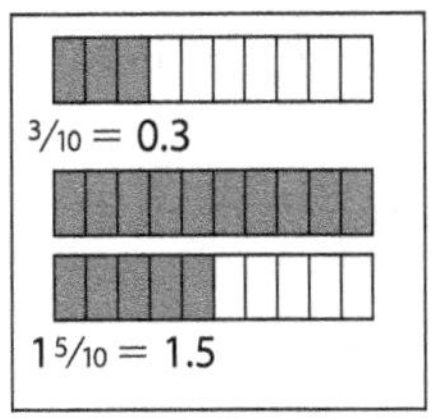

The relationship between tenths and hundreds is seen through creating pictures of various fractions and decimals on 10 × 10 grids. From images created, the relative magnitude of tenths in relation to hundredths is seen. Location of decimals on the number line follows from the pictorial representation.

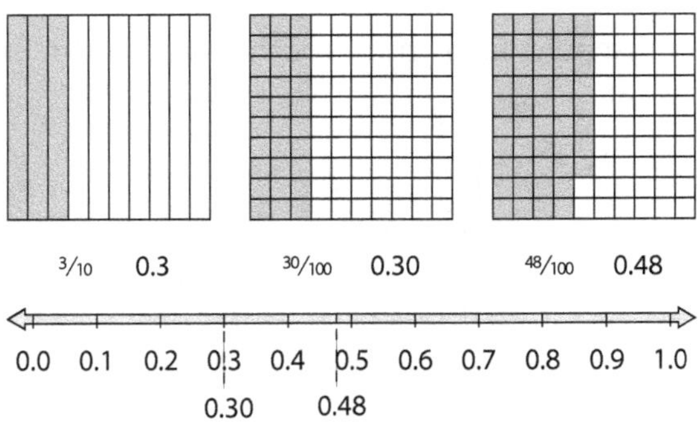

Naming decimals

Many students experience difficulty in reading decimal numbers, in terms of understanding both the size of decimal numbers and the value of the digits within the decimal number. When teachers follow the practice of reading decimal numbers using abbreviated language (e.g. reading 3.4 as 'three point four' rather than 'three and four-tenths'), students are not provided with language support structures in which the meaning of the values of decimal numbers may be emphasised. Pictures of decimal tenths and hundredths provide the link for connecting the symbolic representation to the pictorial representation, and also for naming decimals. The equivalence of three-tenths to thirty-hundredths is seen through overlaying a hundreds grid on the picture of three-tenths. The symbolic representations for these two pictures are 0.3 and 0.30, respectively.

In the symbolic representation, the former decimal indicates that there are three-tenths, and this matches the diagram (see the above grid). The latter decimal in written form indicates that there are 30 hundredths, and this also matches the diagram. The decimal 0.48 depicted on the 10×10 grid shows that this number is made up of four-tenths and eight-hundredths, or forty-eight-hundredths. The appropriate name for this decimal is 'forty-eight-hundredths' rather than 'four-tenths and eight-hundredths', and certainly not 'point four eight'. Appropriate naming of decimals during early explorations ensures that the language links to the symbolic representation and the pictorial image.

Decimals and place value

The symbolic representation of decimal numbers is in accordance with the base 10 numeration system. In Australia and New Zealand, a decimal number is indicated by the presence of a decimal point. (In many European countries, it is indicated with a comma.) The digit to the immediate right of the decimal point is the amount of tenths in that number; the digit to the immediate right of that digit is the amount of hundredths; and so on. One of the most common misconceptions about decimals is thinking that the more digits to the right of the decimal point, the larger the number (Yang & Sianturi 2018). For many students, 0.45 is considered to be of lesser value than 0.2476. This misconception stems from poor understanding of decimal numbers and place value.

Understanding decimals in relation to place value requires understanding of the following three principles:

1. Places within the base 10 numeration system extend in an infinite direction to the right of whole numbers to include decimals.
2. The place value system has symmetry around the ones place, but the value of numbers within each place is not symmetrical.

3. The decimal point apparent in the symbolic representation of decimal numbers indicates the ones place and is not a place of value within the place value system.

Through recording decimals in symbolic form and reading decimals for meaning (0.43 is forty-three-hundredths, not point four three), students' understanding of decimals in terms of places in the base 10 system develops. Naming the place signified by each digit in decimal numbers also builds understanding of the symmetrical nature of the places within the number system. The names of the places follow a similar pattern to that for whole numbers, with decimal places taking on the language pattern of fractions (suffix -th). The ones place is the centre point for the pattern.

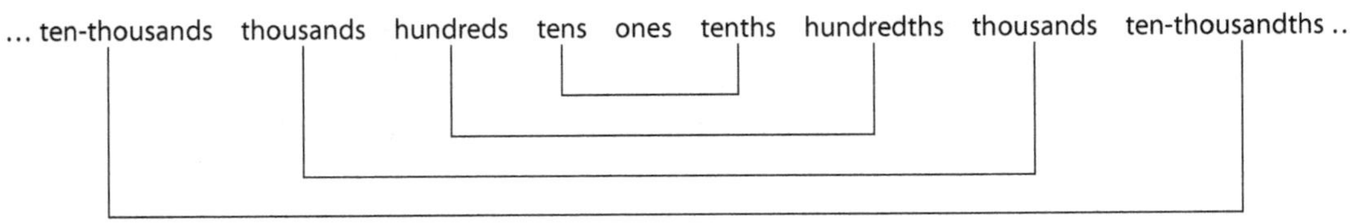

The relative value of decimals that have many digits after the decimal point is consolidated through practice in locating decimal numbers on a number line. Locating decimals on a number line follows from explorations of fractions of denominator 10 or 100, and from counting in decimals. The number 0.5 (five-tenths) is located on a number line halfway between 0 and 1. This requires counting in tenths between 0 and 1. To locate the decimal 0.55 (fifty-five hundredths) means finding the midpoint between 0.5 (fifty-hundredths) and 0.6 (sixty-hundredths). This requires counting in hundredths between 0 and 1. Extending such an investigation requires rewriting the number line in order for each number to be located, and dividing the space between the numbers 0 and 1 into smaller and smaller sections in groups of 10.

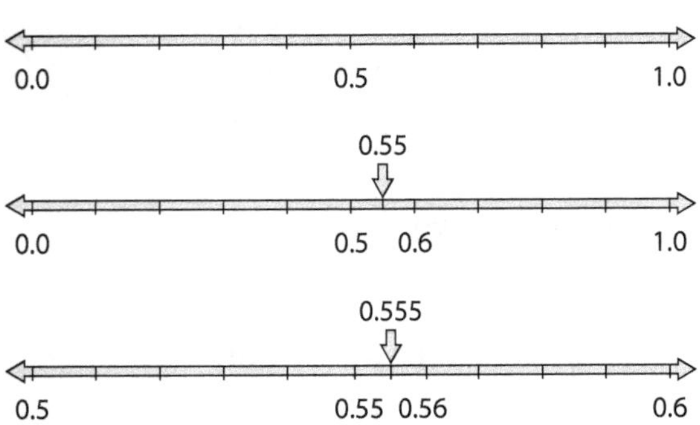

The relationship to 10 of places in our numeration system, established through experiences with whole number study, needs to be reinforced when decimals are introduced. Exploring the effect upon numbers of progressively multiplying and dividing by 10 assists in consolidating this notion. Entering a number on a calculator, and observing what occurs to the calculator display of that number as it is progressively multiplied by 10 shows how digits move across places. Keeping a record of this exploration provides a reference for discussion, particularly when the numbers are recorded in an organised list under place value columns. The number 43, when multiplied by 10, becomes 430. When it is multiplied by 10 again, it becomes 4300. Each digit in the number moves one place to the left, and its value is ten times greater than its previous value. Upon division by 10, the digits move to the right, and eventually the decimal point appears to separate the ones from the tenths. The value of each digit upon successive multiplication or division by 10 is 10 times more or 10 times less than its previous value. Through this investigation, the display of zeros on the calculator can be discussed. A calculator does not show excess zeros beyond the decimal point.

	Th	H	T	O.	t	h	th
			4	3.			
× 10 →		4	3	0.			
× 10 →	4	3	0	0.			
÷ 10 →		4	3	0.			
÷ 10 →			4	3.			
÷ 10 →				4.	3		
÷ 10 →				0.	4	3	
÷ 10 →				0.	0	4	3

Exploring the pattern that occurs when dividing numbers by 10 highlights the place and value of decimal numbers within our numeration system. It emphasises how each place in the numeration system is linked by powers of 10 and that each place to the left is ten times bigger than the place immediately before. Similarly, each place to the right is ten times smaller than the place immediately before. When interpreting decimal numbers, tenths are ten times greater than hundredths, hundredths are ten times greater than thousandths, and so on.

Observation of the movement of digits across the places upon multiplication and division by 10 provides students with an alternative to the commonly held belief that the decimal point moves. It is the *digits* that move—to the left upon multiplication by 10 or powers of 10, and to the right upon division by 10 or powers of 10. The decimal point is always to the immediate right of the ones place and does not move.

Precision and error in rounding decimals

Through experience with decimals, students come to realise that the more digits there are after the decimal point, the more precise the number, but also the increasing insignificance of the value of these digits. When calculating with decimals, it is common for students to be instructed to round their answers to the nearest tenth or hundredth. Knowing the relative value of digits after the decimal point gives meaning to rounding, and also places students in a position to decide the number of places to which a decimal number should be rounded. In the real world, inappropriate rounding of decimals to tenths or hundredths may be detrimental. In medical experiments and operations, inaccurate calculating and measuring may cause risk to patients. In the building industry, machines being calibrated incorrectly could result in damage to equipment. In the classroom, problem-solving and calculations with decimals should be contextualised so that rounding of decimals is appropriate to the situation.

Ratio and proportion

The topics of ratio and proportion are commonly discussed simultaneously, and indeed are conceptually linked. In its simplest form, *ratio* describes a situation in comparative terms; *proportion* is when this comparison is used to describe a related situation in the same comparative terms. For example, if we say that the ratio of boys to girls in a class is two to three, we are comparing the number of boys to the number of girls. When we know that there are 30 children in the class, we know that, proportionally, the number of boys is twelve and the number of girls is eighteen. We are using the base comparison to apply it to the whole situation. In order to understand this relationship, proportional reasoning is used. Proportional thinking and reasoning involve knowing the multiplicative relationship between the base ratio and the proportional situation to which it is applied.

Like all mathematics topics, developing proportional reasoning and thinking skills takes time and is dependent on sound foundations in associated topics, particularly multiplication, and having a rich conceptual understanding of ratio and proportion. Ratio and proportion are two concepts that present a challenge to both teachers and students. Proportional reasoning is fundamental to understanding many topics in the middle-school curriculum, such as the geometry of plane shapes, trigonometry and, particularly, percentages, as well as rate, ratio and proportion applications. As Lesh and colleagues (1988) state, 'Proportional reasoning is the capstone of children's elementary school arithmetic and the cornerstone of all that is to follow' (pp. 93–4).

Traditionally, ratio and proportion were considered topics for upper-primary and secondary grades. However, teachers at all year levels need to provide students with rich experiences and appropriate activities in order for them to develop understanding of these topics, so that students will later be able to apply this knowledge in a meaningful way. Teachers of early primary classes must ensure that the seeds of the concepts are sown in their students' minds in the early years, and as students move through the grades, all teachers must build on and extend this knowledge.

Ratio and fractions

Ratios are commonly described as fractions. In the true mathematical sense, fractions are ratios in that they are rational numbers that can be expressed in the form a/b (provided that b does not equal zero). However, the general meaning of fraction, of being part of a whole (as described in the previous section), interferes with the conceptual meaning of ratio. Continuing with the previous example, it was stated that the ratio of boys to girls in the class is two to three—that is, for every two boys there are three girls. When this is considered in the context of the

TEACHING IDEA

Ratio language through pattern

■■□■■□■■□

For every 2 black there is 1 yellow

●●●■●●●■

For every 3 circles there is 1 square

whole class, there is a base unit of five children, of whom two are boys and three are girls. The boys and girls situation describes the two parts of the whole unit of children. In this sense, the ratio of boys to girls describes a part-to-part relationship, whereas fractions typically represent a part-to-whole situation. In a fraction sense, the class is made up of two-fifths boys and three-fifths girls.

Ratio describes a comparative relationship between two (or more) entities. Ratio can be used to describe part-to-part, whole-to-whole and part-to-whole situations:

- *part-to-part ratio*—e.g. comparing the number of girls in a class to the number of boys in the class
- *whole-to-whole ratio*—e.g. comparing the number of girls in one class to the number of girls in another class
- *part-to-whole ratio*—e.g. comparing the number of girls in the class to the number of children in the class.

Predominantly, ratio is used to describe a part-to-part relationship. The misleading nature of expressing the ratio of boys to girls in fraction form (as $\frac{2}{3}$ in the illustration above) highlights the importance of considering when the symbolic representation for ratio situations is most appropriately introduced to students.

Starting points for ratio and proportion understanding

Through experience and play, learners encounter ratio at an early age. As they get dressed in the morning and are searching for their socks, they know that for every single sock there is another sock; that for every shoe there is another shoe. When they help set the table for dinner, they know that for each person there is a fork and a knife, and for every soup bowl there is a spoon. When playing 'tea parties' with toys, there is one cup and one saucer for each doll. When constructing patterns with blocks, there is one red block for every two blue blocks, or two green blocks for every yellow block. The link between multiplication and division is evident in these examples: if there are five dolls, there are ten cups and saucers; if there are six knives and forks, this is sufficient for three people. Early ratio investigations provide an opportunity for immersing learners in the language of ratio. If it takes three shovels of sand to fill the wheelbarrow, it will take six shovels of sand to fill two wheelbarrows.

Copying, creating and extending patterns with blocks or other material provides a natural environment for ratio language and for posing questions that promote multiplicative thinking, for example:

- What blocks do we need to continue the pattern?
- How many blue blocks for every yellow block?
- How many blue blocks would we need to extend the pattern if we had three yellow blocks?
- How many blue and how many yellow blocks would we need if we had twelve blocks to build the pattern?

Such questions provide a challenge to young minds to consider the base unit of the pattern. Posters of patterns that indicate the base unit of the pattern provide a permanent record of ratio situations.

Guided ratio investigations in the early years can occur through making mixtures. Mixtures that involve edible or drinkable ingredients (such as cake mixtures or fruit juice punch) can be used to appeal to students' sense of taste, and how increasing or decreasing particular elements within various mixtures will alter the taste accordingly. The activity presented in the following Teaching Idea provides opportunities for the promotion of proportional reasoning, through such questions as:

- If perfect punch is made from two cups of orange juice and one cup of lime juice, how many cups of orange juice would be needed to make the ideal punch mix if you had eight cups of lime juice in the mixing container?
- What if it was discovered that the ideal mix for punch was to use one cup of lime juice for every three cups of orange juice? How many cups of each juice would be required to make punch for a party so that the twelve children would each get one cup of punch?

These questions focus on different dimensions of ratio understanding:

- When new mixes are made from a base ratio, they taste the same.
- In any mixture, increasing the number of cups of one mixer requires that the second mixer must also be increased, in accordance with the base ratio, so that the mix tastes the same.
- The two mixers are two parts of the ratio. Adding the parts gives the total number of cups of the base mixture.

TEACHING IDEA

Making punch mixes

Do these two mixes taste the same?

Mix 1:

1 cup lime juice

1 cup orange juice

Mix 2:

2 cups lime juice

2 cups orange juice

Make up the two mixes and pour into their own bowl. Taste them.

- Do the mixes taste the same?
- How many cups of lime juice would be needed if there were 4 cups of orange juice?
- How many cups of each mix is required to mix enough punch for 12 children?
- Make up some of your own mixes.
- Draw pictures of your mixes.
- Make a mix for a party of 24 children. How much of each mixer do you need?

In a mathematical sense, the different dimensions of ratio understanding promoted through mixing activities relate to:

- ratio equivalence (e.g. 1:3 = 2:6)
- part-to-part and the multiplicative structure of ratio
- part-to-part and the relationship to the whole.

In early ratio investigations, the teacher's role is to engage students in discussion, problem-solving and reasoning in a guided sense. Through such explorations, the students can explore the sophisticated topic of ratio and develop proportional reasoning skills through discussion, hypothesising and justification without using mathematical symbolism.

Identifying ratio and non-ratio situations

A further dimension to developing students' knowledge of ratio and proportion is to assist them in distinguishing situations that are proportional and those that are not. There are instances in which situations sound as if they are proportional and yet are not; and, conversely, situations that sound as if they are not proportional and yet are. Students' capacity to make these distinctions is a measure of the sophistication of their proportional reasoning (Van Dooren et al. 2005). For example, if five children sing a song that lasts two minutes, will they sing twice as fast if there are 10 children? If there are seven items of clothing on the line, will the clothes take longer to dry if there are 14 items of clothing? If two cars are racing around a track at the same speed, and the second car starts at the point at which the first car had completed three laps, will the first car have completed 12 laps after the second car has travelled four laps? All these situations sound as if they are proportional, yet they are not. Presenting proportional and non-proportional situations to students creates interesting discussion and promotes students' proportional reasoning development.

Proportional reasoning and numeracy

Many everyday situations and decisions require proportional reasoning. When you are putting detergent into the washing machine, you may use the recommended amount, or you may make a decision dependent on how dirty your clothes are or how fresh you want them to smell. When you are driving in your car and see that your fuel gauge is heading towards empty, you make a decision about when to fill up, how far you have to travel, and how far away is the next fuel station. Because of the real-world nature of so many situations that require proportional reasoning, Dole and colleagues (2015; 2018) contend that proportional reasoning is an essential component of numeracy. Their research shows the value of immersing children in discussions about proportional reasoning that is evident in nature and in other authentic situations (Hilton et al. 2015)—for example, images of large flocks of birds at a lake, big and small animals at a watering hole, or a small bird struggling with a large fish in its mouth. Encouraging students to take photographs of images that require and depict proportional reasoning, such as standing next to a lamp post (how many times taller is the post than the person?) or a carpark full of cars (how many silver cars are there compared to red cars?) fosters and builds awareness of proportional reasoning in the real world.

Promoting language and mental images of ratio and proportion

As students engage in hands-on ratio explorations, a natural context for the use of comparative language to describe ratio situations is provided. The word 'part' is used to label the components in the ratio, and the phrase 'for every . . . there is . . .' is used to describe the relationship between the parts in the ratio. When making mixes, the mix is made up partly of one type of mixer and partly of another type of mixer: 'The mix is part orange juice and part lime juice. For every two cups of orange juice there is one cup of lime juice. For every two parts of orange juice there is one part of lime juice.' Through the use of such language, the word ratio can be introduced in a meaningful sense, and also as a shorthand way of describing the situation: 'The mix of orange juice to lime juice is in the ratio of 2 to 1.' In early ratio explorations, the symbol for ratio (:) should be avoided, but if it is used it should be accompanied with the names of the items being compared (the ratio of orange juice to lime juice is 2:1).

Visual images and mental computation

As students become comfortable with the language of ratio, they need to be provided with activities that promote the development of mental images that assist in mentally carrying out simple ratio and proportion calculations. The three types of ratio calculation students should be able to perform mentally are:

1. generating proportions from simple ratios using multiplication
2. finding simple ratios from proportional situations using division
3. determining missing elements of ratio and proportion situations through part-part-whole concept knowledge.

The following activities provide a sequential progression for assisting mental computational proficiency. The activities utilise coloured counters displayed on ratio tables to promote strong visual images of the process of the calculation. Once students have experienced real-world ratio scenarios, they can then develop this understanding using apps such as Thinking Block Ratios, available at: https://apps.apple.com/au/app/thinking-blocks-ratios/id670448325.

Generating proportions from simple ratios using multiplication

Simple ratio calculation can be performed when a unit ratio is given (i.e. when one of the numbers in the ratio is 1). The unit ratio is the basis for generating equivalent ratios, or proportional amounts. This can be exemplified with concrete materials. Use counters of two different colours (e.g. red and blue, with one red counter and two blue counters) to show a ratio of 1 to 2; that is, for every one red counter, there are two blue counters. Proportions can be generated by adding more counters in the pattern of the original ratio—that for every red counter there are two blue counters. When there are four red counters, there must be eight blue counters. Placing counters in a ratio table, consisting of columns headed with the comparative parts in the ratio, assists in organising the counters and exemplifying the multiplicative nature of proportions. Other ratios can be explored through teacher questions focusing attention on students' prior knowledge of multiplication to generate proportions; for example:

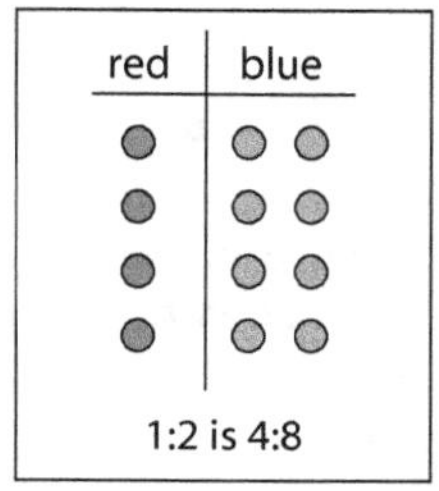

- If red and blue counters are in the ratio of 1:4, how many blue counters will there be if there are 3 red counters?
- If red and blue counters are in the ratio of 5:1, how many red counters will there be if there are three blue counters?

To encourage mental computation, ask students to close their eyes and visualise counters on a ratio table as they calculate proportional amounts from unit ratios. Engage students in discussion about their mental images and strategies, and encourage them to describe their thinking in terms of multiplication. A simple checking exercise will ascertain students' level of comfort with the notion of generating proportions from unit ratios: ask them to mentally complete missing numbers of given proportions (e.g. '1 to 4 is the same as 2 to ?'; '2 to 1 is the same as ? to 10'; '1 to 3 is the same as ? to 12'; and so on).

Exploring ratios that are not unit ratios (i.e. neither of the numbers in the ratio pair has a value of 1) is an extension of exploring unit ratios. As in the preceding activity, counters on a ratio table provide the visual image of the ratio and assist in generation of the proportion (e.g. for the ratio of 2 to 3, there

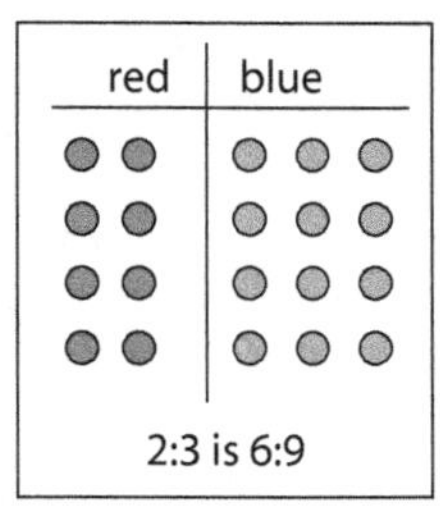

must be 9 blue counters when there are 6 red counters because we multiplied 2 by 3 to give 6 red counters, so we must multiply 3 by 3 to get 9 blue counters).

Simplest form ratio through division

Proficiency with mentally generating proportions from unit ratios lays the foundation for finding *simplest form ratios*; this is actually the reverse of the process of generating proportions from simple ratios. To orientate students to this notion, build on their skill of visualising counters on the ratio table, such as red and blue counters in the ratio of 3:9. Use counters to represent this situation, encouraging students to place the red counters in a single vertical line and to distribute the blue counters evenly between the red. The division process is reinforced through this action. The counters show clearly that for every red counter, there are three blue counters. From a proportion of 3 to 9, the simplest form ratio is 1 to 3.

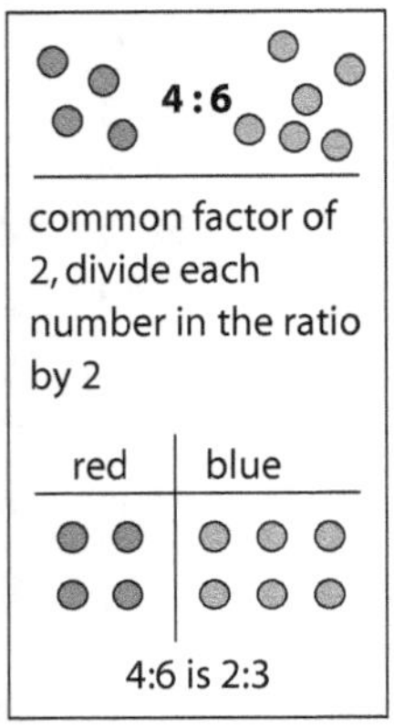

The exploration of many different ratios with concrete materials assists in building students' confidence in the process and familiarity with the language, and should be continued until students can readily calculate simplest form ratio with diminishing reference to the counters. Encourage students to verbalise the situation as a ratio before they separate one group of counters into a vertical line, and to then distribute the other counters equally among the separated counters. Draw on students' number sense before counters are distributed by considering common factors between the two numbers in the ratio. Thinking of ratios in terms of common factors links to the skill of finding simplest form fractions and dealing consistently with all parts in the situation. Encourage mental computation by asking students to close their eyes and visualise the counters on the ratio table as they determine the simplest form ratio.

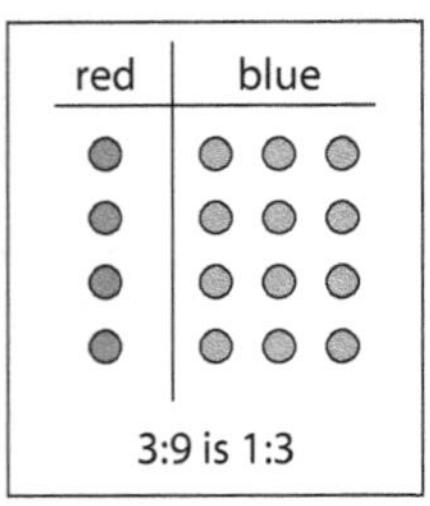

Promoting the part-part-whole structure of ratio situations

In the previous ratio explorations, the focus was on generating equivalent ratios (proportional amounts) or describing a collection of objects as a ratio in its simplest form. This lays the foundation for explicitly promoting the part-part-whole concept of ratio situations and consolidating the language of ratio.

Place red and blue counters on a ratio table to show a ratio of 2:3 (for every two red counters, there must be three blue counters). Focus students' attention on the part-part-whole nature of the situation through direct questioning:

- How many red?
- How many blue?
- How many counters in total?
- If there were a total of ten counters in the ratio of 2:3, how many of each colour would there be? (Four red and six blue.)
- Why? (The ratio 2:3 means five counters in total. If ten counters are used, I need to double each part.)

Explore the part-part-whole nature of other ratios by working through the same sequence of questions (e.g. in a ratio of 2:3, how many counters of each colour are there if there are fifteen counters?; in a ratio of 5:1, how many of each colour are there if there are eighteen counters?). Promote mental computation by encouraging students to determine solutions by visualising the situation.

Problem-solving and ratio calculations

It often happens in ratio instruction that students are provided with problem situations that can be solved easily through multiplication or division. Take the example of finding the amount of fertiliser to be mixed with 15 litres of water in the ratio of two parts fertiliser to three parts water. From the base ratio of three parts water, and comparing it to the 15 litres of water in the given amount, it can be seen that the 3 has been multiplied by 5 to give 15, so the solution is easily acquired by multiplying the 2 by 5 to give 10. When given ratios and amounts are not simply related, the previous strategy does not provide many clues for action to solve particular problems. Consider the example of finding the amount of fertiliser that would be required if 13 litres of water were used. The calculation methods offered to students must lay the foundation for all types of situations that they might encounter, and these methods must be presented in a meaningful way.

Encouraging students to use ratio tables for all ratio explorations provides a solid foundation for promoting the multiplicative thinking necessary for solving 'ugly' ratio problems. As shown in the examples that follow, the *easy* ratios are a process of multiplication and recognition of number patterns and relationships. In dealing with 'ugly' ratios, the process of converting the given ratio into a unit ratio is tracked through the ratio table.

Easy ratio — 'Ugly' ratio

Ratio tables in the vertical sense are shown above. Ratio tables can also be constructed in a horizontal sense. Horizontal ratio tables have been proposed as a means for assisting students to develop mental strategies for solving proportion problems (Dole 2008). Ratio tables encourage the use of number strategies such as halving, doubling, multiplying by 10, and so on. Progressive and simultaneous operation on the given numbers shows how the relationship (ratio) is preserved proportionately. The following example shows how a ratio table can be used to calculate how many seedling plants are packaged into 14 boxes, when each box holds twelve plants:

- *Step 1:* Construct the ratio table and display the given information.

Boxes	1				
Seedlings	12				

- *Step 2:* Select successive operations (e.g. multiplying by 10, halving, doubling) to determine the solution. Use arrows above each number to show the journey to the solution.

		× 10	÷ 2	× 3	− 1
Boxes	1	10	5	15	14
Seedlings	12	120	60	180	168

In this example, the pathway to the solution was to multiply by 10, then divide by 2, then multiply by 3 to get to 15. This is one more than the required number, so we then just subtract 1 box of seedlings (12) to reach the solution of 168.

In the next example, drinks are packaged into fifteen-bottle boxes. How many drinks would there be in sixteen boxes?

- *Step 1:* Construct the ratio table and display the given information.

Boxes	1				
Bottles	15				

- *Step 2:* Select successive operations (e.g., multiplying by 10, halving, doubling) to determine the solution. Use arrows above each number to show the journey to the solution.

		× 10	÷ 2	10 + 5	+ 1
Boxes	1	10	5	15	16
Bottles	15	150	75	225	240

In this example, the pathway to the solution was to multiply by 10, then divide by 2, then add 10 to get to 15. This is one less than the required number, so we then just add one box of drinks (15) to reach the solution of 240.

These two examples show different pathways that can be taken to arrive at the solution. In the first example, after multiplying by 10 and then halving, the next step was to multiply by 3. In the second example, after the amount for 10 and 5 had been determined, these two amounts were added and displayed in the next column (10 + 5 boxes = 15; 150 + 75 bottles = 225). This required looking back at the table to locate previous calculations that assisted in reaching the destination. To use ratio tables effectively, students need lots of practice. They also need support and guidance to ensure that the ratio table is used as an effective tool for ratio calculations. Students need to be reassured that they can extend the ratio table as far as required to arrive at a solution. They also need to be reassured that they do not need to fill in all the cells if they have reached a solution and there are empty cells. They also need to overcome their perception that ratio tables must be constructed 'in order'—that is, that after multiplying by 10, they can readily halve the amount in the next cell and it will not be 'out of order' in the sequence.

Rate

The topic of rate is usually associated with ratio. Many of the procedures associated with ratio calculations also apply to rate situations. Rate describes a relationship between two different measures: cents per litre, dollars per kilogram and metres per second are examples.

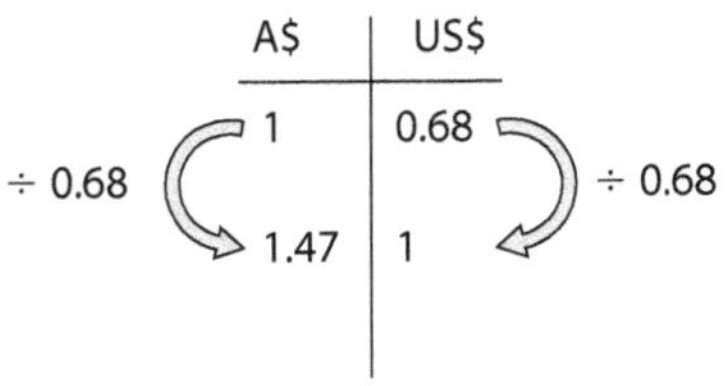

A$–US$ exchange rate mid-2003

Just as ratio tables assist in organising information for ratio investigations, they can also be used to assist in calculation of rates. The use of ratio tables provides a structure for organising the components of the comparative situation and for tracking operations on the numbers during the solution process. With such a structure, problem-solving with rates and ratios becomes a manageable task, and even currency exchange between countries becomes simple to visualise, conceptualise and calculate. For example, a ratio table can assist in conceptualising the exchange rate of US dollars to Australian dollars. As seen in the ratio table depicted, at the mid-2003 rate A$1.00 was equivalent to US$0.68. To make the US dollar the focus of investigation, divide 0.68 by 0.68 to give 1. Correspondingly, the value in the Australian dollar (A$) column on the ratio table must then be divided by 0.68. This gives a total of $1.47. Thus the US$1, when converted to Australian dollars at the current rate of 0.68, is equivalent to A$1.47.

In 2001–02, the Australian dollar compared to the US dollar was approximately 0.56. That is, A$1.00 could purchase 56 cents in the United States. Hence, for every US dollar, you would receive $1.78 in Australian currency at this rate. In 2010, the Australian dollar reached parity with the American dollar. This was great news for Australians travelling to and shopping in the United States. In Australia, in 2010, Converse sneakers were A$90; in the US, they

Time (hr)	0.5	1	1.5
Distance travelled (km)	60	120	180

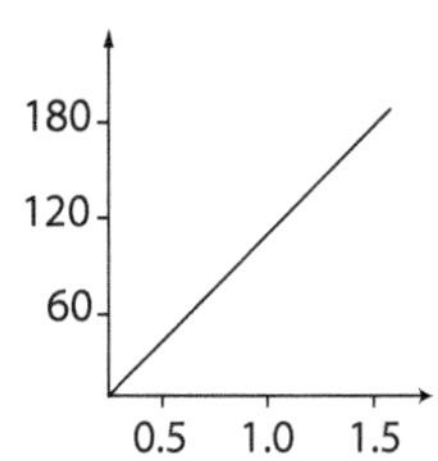

Graph showing speed of a car that travels 60 km in 0.5 hours

were selling for US$60. If the US dollar was at the 2001 rate, the Converse sneakers would be A$107.14, and hence not much cheaper to buy. However, in 2010, the US price meant a great saving to Australians. You can track these calculations by constructing a ratio table in a similar manner to the previous example. For example, if a car travelled 60 kilometres in half an hour, its speed could be calculated as distance (60 km) divided by time (0.5 hour) to give a speed of 120 km/hour. Perhaps the best-known rate is speed. The definition of speed is often given as distance over time and the process of division is used to calculate speed given measures of distance and time. Thinking of rate as merely a division process limits its conceptualisation as a linear relationship between two measures—this can be more appropriately conceptualised in graphical form.

Rate is used frequently in our society, so investigations using rate can easily be linked to the real world. Common rates are found in the purchase of goods in dollars per kilogram, petrol in cents per litre, mobile phone calls in cents per minute (or second) and part-time employment in dollars per hour. Showing rates as a graph reinforces the relationship between the two measures in the rate.

Per cent

Evidence of per cent usage and applications abounds in our society. In shopping, we are confronted with discounts; in banking, we are presented with interest rates; and in media reports, per cent is used to convey information. However, a simplistic view of per cent is often the reason for its misapplication, as the following errors indicate:

- A newspaper article reported the story of a statue picked up for $1 at a garage sale and then being valued at $200, with the headline reading that the lucky statue owner had made a 200 per cent profit.
- Train fares increased 300 per cent, from $3 to $9.
- To encourage sales but also to insure against loss, the store owner increased prices by 25 per cent before advertising a discount of 25 per cent on the new price.

Per cent is a difficult topic to teach and to learn, but it is often taught in a narrow fashion to link only to fractions and decimals (Chen & Rao 2007). A rich conceptual understanding of per cent, and confidence to use and apply per cent, rely not only on knowledge of fractions and decimals but also on ratio understanding and proportional reasoning skills. Students who can operate successfully in the domain of per cent are those with well developed 'per cent sense' (Goos et al. 2017).

Meanings of per cent

Per cent has many meanings and is used for different purposes in different contexts. The multifaceted nature of per cent is seen in the following meanings:

- Per cent is a *fraction:* 35 per cent is 35/100 in the part-whole sense.
- Per cent is a *decimal*: 35 per cent is 0.35.
- Per cent is a *number*: 35 per cent of an amount can be counted and located on a number line.
- Per cent is a *ratio*: There are 35 per cent more girls in Class A than in Class B.
- Per cent is a *proportion*: 70 out of 200 is equivalent to 35 out of 100, which is 35 per cent.

Per cent is a difficult concept because of its many meanings, and the many meanings are masked by a single, concise symbol: %. Per cent instruction needs to promote students' understanding of the various applications of per cent; this is promoted through explicit teaching of per cent as a ratio and a proportion, and through building on and linking to students' prior knowledge of fractions and decimals (Dole 1999; 2000).

Building conceptual understanding of per cent

The basic concept of per cent is its link to a base of 100. Linking per cent to fractions and decimals is an extension of using 10 × 10 grids to represent decimal (and fraction) numbers (as outlined in the section on decimals). As children draw various pictures of fractions with denominators of 100, labelling the diagram as a per cent shows the equivalence of fractions, decimals and per cent. Drawing particular percentages in a highly visual way can also show the link of some common percentages to fractions (e.g. 50 per cent is ½, 25 per cent is ¼, 75 per cent is ¾, 10 per cent is $\frac{1}{10}$) and promote per cent benchmarks (see the following diagram).

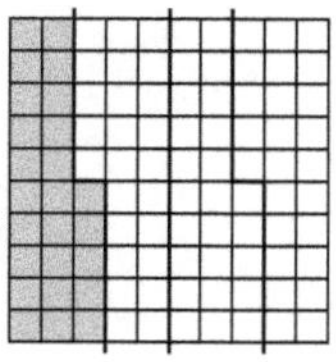

25% as 2 × 10% + 5%

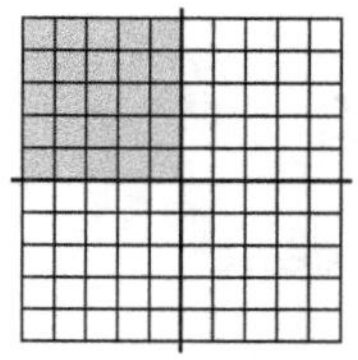

25% as ¼

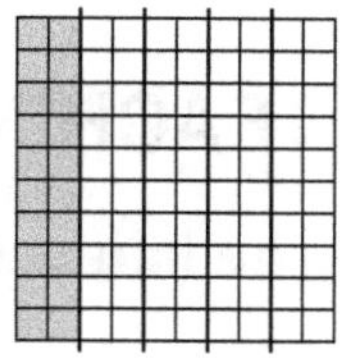

20% as ⅕

The difficulty with using 10 × 10 grids for per cent representation arises when percentages are greater than 1. To draw a picture of 115 per cent, the dilemma is whether to use two grids when one is not enough, which can possibly evoke the misconception of two wholes being required for percentages greater than 1. A better way is to use number lines simultaneously with 10 × 10 grids. For percentages greater than 1, the number line can be extended beyond the 100 per cent mark. Representing percentages on a number line is a means of promoting understanding of per cent as a number.

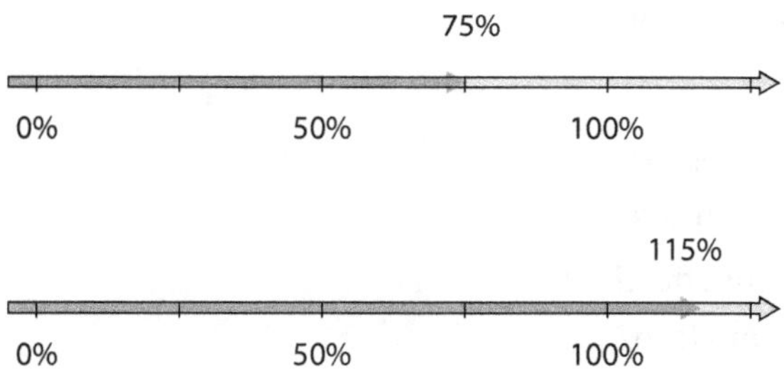

The language of per cent

'Per cent' is frequently used in everyday language and conversation, but quite often its usage does not adhere to the mathematical rules of per cent. This may be a source of confusion for students. It is not uncommon for young children playing sport to be told by the coach that they are expected to put in '110 per cent effort' for the game. If 100 per cent is used to designate the whole, can the whole be more than 100 per cent? Engaging students in discussion of the language of per cent in both colloquial and mathematical terms is a valuable way of determining their intuitive notions of per cent and the extent to which these notions may help or hinder further conceptual development of per cent in a mathematical sense.

TEACHING IDEA

The language of per cent*

In groups, discuss what the following phrases mean and when they might be used.

Can we have:

- 50 per cent class attendance?
- 90 per cent class attendance?
- 100 per cent class attendance?
- 110 per cent class attendance?

Can you be:

- 50 per cent certain?
- 100 per cent certain?
- 110 per cent certain?

How much effort is:

- 50 per cent effort?
- 10 per cent effort?
- 100 per cent effort?
- 110 per cent effort?

What do the following mean?

- 75 per cent full
- 25 per cent full
- 100 per cent full
- 25 per cent empty

*Adapted from Glatzer 1984

Per cent part and complement

To promote a rich understanding of per cent, students need to think of per cent as describing a situation in terms of a base of 100. They also need to see the situation as one part in relation to 100, and the remaining part (the complement) in relation to 100. For example, if a label on a wool sweater states that the sweater is 80 per cent wool, then it is known that 20 per cent of the sweater is made of fibres that are not wool; or if 55 per cent of the people voted for no homework, then 45 per cent of the people voted for homework.

To assist students to visualise per cent situations in terms of the part and the complement, it is useful to represent various situations on 10 × 10 grids. By shading the various parts of the whole on the grid, a visual image is presented, and the part and complement are seen as comprising the whole. Shading 10 × 10 grids is simpler than using a circle graph for representing information in percentages, since the 100 parts in the whole are clearly visible.

Encourage mental computation of simple per cent situations and practise of calculations to 100 by asking students to close their eyes and visualise various per cent part and complement situations. Prior experience of depicting per cent situations on 10 × 10 grids provides a strong visual image for assisting in determining solutions to such questions as:

- If I drank 25 per cent of the milk, what percentage was left?
- Twenty-eight per cent of the Smarties were green; what percentage was not green?
- The jacket was discounted by 25 per cent; what percentage of the original price did I have to pay?
- Rump steak is 82 per cent beef; what percentage is not beef?
- Soap contains 0.5 per cent perfume, 37 per cent pure soap, and the rest is fat. What percentage is fat?
- The human body is about 78 per cent water. What percentage is made up of other matter?

Mental computation of per cent situations

The development of common per cent benchmarks (e.g. 50 per cent is one-half; 25 per cent is one-quarter; 10 per cent is one-tenth) lays the foundation for simple mental computation of finding a percentage of an amount. Per cent/fraction equivalence is useful for supporting mental computation. For example, to find 25 per cent of 80, it is simple to think of 25 per cent as one-quarter and to then divide 80 by 4. Similarly, thinking of 10 per cent as one-tenth assists in mentally dividing the amount by 10 in order to find 10 per cent. Finding a percentage of an amount is the simplest of all per cent calculations, and students need to be provided with opportunities to mentally calculate simple percentages of this type. Use selected examples to support student facility with calculations of percentages (e.g. 25 per cent of 800—think one-quarter of 800; 25 per cent of 650—think half of 650, then half again).

Increases of more than 100 per cent

The three misapplications of per cent presented at the start of this section stem from misunderstanding of per cent increases of more than 100 per cent. The difficulty in conceptualising and checking calculations of increases of more than 100 per cent lies in the potential confusion caused by the fact that the situation can be considered from both an additive and a multiplicative perspective. The first example given at the beginning of this section describes incorrect reporting in a newspaper article of a statue purchased for $1 and then valued at $200, and this being described as a 200 per cent increase. The confusion is that the original whole amount ($1) has not been considered as 100 per cent, but has possibly been considered as a value of 1. Since the original amount of $1 is equivalent to 100 per cent, an increase of 100 per cent would mean doubling the original amount to give a new whole of $2. A 200 per cent increase is actually three times the original amount, which in this case would be $3. By considering percentage increases progressively in this way, the magnitude of the actual percentage increase to reach $200 becomes apparent. The actual percentage increase from $1 to $200 is 19 900 per cent.

Considering 100 per cent and 200 per cent increases, as in the example above, the additive and multiplicative interpretations of per cent increases can be seen. If the statue had been purchased for $1 and resold for $2, the per cent increase from an additive interpretation would have been 100 per cent. That is, 100 per cent has been added to the original whole. In a multiplicative sense, the new price is 200 per cent of its original price. In the train fare example given at the start of this section, an increase from $3 to $9 means that the fare has been increased from the original price additively by 200 per cent, or multiplicatively by 300 per cent. The subtle differences between interpreting increase situations in a multiplicative or an additive sense help to explain how simple errors can be made.

TEACHING IDEA

Exploring increases of 100 per cent

- Take a group of counters.
- Encase in the hands and state, 'This is one whole.'
- Increase the group by 100 per cent.
- State how many counters there are in the new group. State this additively, as a per cent increase.
- State this multiplicatively as the number of times bigger the new group is than the original.

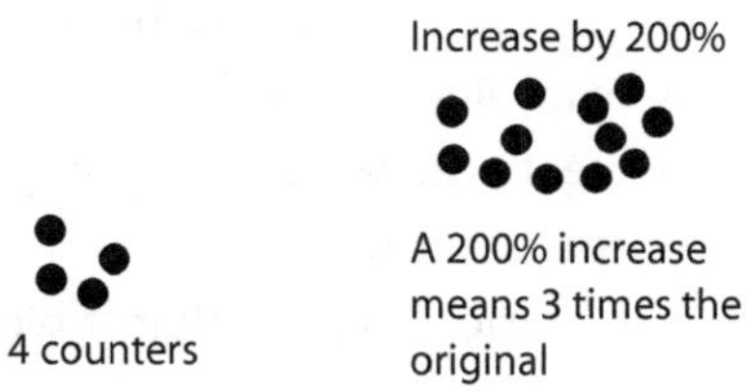

The multiplicative and additive interpretations of percentage increases of 100 per cent and other multiples of 100 can be assisted through explorations with concrete materials. Take a small group of counters and increase the number by 100 per cent, by adding the same amount of counters as the original group. The size of this group is then compared to the original amount (see the previous Teaching Idea). An increase of 100 per cent means the new size of the group is twice that of the original amount. When the group is increased 200 per cent, the increase is three times the original. Through such explorations, the teacher's role is to encourage students to generalise what happens as a collection is increased by an amount of more than 100 per cent or by multiples of 100 per cent. Draw students' attention to the fact that, for example, a 200 per cent increase is not the same as multiplying by 2. Such investigations lay the foundation for visualising mental calculations of amounts greater than 100 per cent. Ask students to close their eyes and visualise increasing groups by various percentages (e.g. 4 counters, increase 100 per cent; 50 counters, increase 100 per cent; 800 counters, increase 300 per cent). Always encourage students to discuss their solution strategies.

Per cent increases and decreases

For exploring per cent increases of more than 100 per cent, the language foundation and multiplicative structure are emphasised in order to use the language of per cent to describe situations appropriately. Refer to the example given at the start of this section of the shop owner increasing prices before a sale by 25 per cent and then giving a 25 per cent discount. Does the shop owner win, lose or break even in this situation? In fact, by employing this strategy, the shop owner loses. Modelling this situation with counters exemplifies why this is so. Take a set of eight counters. Increase the group by 25 per cent (i.e. add two counters). The new number of counters is ten. Therefore, the new *whole* totals ten. How many counters must be removed from this group to take it back to its original amount? Two. What fraction of the new total is this? One-fifth. What percentage is this? It is 20 per cent (see the accompanying diagram).

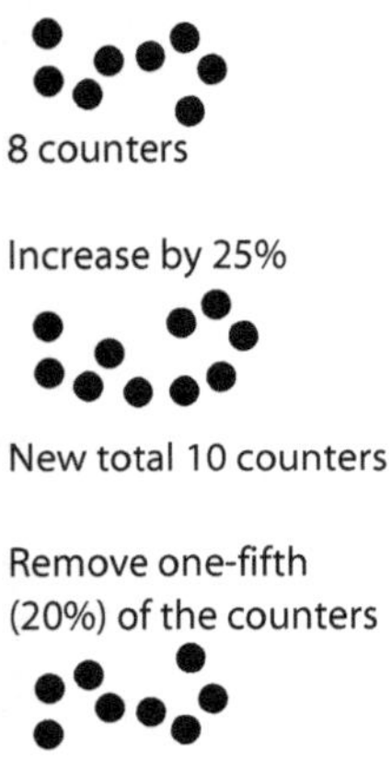

From this demonstration, we see that if we increase something by 25 per cent, we actually need to reduce it by 20 per cent to get back to the original amount. Relating this back to the shop owner, a 25 per cent discount after increasing the original price by 25 per cent means that the shop owner is selling the product for less than the price before the sale. The shop owner would not have a hope of breaking even or making a profit.

Through modelling with counters, the following relationships can be discovered:

- An increase of 50 per cent means a reduction of 33⅓ per cent to get back to the original.
- An increase of 25 per cent means a reduction of 20 per cent to get back to the original.
- An increase of 33⅓ per cent means a reduction of 25 per cent to get back to the original.
- An increase of 100 per cent means a reduction of 50 per cent to get back to the original.

Teaching rational number in the middle years

Rational number is situated squarely in the middle-years mathematics curriculum. In fact, the focus of number in the middle years is on promoting rational number knowledge. In the early middle years, decimal and common fraction understanding is the focus. This is extended gradually to ratio and percentage as students progress to the upper-middle years. The study of rational number is difficult for students, and students will exhibit varying levels of understanding in the middle years of schooling. In this chapter, foundations for rational number have been provided, enabling teachers to look back to activities for building conceptual understanding when students experience difficulty. Using concrete materials and structuring learning experiences will support students at any year level, and strategies of these types have been advocated in this chapter.

Summary

For many students, rational number topics pose great difficulty. It is important to provide opportunities for students to engage with the big ideas of rational number topics so that they can both see connections between these topics and understand key ideas within each topic. Concrete materials need to be selected carefully on the basis of their value in accurately representing key concepts for rational number understanding, and the associated language must become a natural part of students' dialogue. Confusion with rational number topics often occurs as a result of rushing to symbolic representation and manipulation of numbers before a rich conceptual framework has been developed sufficiently. Rational number study needs to be an integral part of early mathematics classrooms in an exploratory sense, so that a solid foundation is laid for more formal investigations in the middle and later years.

Review questions

11.1 Describe rational numbers in terms of their location on a number line. Outline how rational numbers extend students' conceptualisation of numbers on a number line.

11.2 Why is decimal knowledge only partly dependent on fraction knowledge?

11.3 Describe how a fraction meaning may interfere with developing ratio meaning.

11.4 Outline the multiplicative structure of ratio and percentage increase situations.

Further reading

Carpenter, T.P., Fennema, E. & Romberg, T.A. (2012). *Rational numbers: An integration of research.* New York, NY: Routledge.

Condon, C. & Hilton, S. (1999). Decimal dilemmas. *Australian Primary Mathematics Classroom, 4*(3), 26–31.

Cramer, K. & Bezuk, N. (1991). Multiplication of fractions: Teaching for understanding. *Arithmetic Teacher, 39*(3), 34–7.

Eames, C. & Barker, M. (2011). Understanding student learning in environmental education in Aotearoa New Zealand. *Australian Journal of Environmental Education, 27*(1), 186–91.

Empson, S. (2001). Equal sharing and the roots of fraction equivalence. *Teaching Children Mathematics*, 7(7), 421–5.

Graeber, A.O. & Baker, K.M. (1992). Little into big is the way it always is. *Arithmetic Teacher, 39*(8), 18–21.

Haubner, M.A. (1992). Percents: Developing meaning through models. *Arithmetic Teacher, 40*(4), 232–4.

Steffe, L.P. & Olive, J. (1991). The problem of fractions in the elementary school. *Arithmetic Teacher, 38*(9), 22–4.

Thompson, C.S. & Walker, V. (1996). Connecting decimals and other mathematical content. *Teaching Children Mathematics, 2*(8), 496–502.

Yang, D. & Reys, R. (2001). One fraction problem, many solution paths. *Mathematics Teaching in the Middle School*, 7(3), 164–6.

References

Chen, H. & Rao, A.R. (2007). When two plus two is not equal to four: Errors in processing multiple percentage changes. *Journal of Consumer Research, 34*(3), 327–40.

Dole, S. (1999). Successful percent problem solving for Year 8 students using the proportional number line method. In J.M. Truran and K.M. Truran (eds), *Making the difference: Proceedings of the 22nd Annual Conference of the Mathematics Education Research Group of Australasia.* Adelaide: MERGA.

Dole, S. (2000). Promoting percent as a proportion in eighth-grade mathematics. *School Science and Mathematics, 100*(7), 380–9.

Dole, S. (2003). Applying psychological theory to helping students overcome learned difficulties in mathematics: An alternative approach to intervention. *School Psychology International, 24*(1), 95–114.

Dole, S. (2008). Ratio tables to promote proportional reasonings in the primary classroom. *Australian Primary Mathematics Classroom, 13*(2), 18–22.

Dole, S., Goos, M., Hilton, A. & Hilton, G. (2018). Authentic numeracy context for proportional reasoning—the case of the seven summits. In H. Hunter, P. Perger and L. Darragh (eds), *Making waves, opening spaces: Proceedings of the 41st Annual Conference of the Mathematics Education Research Group of Australasia* (pp. 59–62). Adeladie: MERGA.

Dole, S., Hilton, A. & Hilton, G. (2015). Proportional reasoning as essential numeracy. In M. Marshman, V. Geiger and A. Bennison (eds), *Mathematics education in the margins: Proceedings of the 38th Annual Conference of the Mathematics Education Research Group of Australasia* (pp. 189–96). Adeladie: MERGA.

Glatzer, D.J. (1984). Teaching percentage: Ideas and suggestions. *Arithmetic Teacher, 31*, 24–6.

Goos, M., Vale, C. & Stillman, G. (2017). *Teaching secondary school mathematics: Research and practice for the 21st century*. Crows Nest: Allen & Unwin.

Hilton, A., Dole, S., Hilton, G. & Goos, M. (2015). Using photographic images to enhance conceptual development in situations of proportion. *Australian Primary Mathematics Classroom, 20*(1), 3–9.

Lesh, R., Post, T. & Behr, M. (1988). Proportional reasoning. In J. Hiebert and M. Behr (eds), *Number concepts and operations in the middle grades* (pp. 93–118). Hillsdale, NJ: Lawrence Erlbaum.

Ministry of Education, New Zealand. (2008). *The number framework*. Wellington: Ministry of Education.

Van Dooren, W., De Bock, D., Hessels, A., Janssens, D. & Verschaffel, L. (2005). Not everything is proportional: Effects of age and problem type on propensities for overgeneralization. *Cognition and Instruction, 23*(1), 57–86.

Yang, D.C. & Sianturi, I.A.J. (2018). Sixth grade students' performance, misconceptions, and confidence when judging the reasonableness of computational results. *International Journal of Science and Mathematics Education*, 1–22. https://doi.org/10.1007/s10763-018-09941-4.

CHAPTER 12

PATTERNS AND ALGEBRA

For many students, the study of algebra is associated with a turning point—or, more precisely, a turning-off point—in the study of school mathematics. They leave school with vague feelings of uneasiness associated with *x* and *y*, and trying to make sense of something they feel is quite meaningless. Yet this should not be the case, since understanding of algebra and algebraic thinking lead on from early learning about numbers and patterns. Students' difficulties with algebra are often associated with problems in arithmetic and computational understanding (Kieran 2018).

Algebra is often referred to as generalised arithmetic, whereby students take their early learning about number and operations and move to seeing generalisations. Exploring patterns is key. Patterning has long been, and continues to be, a main element of the mathematics curriculum in the early years (Lüken & Kampmann 2018), and its importance as a foundation for algebra is now being increasingly recognised and acknowledged (Maj-Tatsis & Tatsis 2018). Playing with numbers, building understanding of how numbers work and noticing patterns in the early years lay the foundation for algebraic understanding (Blanton et al. 2018; Carraher & Schliemann 2018).

In many cases, students' introduction to formal algebra is through endless textbook exercises involving solving equations and inequalities, finding the value of *x*, collecting like terms and substituting values for unknowns. Although this is part of what comprises algebraic knowledge, if taught without meaning, there is little chance that students will see algebra as useful (Kieren 2018). Algebra assists in solving real problems, and is therefore a valuable numeracy tool.

Learning algebra is important from a perspective of equity, because algebra is regarded as a key to mathematical literacy, which in turn is the key to economic access and citizenship (Bond 2003). Providing activities that involve simple patterning entails students thinking logically, and this, in turn, supports students' general learning of mathematics. Algebraic thinking is supported through the work of patterning and relationships; thus mathematics education in primary school needs to have a strong emphasis on patterning in order to prepare students for algebraic thinking. For many students in the middle years of schooling, experiences with algebra are problematic because they have not had a good foundation in early algebra.

Patterning

There are three distinct aspects of the teaching of patterning: repeating patterns, growing patterns and number patterns. Early experiences should be ones that students can understand easily, and teachers should organise learning experiences that progressively extend students' thinking. It is important to allow students time to justify, explain or extend their patterns, since there may be other ways to see a pattern than those envisaged by the teacher. Observing students working with patterns can provide a basis for understanding their thinking, and so allow the teacher to better plan for teaching. Blume and Heckman (2000) note that students are more likely to be able to continue a pattern than to insert a missing element. It is important for students to experience both linear patterns—that is, patterns back and forth in one direction only, as well as patterns in two directions, such as those in the accompanying picture.

Early experiences with patterning in two directions

When teaching patterning, it is important to focus on students developing appropriate language relating to patterns, so that they will become familiar with identifying particular elements within a pattern—for example, what will the fifth element be?

Repeating patterns

Repeating patterns are usually introduced in a linear form. Once students are familiar with this format, they can work with formats in which the repetition goes in more than one direction. Study of patterns can often involve geometric patterns and patterns in cultural activities, such as quilting, mat weaving, and so on.

Early pattern work will involve patterns of various colours or shapes. As a general rule, early work should involve patterning where the only attribute is the focus of the pattern. Blocks that vary *only* in colour or shape are very useful for this early work. Use of Unifix blocks can be a good starting point, since these blocks are identical except for colour. This allows students to focus on the changes in colour. Experiences with patterning involve students:

- copying a pattern
- describing a pattern
- copying and extending a pattern
- making and describing a pattern.

The cognitive demands placed on students with these patterning activities increase, so early work should involve considerable copying of patterns and building towards creating patterns. Patterns may initially involve only one of each item (red, green, blue, red, green, blue), but later they will involve multiples of an item (red, red, green, red, red, green).

Early experiences with patterns will normally consist of only two items involving one attribute. These are commonly referred to as *AB patterns*. Typically, this could be of a shape (square, circle, square, circle) or other attributes, as shown in the picture. Various items commonly found in classrooms can be used for this type of patterning.

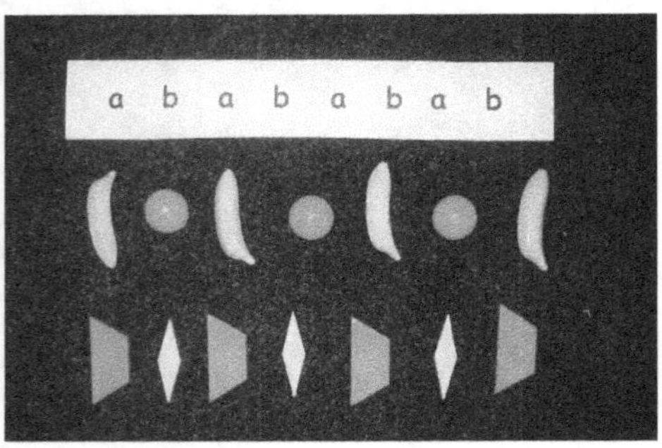

Simple AB patterns

Early experiences in patterns can vary in one attribute such as colour, shape or size, as in the next diagram. The shape and size patterns here are simple ABC patterns. As students become confident in continuing or creating patterns of this format, activities can focus on making the patterns more complex by using formats such as ABB or ABBC. While geometric shapes are often the basis of early patterning, activities can also involve the use of alphanumeric-symbolic forms, pictures, and so on.

Colour

Shape

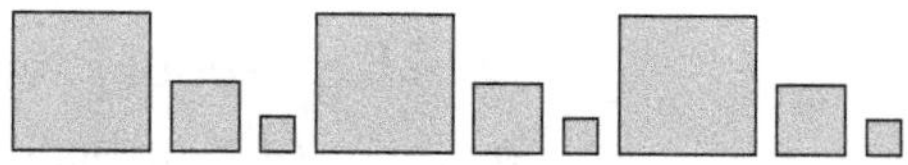

Size

Once students are familiar with repeating patterns where there is a change in one attribute, teachers can introduce changes to the other attributes. In the example below, both colour and shape have been changed to create a more complex pattern, which includes first-order and second-order patterns—that is, a smaller pattern is contained within a larger pattern. Here, the first-order pattern is related to the shapes—square, square, circle. This is contained within a larger pattern of colour—black, grey, white, shade. That is, the first-order pattern of shape is contained within the second-order pattern of colour. This is fairly complex, and would be introduced later in the process of patterning.

Growing patterns

Rather than a sequence of patterns that is repeated over and over again, a growing pattern involves an increase or growth in the subsequent sections of the sequence when compared with the previous sections. An example of a growing pattern is AB, AAB, AAAB. Here, the number of Bs remains unchanged, but the number of As increases by one in each subsequent section. While there is a change in each section, the change is constant (increases by one)—that is, the first element in the pattern is growing (or increasing) by one each time while the second element repeats. Initial activities should involve only one element of the pattern increasing, but as students become confident with growth in one element, growth in more than one element can occur.

Growing patterns can also be two dimensional. This can involve the use of tiles so that the pattern can be seen to grow in two directions.

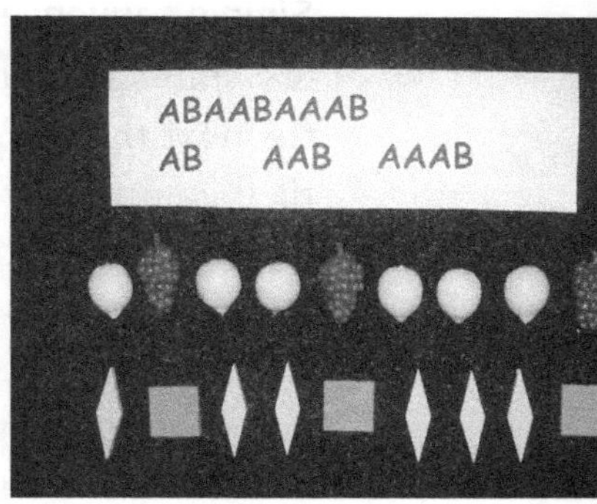

Growing patterns

Number patterns

Number patterns, like growing patterns, are about describing changes. For example, in the number sequence 2, 4, 6, what number comes next? Students need to identify the change that is occurring (here, adding 2) or, in another example, notice that when subsequent numbers are compared to the first, the second has increased by 2 and the third by 4, so that the next will increase by 6; thus, the number after 6 is 8.

Students should come to realise that when working with patterns in sequences of numbers, knowing the first three terms only might not be sufficient to develop a theory to explain the pattern. In some cases, more than three terms might be required. Also, as indicated in the picture, there might be several ways to describe a particular pattern.

Simple sequences of numbers

Students should be provided with a variety of activities, such as finding the next three terms in the sequence 2, 4, 6, 8 . . . , or finding terms at the beginning of a sequence such as . . . 31, 38, 45, 52. Students should be encouraged to describe in words the sequence and the change from one term to the next. The teacher's role is to explicitly draw students' attention to making generalisations about number patterns. When students are adept at counting in twos, ask them to look at the relationship between the number and its solution:

- What is the first number? Two.
- What is the second number? Four.
- What is the third number? Six.
- What do you notice about the position of the number in the sequence and its solution? It is double.
- So . . . what is the tenth number in the sequence? 20.
- What is the hundredth number? 200.

In simple algebraic terms, this means that the general rule is $2n$ (where n is an integer). This is actually an exploration of even numbers (see below).

Odd and even numbers

The earliest experiences with number theory arise through the study of odd and even numbers. This is often taught through the use of grids (or egg cartons) where students can see what happens when odd and even numbers of items are placed into a two-grid (see diagram).

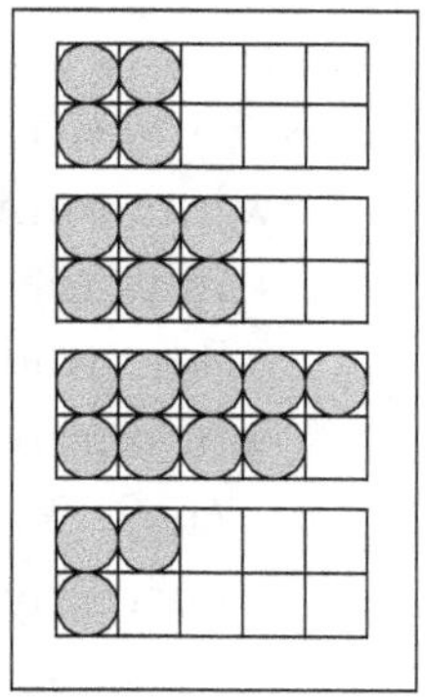

Opportunities should be provided for students to develop their own theories about odd and even numbers, to articulate what they see and to develop a sense of what is happening. They will also need opportunities to determine for themselves the following patterns: even numbers are integers that end with 0, 2, 4, 6 or 8, and odd numbers are integers that end in 1, 3, 5, 7 or 9. Like developing the generalisation that $2n$ is the algebraic shorthand for even numbers, a generalisation for odd numbers can occur by drawing students' attention to the position in the sequence and its solution:

- What is the first number? One.
- What is the second number? Three.
- What is the third number? Five.
- What do you notice about the position of the number in the sequence and its solution? It is double minus 1.
- So . . . what is the tenth number in the sequence? Double 10 minus 1, which is 19.
- What is the hundredth number? 199.

In algebraic terms, this means that the general rule for an odd number is $2n - 1$ (where n is an integer). As with all scaffolding, good questions are important for students to develop robust understandings. Another good question to pose is whether zero is even or odd.

Rich discussion can ensue from such questions. By the way, zero is an even number. The definition of an even number is that it is an integer exactly divisible by 2. Multiplying and dividing numbers by 2 and exploring patterns will lead to healthy discussion on why zero is even and, further, whether odd and even numbers can be negative numbers (and, yes, they can).

Patterns involving two related sequences of numbers (relationships)

Mathematics can involve finding and describing relationships between items. Early work with relationships focuses on simple experiences, such as how the number of bicycles is related to the number of wheels or how the number of hands is related to the number of fingers. The study of relationships should link closely to other aspects of number study—for example, it should involve the operations (1, 2, 3, 4) currently being studied, and numbers in the range currently being studied.

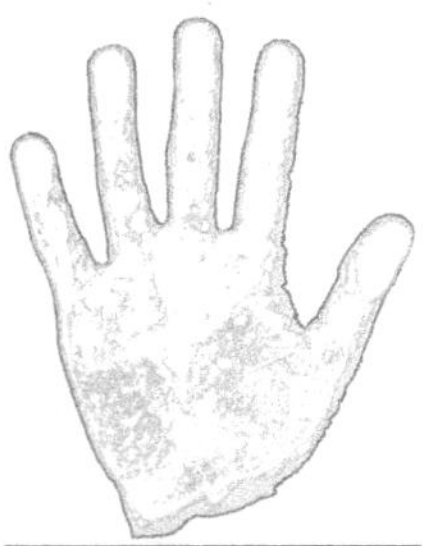

1	2	3	4	n
5	10	15	20	$5n$

At some point, students need to develop effective ways to record their work. Tables or grids can be used, as shown in the diagram. Terms in the upper row increase by 1, and students determine the constant increase in the lower row (in this case, 5). A more complex response is to answer in the form of a generalisation—that is, to describe a rule for determining terms in the lower row from the corresponding term above. In this case, the generalisation is that the number is multiplied by 5 ($5n$). By developing a rule in this way, students can predict any term in the sequence. Consider the task of working out the one-hundredth term in the sequence. This is made much easier by using the generalisation or rule.

Function machines

'Guess my rule' is an activity used with the aid of a function machine. To construct a function machine, simply draw a square shape on the board, with a 'funnel' shape going into the square on one side and a funnel coming out of the square on the other side. Think of a rule that can be applied to a number (e.g. double the number, then add 1). Students volunteer a number, and it is written on the whiteboard near the incoming funnel of the function machine. Once the rule has been applied, the outgoing number is written near the outgoing funnel of the function machine. For example, 7 goes in, 15 comes out. After several numbers going in and coming out, ask students to volunteer what they think the rule is. Again, consider that more than one rule may be correct. There are many interactive websites that contain animated function machines, which can be used with the whole class via an IWB or at single computers for small-group or individual work. Just Google function machines and browse the range of options available—for example, www.mathplayground.com/functionmachine.html. An even easier approach is to provide students with simple tables of values, with columns headed 'input' and 'output' for exploration and analysis. Ensure students also have the opportunity to make up their own rules for the whole class to determine.

Function machines and the types of pattern exploration described above can provide a context for introducing students to graphing (or plotting) pairs of numbers on a grid. In this way, students can be introduced to displaying the function rule geometrically. For example, consider the simple function machine in the diagram, where the function (or change) is add 3. In this example, a number—say, 5 (the input)—comes into the function machine, the change (+ 3) is made and the number 8 (the output) exits from the machine.

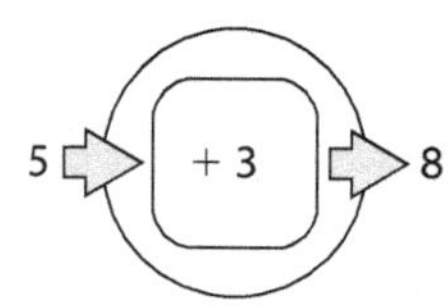

+ 3	1	2	3	4	5	6
	4	5	6	7	8	9

A table showing the input numbers 1 to 6 and their corresponding output numbers can then be constructed. The numbers can then be plotted on a grid showing the horizontal numbers for 1 to 6 and the vertical numbers to 9. When the numbers are plotted, a straight line results.

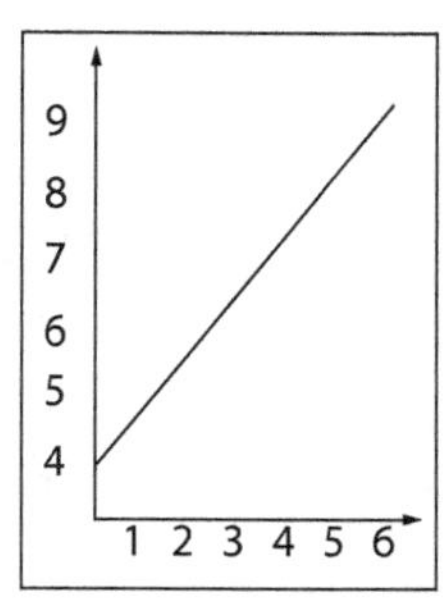

The plotted line (graph) can then be used to determine other output values for given input values. For example, if the input is 7, what is the output? If the graph is extended over a longer range of values, other output numbers can be read off. Students can also work in this way with numbers involving decimals (or fractions). For example, what input would give an output of 7.5? The student's task is to determine this from the graph.

Using approaches such as these, teachers can informally introduce ideas and language relating to algebra and functions. Providing students with variety in representing patterns and relationships to include tables, graphs, words and symbols is an important part of algebra knowledge (Carraher et al. 2016; Kieran 2007). A second digital resource to support student understanding of functions is available from the following website: www.nrich.maths.org/2727.

Technology and algebraic thinking

As has been noted, there are many interesting ways in which digital technology can be used to support the development of students' mathematical thinking. One that is very relevant to number patterns and algebra involves the use of spreadsheet software (Calder 2004). A spreadsheet is a computer program that will do calculations automatically according to formulae entered into the computer, based on a table of numbers arranged in rows and columns. Thus, in contexts such as the study of interest rates, foreign currencies or results of sports competitions, students can use spreadsheets to do the routine aspects of calculation, and thereby focus on thinking about the results of the calculations. Spreadsheets are also an ideal tool for undertaking activities involving function machines or related sequences of numbers (see above).

One type of activity can involve students reflecting on two related columns of numbers in order to determine a number pattern and to study or work out the formula involved. This process encourages students to develop a general rule, articulate it in a language applicable to the software and then carry out the necessary tasks to program the spreadsheet. A teaching tip: you might need to hide the formula bar so that students can't just hover the mouse over it to see the formula. Most spreadsheet programs are simple to use, and most are appropriate for upper-primary students. For example, if students were converting foreign currencies into the local currency, they could fill in the first cell, work out the multiplier for the second cell and then, using the appropriate spreadsheet language, fill in the commands for the program being used, completing the third cell using the 'fill' mechanism. This process encourages students to think algebraically by developing and articulating the general rule needed. In addition to spreadsheets, there are a number of algebra apps that support student learning. One we recommend is found here: www.venturaes.com/iosapps/algebra_concepts.html.

Number, arithmetic and algebra

Through number study in the early years, students are working with powerful algebraic ideas, but they often are not aware of this. When learning basic facts, for example, students are introduced to the commutative law of arithmetic—that is, when they learn one basic fact, they are actually learning two, since they see that, for example, 7 + 5 gives the same total as 5 + 7. When they are performing mental computation, the associative law of arithmetic is often used. For example, when adding 9 + 7 + 3, it is probably easier to add 7 and 3 first, then add 9, rather than to begin at the start and add 9 and 7, then 3. These understandings about numbers and how they work are important foundations for algebra. This does not mean that explicit teaching of the laws of arithmetic is required, but rather that students need to be immersed in an environment in which rich number explorations and investigations are a feature of number study.

Equality

Understanding of the equals sign is a major foundation of algebra. Yet misconceptions about the equals sign can interfere with the development of flexible thinking about numbers and, hence, about algebraic thinking. When engaging students in early computation activities, the use of language must be considered carefully. For example, number sentences such as 3 + 4 = 7 may be said as '3 and 4 makes 7'. The concept of 'making' results in confusion because it does not necessarily highlight the idea of equivalence. For example, when posing the question, 'Is this sentence correct: 4 + 5 = 2 + 7?', students might regard it as incorrect, since it does not conform to their expectations. Commonly, 5 – 2 = 3 is regarded as back to front, or it is believed that the answer is 5. Students are unaware of the relationship between 5 and 2 and 3, or that this involves equality. Such misunderstandings are a result of narrow teaching and lack of opportunity to explore numbers and the ways in which they interrelate. Balance scales or a number beam balance can be used to convey the idea of the equals sign as representing equivalence. Students place blocks on each side of the beam to make it balance. They then remove blocks, realising that they must remove an equivalent number of blocks from the other side to keep the beam in balance. This supports later algebraic activities of solving for unknowns. As with all manipulatives, the value in the balance beam relates to how it is used and whether students are able to abstract the notion of equality from the balancing process. A worthwhile digital resource to support the notion of equal as a form of balance is the Animal See Saw (www.counton.org/games/animal-see-saw/index.html). When equivalent number situations are symbolised, they should be presented in various modes. For example, when considering the equivalence relationship of $3 + 4 = 7$, discussion about why $7 = 4 + 3$ is also correct should occur. Such simple and repeated exposure to representation of numerical statements of equivalence can promote students' understanding of mathematical equivalence (McNeil et al. 2015).

Arithmetic number study

Some important terms associated with the four basic operations of arithmetic are presented in the table below.

Operation	Key terms	Example	Application of terms
Addition	Addend, sum	$6 + 4 = 10$	Addends: 6 and 4 Sum: 10
Subtraction	Subtrahend, difference, minuend	$15 - 12 = 3$	Subtrahend: 12 Difference: 3 Minuend: 15
Multiplication	Factor, product (multiplier, multiplicand)	$8 \times 4 = 32$	Factors: 8 and 4 Product: 32 Multiplier: 4 Multiplicand: 8
Division	Divisor, dividend, quotient, remainder	$43 \div 8 = 5$ r 3	Divisor: 8 Dividend: 43 Quotient: 5 Remainder: 3

Laws of arithmetic

A mental computation program assists students to become familiar with the laws of arithmetic. Exploring strategies for mental computation supports flexible thinking about number and understanding of number relationships. The commutative law is learned through spin-arounds when learning basic facts (i.e. 9 + 3 = 3 + 9).

Students are meeting the inverse law when they learn subtraction and division facts. Once addition and multiplication facts are known, students develop subtraction and division facts by using 'think addition' and 'think multiplication', respectively. When students learn addition of zero or multiplication by 1, they are learning the identity law. The laws of arithmetic are provided in the table below.

Law	Addition	Multiplication
Commutative	a + b = b + a (9 + 3 = 3 + 9)	a × b = b × a (9 × 3 = 3 × 9)
Associative	(a + b) + c = a + (b + c) (9 + 7) + 3 = 9 + (7 + 3)	(a × b) × c = a × (b × c) (4 × 7) × 5 = 4 × (7 × 5)
Distributive	a × (b + c) = a × b + a × c 74 × 102 = 74 × 100 + 74 × 2	(a + b) × c = a × c + b × c 102 × 56 = 100 × 56 + 2 × 56
Identity	a + 0 = a 5 + 0 = 5	a × 1 = a 6 × 1 = 6
Inverse	a + (−a) = 0 3 − 3 = 0	a × ⅟a = 1 4 × ¼ = 1

Factors and multiples

A *factor* is a whole number that can be divided equally into another number. A *multiple* is a whole number that is created by multiplying the original number by a whole number. A good way to discuss some of the language of factors and multiples is through the use of examples: 6 is a multiple of 2; 20 is a multiple of 5; 100 is a multiple of 10. The general rule for multiples is that *a* is a multiple of *b* if *a* can be obtained by multiplying *b* by a whole number. Factors can be explored in a similar way: 4 is a factor of 12, 8 is a factor of 16, and so on. The general rule for factors is that *a* is a factor of *b* if *b* can be obtained by multiplying *a* by a whole number.

Prime and composite numbers

Through the study of factors and multiples, students come to see that some numbers in the 1–100 range have a relatively large number of factors. For example, 36 has nine factors—1, 2, 3, 4, 6, 9, 12, 18 and 36—and 90 has 12 factors—1, 2, 3, 5, 6, 9, 10, 15, 18, 30, 45 and 90. They will also discover that there are some numbers that have no more than two different factors—that is, 1 and the number itself (the number 1 is not considered prime because its only two factors are the same). For example, 13 has 1 and 13 as its only factors. Numbers with only two factors are referred to as *prime numbers*; numbers with more than two factors are referred to as *composite numbers*.

Number explorations

Geometric numbers

Exploring geometric numbers and trying to find the general way to describe the pattern is another way to promote algebraic reasoning and flexible number knowledge. Using concrete materials and constructing tables of values are the starting points. Use counters to create square numbers, triangular numbers and rectangular numbers, and keep a record (using an organised list or table) of the number of counters required to construct each number in the sequence. Look for a relationship between the position of the number and the number of counters required. Try to determine the tenth, and then the hundredth number in each sequence.

The table below shows the number of counters required to construct the first five square, triangular and rectangular numbers.

Position	1	2	3	4	5
Square number	1	4	9	16	25
Triangular number	1	3	6	10	15
Rectangular number	2	6	12	20	30

Palindromic numbers

Palindromic numbers are numbers that read the same from left to right and right to left. Examples of these are 121, 66, 147 and 741. Students should be familiar with palindromes from their language studies, and they are often excited at seeing numbers working in a similar way.

Palindromic numbers can be obtained via an activity in which a number is written down, and the reverse of that number is written below it. The numbers are then added together. If their sum is a palindromic number, it is said to be a first-order palindromic number. If the result is not a palindromic number, the cycle is repeated until a palindromic number is obtained. For example, starting with 728, this is reversed to 827 and the two numbers are added. The sum is 1555 (which is not palindromic), so the cycle is repeated: 5551 + 1555 = 7106, which, again, is not palindromic. The third cycle results in 13 123, the fourth in 45 254, the latter of which is a palindromic number. For any number, repeating this cycle of operations will eventually produce a palindromic number. Later students should be encouraged to determine why this cycle works mathematically.

The importance of algebra in the primary school

MacGregor and Stacey (1999) argue that the roots of algebra learning arise in mathematics in the early years when students begin to notice regularities in calculations with numbers. They discuss five aspects of number work in the primary years that are important as a basis for algebra: understanding equality, using a wide range of numbers, understanding important properties of numbers, recognising the operations, and describing patterns and functions.

Understanding equality relates to understanding number sentences and algebraic equations, and, in particular, the meaning of the equals sign in number sentences and algebraic equations. A wide range of numbers is used because of the importance of students working with fractions and decimals, as well as with whole numbers; numbers in the forms of fractions and decimals can occur frequently in algebra.

An example of an important property of numbers that needs to be understood as a basis for algebra is the following: if you subtract one less from a number, then the answer is one more. For example, 2475 – (357 – 1) = (2475 – 357) + 1. Understandings like this are necessary when algebraic equations are used to solve problems.

Recognising the operations relates to the need for students to develop a full understanding of addition, subtraction, multiplication and division as a basis for using algebra in problem-solving. This is because algebra focuses explicitly on *properties* of operations, and on *relations* among operations, rather than on merely *carrying out* operations, which is the main focus in arithmetic. Important examples of properties of operations (i.e. addition, subtraction, multiplication and division) are the commutative, associative and distributive properties (also known as 'principles' or 'laws'). Algebra at the primary-school level does not include a formal study of these properties. Rather, the focus is on developing intuitive understandings of these properties in working with numbers. In relation to patterns and functions, the focus of this chapter has been the importance of opportunities for students to explore patterns, find general rules and make generalisations.

Teaching algebra in the middle years

For many students, the formal introduction to algebra topics occurs in the middle years, when they start secondary-school mathematics. And, for many students, this formal introduction to algebra is the turning point at which mathematics starts to no longer make sense. To support students in the middle years, early patterning experiences in geometry and number (as described in this chapter) are essential. In the middle years, the task is to build on this knowledge to support meaningful understanding of symbol manipulation exercises and other algebraic activities.

Algebra is a major tool for supporting problem-solving in the real world. Consider Shannon, who makes cupcakes every week to sell in the tuckshop during the netball matches played on Saturdays. Shannon has to price the cupcakes reasonably, but he also needs to ensure that his effort is rewarded through the making of a profit. How can algebra help here? Shannon calculates the price of all the ingredients and determines that the flour, sugar, milk, vanilla, eggs, icing and any other ingredients cost approximately $1.25 per cupcake (fancy ones cost a bit more). He needs to pay the tuckshop a weekly fee of $15 to be able to display his cupcakes. He has undertaken some market research and noted that some coffee shops charge $4.50 for cupcakes—sometimes higher. He decides to cost his cupcakes at $4.20. Shannon can readily calculate the number of cupcakes he needs to sell each Saturday to break even once he has constructed an equation for the situation ($\$4.20x - \$15$). In addition to generating an equation, Shannon could also create a function table with inputs and outputs to determine a solution. Providing students with real-world problems such as this will support their capacity to use and apply algebra in their lives, as well as the development of beliefs about the value of mathematics.

Review questions

12.1 Outline the process you might undertake to teach patterning to students.

12.2 In your staffroom, there has been heated debate about the place of algebra in the primary school. Outline the position you would take in relation to teaching algebra, and justify that position.

12.3 Equality can often be a difficult concept for students to grasp. Outline some of the ways in which current teaching methods contribute to the difficulties encountered by students, and devise three strategies you could use to support students to develop this important mathematical idea.

12.4 Outline some approaches to teaching students about the commutative property in the contexts of addition, subtraction, multiplication and division. Some of your approaches might involve the use of concrete materials.

12.5 Determine some approaches to teaching about prime, composite, square and triangular numbers using two-dimensional shapes.

12.6 Technology is important in the development of algebraic thinking. Devise some relevant strategies that could be implemented in the classroom.

Further reading

Hitt, F., Saboya, M. & Zavala, C.C. (2016). An arithmetic-algebraic work space for the promotion of arithmetic and algebraic thinking: Triangular numbers. *ZDM: The International Journal on Mathematics Education*, *48*(6), 775–91.

Kaput, J.J., Carraher, D.W. & Blanton, M.L. (eds). (2008). *Algebra in the early grades*. New York: Taylor and Francis.

Knuth, E., Stephens, A., Blanton, M. & Gardiner, A. (2016). Build an early foundation for algebra success. *Phi Delta Kappan*, *97*(6), 65–8.

Lamb, L.L., Bishop, J.P., Philipp, R.A., Whitacre, I. & Schappelle, B.P. (2018). A cross-sectional investigation of students' reasoning about integer addition and subtraction: Ways of reasoning, problem types, and flexibility. *Journal for Research in Mathematics Education*, *49*(5), 575–613.

References

Blanton, M., Brizuela, B., Stephens, A., Knuth, E., Isler, I., Murphy Gardiner, A., Stroud, R., Fonger, N. & Stylianou, D. (2018). Implementing a framework for early algebra. In C. Kieran (ed.), *Teaching and learning algebraic thinking with 5- to 12-year-olds* (pp. 27–49). Cham: Springer.

Blume, G.W. & Heckman, P.E. (2000). Algebra and functions. In E. Silver and M.J. Kenney (eds), *Results of the 7th mathematics assessment of the national assessment of education progress*. Reston, VA: NCTM.

Bond, M. (2003). Powers to the people: Interview with Robert Moses. *New Scientist,* 19 April.

Calder, N.S. (2004). Shaping understanding: How does investigating in a spreadsheet environment affect the conversations of initial training students doing mathematics? Mathematics education for the third millennium. *Towards 2010*, 143–50.

Carraher, D. & Schliemann, A. (2018). Cultivating early algebraic thinking. In C. Kieran (ed.), *Teaching and learning algebraic thinking with 5- to 12-year-olds* (pp. 107–38). Cham: Springer.

Carraher, D., Schliemann, A., Brizuela, B. & Earnest, D. (2016). Arithmetic and algebra in early mathematics education. In E.A. Silver and P.A. Kenney (eds), *More lessons learned from research, vol. 2: Helping all students understand important mathematics* (pp. 109–22). Reston, VA: NCTM.

Kieran, C. (2007). Learning and teaching algebra in the middle school through college levels: Building meaning for symbols and their manipulation. In F. Lester (ed.), *Second handbook of research in mathematics teaching and learning* (pp. 707–804). Charlotte, NC: Information Age.

Kieran, C. (2018). Seeking, using, and expressing structure in numbers and numerical operations: A fundamental path to developing early algebraic thinking. In C. Kieran (ed.), *Teaching and learning algebraic thinking with 5- to 12-year-olds* (pp. 79–105). Cham: Springer.

Lüken, M.M. & Kampmann, R. (2018). The influence of fostering children's patterning abilities on their arithmetic skills in Grade 1. In I. Elia, J. Mulligan, A. Anderson, A. Baccaglini-Frank and C. Benz (eds), *Contemporary research and perspectives on early childhood mathematics education* (pp. 55–66). Cham: Springer.

Maj-Tatsis, B. & Tatsis, K. (2018). The use of variables in a patterning activity: Counting dots. *Center for Educational Policy Studies Journal, 8*(2), 55–70.

MacGregor, M. & Stacey, K. (1999). A flying start to algebra. *Teaching Children Mathematics, 6*(2), 78–85.

McNeil, N., Fyfe, E. & Dunwiddle, A. (2015). Arithmetic practice can be modified to promote understanding of mathematical equivalence. *Journal of Educational Psychology, 107*(2), 423–36.

CHAPTER 13

MEASUREMENT

What is measurement?

Measurement involves the assignment of a numerical value to an attribute of an object or event. While there is a degree of arbitrariness to what is included in this content area and the way it is organised (whether around content or processes), the focus of the strand is the measurement of some attribute. In some cases, probability is seen as the measurement of chance and will be assigned to this strand, whereas in other documents it will remain as its own strand. Similarly, measuring angle is the measurement of the turning attribute and could be part of this strand. However, in some documents it is seen to be an integral part of geometry and is located in the space strand. Many curriculum documents (e.g. the Australian Curriculum: Mathematics) demonstrate the link between geometry and measurement by presenting them as two parts of the one strand.

Measurement has an important role to play in the mathematics curriculum. The National Council of Teachers of Mathematics (NCTM 1989) notes that 'measurement is of central importance to the curriculum because of its power to help children see that mathematics is useful in everyday life and to help them develop many mathematical concepts and skills' (p. 51). The NCTM further argues that measurement pervades 'so many aspects of everyday life . . . [it] offers the opportunity for learning and applying other mathematics, including number operations, geometric ideas, statistical concepts and notions of function. It highlights connections within mathematics and between mathematics and areas outside of mathematics, such as social studies, science, art and physical education' (NCTM 2000, p. 44).

To consider the importance of measurement, stop and think about when you have used mathematics in the past week. You will likely notice that it was usually in relation to measurement—how far you walked, how much money you spent, how many drinks you had during the day, how much petrol was needed to fill the car. Since measurement tends to be the strand of mathematics that is used most in everyday life, it should be taught in a very practical manner, making it highly enjoyable for students (and teachers).

For teachers, the measurement strand is useful for linking with other curriculum strands, thus making its teaching highly integrative. It links strongly with the number strand (calculating area involves multiplication); using analogue clocks reinforces fraction work; conversion of units of measures uses number work; and so on. Similarly, links can be made across curriculum areas—sporting events such as running (time) or jumping (length), international travel in social studies (timelines, geometry on a sphere), writing reports for science and English (using measurements of temperature and calculations, and representation of data). Similarly, within mathematics, many of the topics that are taught in other strands have a strong connection with measurement. For example, the use of the number line links with length, multiplication grids link with rectangular areas and measuring to the nearest unit often involves rounding.

Aside from the mathematical aspects, one of the more powerful aspects of teaching the measurement strand is its pedagogical benefit. The teaching of measurement frequently involves hands-on work, and students enjoy the change of pace in these lessons. Often, students do not even see what is being taught as mathematics (particularly where they have a perception of mathematics as being a boring, solitary activity involving things such as completing worksheets). Organising a class party at which students have to budget (money, operations), work out best buys (ratio), cook food (volume, mass), shop (money), fundraise and plan (reasoning and problem-solving) creates a very positive atmosphere in the classroom as students work collaboratively on realistic tasks with meaningful ends. The social dynamics of the class improve, students enjoy the tasks and they learn considerable mathematics without necessarily being conscious of it. Measurement gives students the opportunity to apply their knowledge and skills in a meaningful and enjoyable manner.

In summary, the measurement strand is an integral part of the mathematics curriculum for several reasons, including that it:

- applies to the everyday experiences of students
- provides application to the world beyond school
- supports other areas of the mathematics curriculum
- supports other areas of the curriculum
- involves active, hands-on learning experiences
- links strongly to problem-solving.

Sub-strands of measurement

In most curriculum documents, the recurring themes are centred around eight common concepts: length, area, volume and capacity, mass, angle, temperature, value (money) and time. There is variation, as noted earlier, in how and where these sub-strands are located in any particular curriculum.

Length

Length is the most easily perceived of the measurement attributes, and refers to how long or how far something is. This length can be along a single plane or through two dimensions (such as around the circumference of a bottle). 'Length' refers to the attribute, while other terms such as 'width' or 'height' refer to specific contexts for length. In most cases, students come to school with some concept of length and the language associated with it. Many students have misconceptions about length, such as those illustrated in Piaget's conservation of length tasks, but, as they progress through school and their lives beyond school, their experiences in most cases allow them to construct more rigid concepts. Frequently, length is seen to be an attribute of something long and thin, such as a piece of string or a line, but over time this conception gives way to other conceptions, including the distance between two points—for example, home and school—or the circumference of circular objects such as bottles or cans. Distance becomes another aspect of length; it is often more difficult for students, so it tends to be

introduced later in schooling. Distance has a specific language (such as 'near' and 'far').

Area

Area is a two-dimensional concept that is related to the region enclosed within a plane shape. Students usually begin exploring area by investigating the amount of surface area. They do this through a process of 'covering'. Teaching methods in the past frequently relied on teaching rote formulae (A of a rectangle = L × W), but these resulted in many students confusing area and perimeter. It is now recognised that students should be taught so that they have an understanding of the concept and develop intuitive ways to calculate area so that formulae become simply a formal recognition of the work they have been developing progressively.

Volume and capacity

Volume is the amount of space taken up by an object. *Capacity* is the amount an object can hold. Often, but not always, volume can be seen as a solid measure and capacity as a liquid (or fluid) measure. By working in this sub-strand, students learn the relationship between cubic centimetres and cubic metres, and between millilitres and litres.

Mass

Mass and weight are often confused: while they are very different mathematically and scientifically, for most everyday experiences they are the same, so it is not uncommon for the terms to be used interchangeably (even by teachers). It is most common to hear people referring to the process of 'weighing' when they are, in fact, measuring mass. *Mass* refers to the amount of inertia required to stop an object in motion or start an object into motion. Mass also relates to substance and is 'constant' but weight varies. Although larger things often have more mass, it is important to realise that comparisons of mass are not directly related to volume—that is, a small object (marble) may have more mass than a larger item (air-filled balloon). Mass is often identified through the process of hefting (lifting). For example, it requires far greater force to move a greater mass than a smaller mass. Similarly, if the same amount of force is applied to two objects of differing mass, the smaller object will move further than the larger one. In contrast, weight is the force that gravity exerts on an object. Thus, while mass will remain constant, weight varies depending on where an object is located. For example, an object's weight is less on the moon than it is on Earth but its mass is the same.

Angle

Angle refers to the amount of turn or rotation. Differentiating between these two processes is an important realisation teachers need to make in their planning (Kontorovich & Zazkis 2016). Most frequently, angle is taught as a static measure, such as the distance between two rays emanating from the same point (see diagram A). It is often seen in activities in which students are expected to compare the sizes of two-dimensional corners in a shape. However, mathematically it is more correct to refer to the amount of *rotation*. Because angles have been taught as the 'area' between two rays and measured with a protractor, students frequently construct the misconception that angles are less than 180°, so that when they are required to measure rotations greater than 180° they become confused, because they do not see this as an angle (see diagram B). This sub-strand is often located within the geometry or space strands of the curriculum.

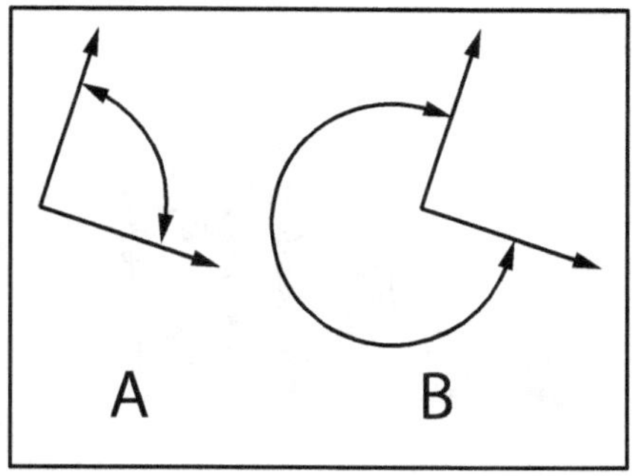

Interior angle Exterior angle

Temperature

Temperature is used to measure how hot or cold things are. In most cases, the difference between hot and cold is apparent. Certain circumstances can cause misperceptions of temperature to occur, creating the need for formal measurement with a thermometer. Different scales have been designed to measure temperature, where arbitrary units have been assigned to key experiences—the points at which water boils and freezes. The Celsius system uses the arbitrary measures of zero for water freezing and 100° for water boiling. Other systems use different scales for these key points: Fahrenheit, for example, uses 32° for freezing and 212° for boiling. The Celsius system has created a need for negative numbers to identify temperatures below freezing. Depending on students' out-of-school experiences, temperature can provide the ideal learning situation for the introduction to negative (or directed) numbers. It can also be used for linking with other curriculum areas (Moore 1999), particularly science (McHugh et al. 2017).

Value (money)

The sub-strand of money often results in the misperception that money itself is the item of measurement. More correctly, money is the unit of measure of a particular concept, in this case *value* (or *cost*). In most Western countries, money is used to measure the value or cost of something. Unlike other measures, value can be very flexible, as can be seen in the fluctuations in currency exchange rates. The cost of an item changes according to how much it is 'valued' at a particular point in time. For example, people are often prepared to pay more for something that they particularly want (such as a house at auction or collector's cards) than it would have been worth to them had they not been so keen to purchase it.

Money is frequently found in the measurement strand, although in some curricula it is found in the number strand. The justification for inclusion in the number strand is money's close relationship to number, particularly to decimals. Students in many countries encounter money as whole dollars and part cents, which means that they encounter decimal fractions in their everyday experiences; thus, from a teaching perspective, decimal fractions are easier to teach. In such countries, students in the first few years of schooling deal with numbers up to 100 and then move into decimal fractions. In countries such as Brazil, Thailand and (prior to the introduction of the euro) Portugal and Italy, where the denominations of banknotes are mostly in multiples of 1000, by the third year of schooling students are dealing with numbers in the thousands, but they do not work with decimal fractions; instead, common fractions are used as their experience of part number. Thus, the monetary system of a nation can impact the ways in which students encounter number study.

Financial literacy programs that seek to teach students important life skills around money handling have become commonplace in the mathematics curriculum. Part of the rationale for inclusion of financial literacy within schools is that it increases people's level of empowerment if they are able to understand and manage money effectively (Davies 2015). While there is not a lot of evidence to show their effectiveness (Mandell & Schmid Klein 2009), there has been some research to show that particular pedagogical approaches have had more effect than others.

For school-age students, experiential learning has had the greatest impact (Amagir et al. 2018). Numerous online and corporate-sponsored packages can assist in the teaching of financial literacy. Of course, corporate-sponsored packages need to be used thoughtfully by teachers.

Time

Time is often the most difficult concept of measurement for students to understand. Unlike the other attributes, time cannot be measured in a perceptual way. It is an historical artefact, and hence is open to different interpretations (Poole 1998). In most Western cultures, time is measured through mechanical means, such as a clock, and seasons are measured by the calendar. In other cultures, qualitative experiences, rather than a calendar, may be used to identify the seasons, such as when particular forms of animal life or vegetation appear. Thus, the seasons can vary in length depending on these qualitative experiences (Harris 1991), and it is not uncommon for people to talk about summer being 'late this year' or being 'very short'.

Two aspects of time appear in the curriculum. The first is the passage of time, where the amount of time from one point to the other is measured as, for example, minutes, days or years. The second is a point in time that is used to identify a particular moment in time—such as Monday, January or winter. The passage of time is often difficult for students to comprehend due to their qualitatively different experiences of time—compare running for one minute with waiting for one minute.

Teaching measurement

Good teaching requires the teacher to identify common patterns in students' thinking and then move that thinking forward to the goal identified by the teacher and/or the curriculum. Often, this approach is referred to as *discovery learning*, but whether students 'discover' anything new is dependent on two key attributes: whether the student is able to construct the desired understandings planned for the lesson, and whether the teacher is able to organise the learning environment so that the student is able to construct the intended learnings. Rather than refer to this approach as 'discovery'—the discovery has been planned by the teacher—leading students to learn desired knowledge can be framed within Vygotskian constructs, in particular using the notions of scaffolding and the zone of proximal development (see Chapter 2). Such approaches are particularly desirable within the teaching of measurement. It has been found that most teachers tend to use a constructivist (as opposed to a traditional, teacher-directed) approach to teaching measurement (Lui & Bonner 2016).

Most 'traditional teaching' of measurement has been focused on the rote memorisation of formulae that are rarely understood and thus become mere tools with which to operate. Such approaches result in students constructing misconceptions and decontextualised knowledge that have little meaning. Rather than focus on the teaching of abstract concepts—such as the gradation on the ruler or a formula for area or volume—a scaffolding approach requires the teacher to identify students' current patterns of thinking and to then create learning environments that support the development of more complex understandings.

It is important, when planning a lesson or sequence of lessons (a unit) in measurement, that the intended learning outcomes are articulated clearly, so that subsequent activities can be identified as linking into the learning outcomes. Teachers being able to articulate what they want students to learn is fundamental. Of course, students may learn more than is intended, but there is an expectation that there is some particular learning that is key to the teaching. By appreciating what is prerequisite knowledge and what comes later, teachers are able to plan to include all the students in the class. Hence, models of how students learn measurement are critical, since they allow teachers to plan for students who are not grasping the concept, as well as for students who are well beyond the understandings of the majority.

A number of key considerations should underpin work in measurement:

- Students should measure frequently and often. The exercises they undertake should be based on real-world tasks rather than on exercises taken from textbooks. This gives students an appreciation for why we measure and why it is important to measure.
- Students should be actively involved in measuring activities, rather than passively observing them. The activities should encourage discussion so that students can refine their measurements and skills of measuring. Insofar as planning is concerned, this means ensuring

adequate resources are available for students (as well as having a high level of tolerance for noise!).

- Links should be made to other areas of mathematics and to other curriculum areas.

Conceptual development model of planning in measurement

A model commonly used in planning for teaching of any measurement topic delineates five phases:

1. identifying the attribute
2. comparing and ordering
3. informal measurement using non-standard units
4. formal measurement using standard units
5. application to problem-solving contexts.

The basic premise of this model is the underlying complexity of measurement concepts—much like the development of number and operations. The model follows a logical path from simple recognition of attributes through to formal measurements and the application of formulae in problem-solving, providing a structure for planning learning experiences for students. The model is not prescriptive or year- or age-specific, but it provides an outline for a general process through which the teaching of measurement can occur. While the model goes some way towards describing the learning process for many students, learning is highly complex and cannot be accounted for adequately by this simple sequence.

The model, therefore, should be seen only as a guide for planning; teachers also need to recognise that students do not always follow its logic, since prior learning and experiences cannot always be accounted for adequately. As students enter the classroom with a wide range of backgrounds and experiences, the latent learning that has occurred outside school can result in some students having developed considerable intuitive understandings of measurement concepts and skills, while others still have very little understanding. Where 'anomalies' occur, teachers should be ready with alternatives for students.

Identifying the attribute

In order for students to be able to undertake learning in any particular measurement strand, they need to be able to recognise the attribute

Free-play experience with volume to identify the attribute

that is under investigation. If one considers the teaching of length, for example, there are often other distractors in the activity, yet students are expected to identify the length attribute over all other attributes, such as colour, texture, width, purpose, and so on. Unless the students can isolate the attribute being taught from all other attributes, it is unlikely that they will be able to understand any further developments of the concept. For example, if a number of items, such as a ball, a tennis racquet and a fishing line, are used when talking about length, the concept may be very difficult to grasp for a student who is considering the recreational aspects of the items, or their hardness, or their colour, or any other attribute. It would be best to use only very similar items (e.g. pencils) that differed in length. In the very early stages, it would be best to use the same item in different lengths—all red straws or the same type of pencil—in order to reduce distraction. That said, research from the Early Numeracy Research Project in Victoria (Clarke et al. 2000) has shown that many students come to school with a reasonable understanding of the length concept. The same cannot be assumed for other aspects of measurement. For the early-years experiences, play is very important, as is the use of children's storytelling (Chigeza & Sorin 2016).

Early-years experience comparing and ordering volume

Comparing and ordering (using direct and indirect comparisons)

Once students have been able to identify the attribute, they need experiences to compare and order (or seriate) objects according to that attribute. Beginning with two objects, students can talk about which is taller/

shorter (for length), hotter/colder (for temperature), and so on—again, depending on the attribute to be compared. More items are introduced progressively, and students order according to the attribute. Comparisons are initially made directly, and then through an intermediary tool (indirect comparison). The use of indirect measures creates a need for informal measurement.

Informal measurement using non-standard units

Initially, everyday experiences form the basis for measurement—handprints for area, cups for volume, steps for length—in order to develop principles for effective measuring skills, and in preparation for the introduction of formal units. Using informal measurement tools, students have access to experiences that allow them to recognise that different units can be used for different contexts—for example, while it is effective to use paperclips to measure the length of a book, a paperclip would not be a good unit for measuring the length of a room. Such experiences prepare students for the arbitrariness of formal units of measurement.

Similarly, experiences with informal units create a need for common formal units—for example, when two groups of students measure the same item with different units, they get different results. By showing that the items are the same length but have returned different scores for their recorded lengths, it becomes apparent that common units of length are needed when comparing items. Digital tools have also been shown to enhance the learning of measurement using non-standard units (Aladé et al. 2016). Early-years learners can learn about measurement through virtual manipulation of objects. Interactivity was seen as a valuable part of the learning process.

The use of non-standard units must be developed over time, and there are six concepts that are foundational to measuring with non-standard units. Teachers should recognise these very different concepts and plan their lessons to ensure that a range of activities is undertaken in order that such skills are developed:

1. Measuring involves both a number and a unit of measure. Students must be able to count how many units and include the unit of measure—for example, five paperclips, fourteen steps, six hand spans.
2. Measurements can be compared easily if the units of measure are the same—that is, items can only be compared if the same unit has been used.

Young children often rely on just the number to compare items—six paperclips may be thought of as being bigger than three rulers.

3. Depending on the context, one unit may be more appropriate than another. Measuring the capacity of a bucket with cupfuls of sand may be effective, but cupfuls of sand would not be a good unit of measure for the capacity of the sandpit.
4. The size of the unit of measure is inversely proportional to the number of units. When using small units to measure an item, the count will be higher than when using large units for measuring the same item.
5. Communication of measurement is most effective when standard units of measure are used. Non-standard units of measure do not communicate measurement effectively—books, paperclips, blocks, steps, cupfuls, and so on can vary in size, so communication is not effective. Use of standard units of measure, such as centimetres, grams, millilitres, and so on, allows all parties to appreciate the magnitude of the objects being measured.
6. More exact measurements can be undertaken with smaller units. Most measurements are estimations: 'about 6 cm', 'between 5 and 6 kg', and so on. When smaller units are used, greater accuracy can occur—'less than a metre' is not as accurate as 'between 60 and 70 cm'.

Formal measurement using standard units of measure

Having created a need for standard units of measure and developed principles for measuring, students are introduced to standard units of measure for the various attributes. In concert with the standard units being used, students need to develop the appropriate language for such measurement, along with the correct symbolic representation for units of measure (such as cm^2). Estimation of quantities is also central.

The experiences that the teacher organises for students should ensure that students develop a sense of how big the units of measure are—for example, how heavy a kilogram is, how much a litre is, how long a metre is, and so on. These skills are important for estimation. Similarly, there are key referents that students should also develop. By knowing what a two-litre container of milk looks like, a student is able to estimate the capacity of another item by comparing it with this item. Similar referents occur for height: 'I am 1.5 metres tall, and you are a bit shorter than me, so I think you are about 1.45 metres'; 'My dog weighs about 15 kg and that dog is about the same, so it is probably about 15 kg, too.'

Application to problem-solving contexts

Once students have developed, and are confident with, formal units of measure, they can be introduced to applications of the units to enable the development of formulae. In subsequent years, particularly in secondary school, the development of formulae with specified values gives way to the use of variables. This stage should not be seen as entirely separate from the others, as if problem-solving should only be undertaken once the concept is fully and formally developed. Many of the earlier stages should include problem-solving in their planning. This stage tends to focus very heavily on problem-solving in terms of teaching methods.

The value of this model of concept development is that it progresses through the primary-school years. It allows teachers to develop an understanding of what should have happened in earlier years and what is to come later. By having a deeper understanding of where things come from and where they are going, teachers can focus on their own role in developing measuring concepts and skills. For example, it is possible to think that teaching area is about the formula 'length × width', where this is the memory of the teacher's experience. However, using the model just discussed, it becomes clear that considerable work is undertaken before the formula is introduced. Indeed, in a model such as this, the formula would likely not be taught; rather, the sequence of activities leading up to introducing the formula should mean that students 'discover' it for themselves.

Discovering or 'inventing' the formula means the formula becomes personally meaningful to students, and hence is more likely to be retained (and less likely to be confused with the formula for perimeter). Through a process of progressive scaffolding, the learning experiences provided by the teacher create an environment in which the formula appears as a 'natural' development from the experiences provided.

Limitations of the concept development model

As with any model that is being used in the complex processes of teaching and learning, teachers should have some sense of the concept development model's limitations. Such critique is necessary because, while the model serves a purpose, it often reflects the ethos of various stakeholders. Seeing the model as unproblematic risks its becoming dogma and being embraced without the recognition that, in some cases, it may not work. In some cases, it can put students at risk of failing, and hence put the teacher at risk of criticism for not understanding the intricacies of the learning processes of some students.

This model has as its root a notion that learning occurs in a sequential manner, suggesting linearity in development that may hold true for a significant number of students. However, learning is a highly complex process, one that is often far from linear, and the linear model may deny the complexity of thinking and development in students whose experiences may be very different from those of the other students in the class. Imagine a student who has spent considerable time with her parent, who is a tiler. This student's spatial awareness, tessellation knowledge and concept of area could be expected to be significantly higher than that of other students in the same class, but lacking the formal understandings expected in such a sequence.

To think of the sequential process of this model as 'natural' is highly problematic. While teachers may confirm that this is the learning sequence they observe in their classrooms, it may not necessarily represent the 'natural' learning process. It could be that the progression that is noted is a reflection of teachers' experiences in schools where measurement is concerned. If teachers are organising learning experiences in line with this sequence, it makes sense that observation of student behaviour will reflect their experiences. If the scaffolding provided by the teacher supports particular outcomes, then it is hardly a natural process. Consider a case in which students are provided with experiences with blocks in which they have to count the blocks in order to see how long an item is, and then provided a situation in which their counting is ineffective—such as some students using the blocks to count the length of a book, while others use paperclips—so that the students will recognise the need for a common unit to measure and compare with. This would not represent a natural movement into the next phase of the model. Rather, the teacher has contrived a situation (either intentionally or unintentionally) whereby students have come to recognise the need for the common units.

The five stages noted in this model can be condensed, since there are overlaps between the various levels. Rather than seeing them as discrete skills, they should be viewed as existing along a continuum. For example, identification of an attribute is often undertaken in concert with comparison of the attribute against others in order to identify it. Identifying mass is difficult without comparing one object against another—which is heavier, which is lighter—while keeping the focus on mass rather than on volume or length. Similarly, the introduction to informal units links closely with indirect comparison. Students may decide to see 'how many rulers long' each of two items is because the ruler is the most convenient item for them. That is, the process of indirect comparison has led to the use of informal or non-standard units of measure. The same process can occur as students move into formal units of measure—when they are measuring how many rulers long a desk is, a need for part measures becomes apparent, so they note that the desk is three rulers and 22 cm long, indicating that the table ended at the 22 cm mark.

Despite the limitations of the process, and the fact that the model is more useful as a planning tool than as a reflection of learning, the concept development model is helpful in planning learning experiences for students. Since length is the most commonly taught measurement attribute, and the basis of many other attributes—area and volume, for instance—considerable space is devoted to this topic to illustrate in detail how the concept is developed through the five stages. Subsequent sub-strands are dealt with in less detail.

Length

As noted earlier, length is the most easily perceived attribute, and most children come to school with some understanding of length and its associated language. There are many strategies to support the learning of length (Lee & Francis 2016), but it is recognised that the use of manipulatives and resources (physical and digital) is important in building conceptual understanding of length. Below are some key points about the teaching of length.

Teaching strategies

- Early experiences with length should focus on the language of length, including classifying objects according to whether they are short or long, thick or thin.
- Two key skills need to be developed: direct comparison, where items are compared directly against each other (e.g. placing two pencils or children side by side) to see which one is longer/taller or shorter, and indirect comparison, where an intermediary device is used. Indirect comparison might consist of using a body part—such as 'the door is ten hand spans long' and 'the desk is five hand spans long', so the door is longer than the desk. Indirect comparison is needed when

measurement of two objects against each other is not possible. It is also the precursor to developing the need for intermediary devices, thus eventually leading to formal tools for measuring.

- Comparing and ordering lengths can be undertaken through the following sequence:
  1. Initially, ask students to copy a sequence that has been constructed.
  2. Ask students to place an item correctly at one of the ends of the sequence.
  3. Ask students to place an item correctly in the given sequence (somewhere other than an end).
  4. Order three or more items. Once students can consistently order three items, more items can be introduced progressively.
- Students need to be able to make sure that informal units are lined up in a straight line, that there are no gaps or overlaps between units, that the starting point is level with the item being measured and that counting of the items is accurate. These comprise a new set of skills to be learned. Blocks that link together, such as Unifix or interlocking blocks, are very useful for early experiences with non-standard units, since they hold together in a relatively stable manner. Teachers should ensure that there are sufficient blocks for all students when undertaking these activities.
- Students place a number of units along the item to be measured and then take one of the earlier units and place it in the line, counting as they go (iteration). Some students will prefer to use marks to keep count of their work. This is a complex skill and should not be introduced too early; if students are having difficulty, leave it until a later stage. It is a skill that is useful later on, when students need to measure lengths that are longer than the instrument they are using, such as a ruler.
- The history of length can be implemented—for example, the history of the cubit, the foot, a span and a digit.
- Students should develop a sense of the size of a centimetre and metre. A commonly used referent for primary-school students is the student's and other significant people's height(s).

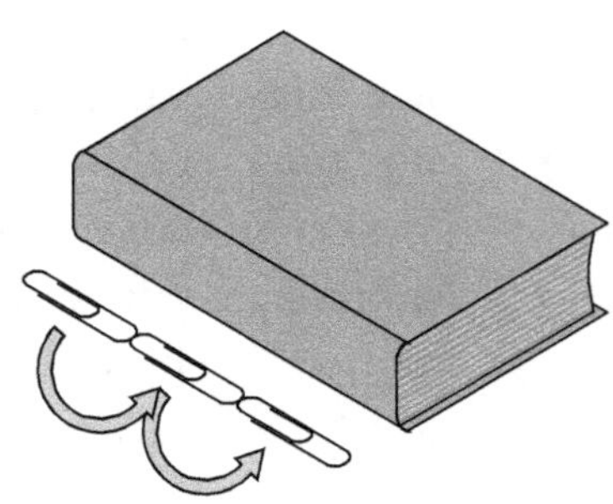

Using paperclips to measure the length of a book

- A kilometre is more difficult to gain a sense of because it is such a large quantity and can be experienced in qualitatively different ways—for example, driving to school is quite different from running the same distance.
- A common mistake that students make when using an instrument such as a ruler is not lining up the item to be measured with the 0 point on the ruler. They may either use the start of the ruler (the section that precedes 0) or line the item up at the 1 cm mark. When introducing standard units, it is best to use a ruler that has only centimetre gradations on it (rather than ones with millimetre marks). This allows the student to focus on a single unit of measure—a centimetre. Once they are competent with this instrument, they can use rulers with millimetre divisions.
- Having developed effective measurement principles, problem contexts can now be developed. This often involves measuring and calculating perimeter (the length around an object). A difficulty that commonly arises in the teaching of perimeter is that students can become very confused about the difference between area and perimeter. Newer versions of some tablet software come with a length-measuring app, whereby a cursor is positioned on one end of an object and then repositioned at the other end, with a measurement then being provided. This tool should only be used once students have developed the physical skill of measuring.

Area

Area is often taught through a process of multiplication and formulaic thinking. However, it has been found that the use of manipulatives, rather than the use of a formula, enables richer understandings of area to be developed (Huang 2017). Teachers need to plan activities that allow students to identify area as being the amount of covering over a surface, along with the appropriate language of area (Tossavainen et al. 2017). Activities could include covering flat objects such as tables, mats, books, and so forth with paper, tiles, base 10 block flats or other objects. Pasting activities using standard-sized pieces of paper (such as confetti for small objects) give students a sense of which is bigger (i.e. they may need several pieces of confetti to cover the one original shape). Leaving no gaps when covering areas with shapes is also very important to promote conceptual understanding of area measurements, and provides the foundation for the understanding of standard units for area. It is also important that non-standard shapes (e.g. kidney bean shapes, stars, etc.) are used for area measuring experiences; otherwise, students may develop the misunderstanding that only mathematical shapes (squares, rectangles, triangles, etc.) have area.

After encouraging students to experience covering with units that are congruent (same size and shape), progress to finding the area of shapes, of different size or shape, in square centimetres by either ruling grid lines on the shape or placing a transparent grid over the object to count the number of square centimetres and therefore determine the area of the shape in square centimetres. Using this language assists children in developing meaning for the symbol cm^2. Measuring larger areas (for example, floor spaces) with square metres, or having students make a square metre from newspaper and placing it on the floor, assists in developing the concept of size of a square metre and assists in building estimation skills (see Teaching Idea).

TEACHING IDEA

Take an A4 sheet of paper and cut it diagonally. Reorganise the shape so that it looks like the second shape here.

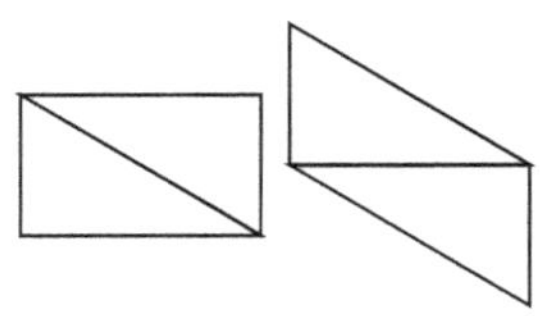

- Which is bigger in area? Justify your answer.

TEACHING IDEA

Take an A4 sheet of paper. Cut it into four or so pieces using straight cuts (or use a tangram). Create new shapes. Allow students to predict and explore the amount of area in the new shapes and the new perimeters.

- Create a table of results.
- What did you notice?

Teaching strategies

- Grid paper can be a useful tool for implementing formal units of measure. Centimetre-square paper can be used for drawing or tracing shapes and then counting the area. The grid paper can also be copied onto transparencies so that students can overlay it on shapes and count the number of squares. It is useful to begin with whole numbers and build progressively to more complex numbers and shapes (including circles).
- Use large sheets of newspaper to make square metres so that larger areas can be covered.
- Create situations in which students can 'discover' formulae for various shapes. Allowing them to discover a formula will make it more meaningful than if it has been rote learned. Once students have developed a strong conceptual understanding of area, software such as GeoGebra can be used to reinforce their understanding (www.geogebra.org/m/vDxUbcTT).
- As before, personal referents should be developed for area; for example, a small fingernail might represent 1 cm^2.
- Large areas, such as hectares and large land or sea areas, are difficult to conceptualise. World mapping exercises can be fun and illuminating for students. Compare the area of Australia with that of the United States and the United Kingdom. Students often think that countries they hear a lot about must be big in terms of their areas. It is often very surprising to learn that Australia and Europe are almost the same size; likewise, the United States and Australia are of similar areas. Compare how many times the United Kingdom fits into countries such as New Zealand, Canada, Australia or Kenya. These activities also have application to social studies. Software such as Google Earth provides useful resources for helping children conceptualise large areas.
- Teaching area using tessellation (which links with the space strand) is useful, particularly when regular shapes (rectangles, squares) are used, which allows the student to build rows of blocks, thus leading towards the development of the formula. This process links with number study, since students need to have well-developed counting skills in order to move towards developing an intuitive understanding of the formula. Skip counting and multiplication facts are almost prerequisite knowledge.

TEACHING IDEA

Using Tasmania as a unit of one, estimate how many times it will fit into each of the other states/territories.

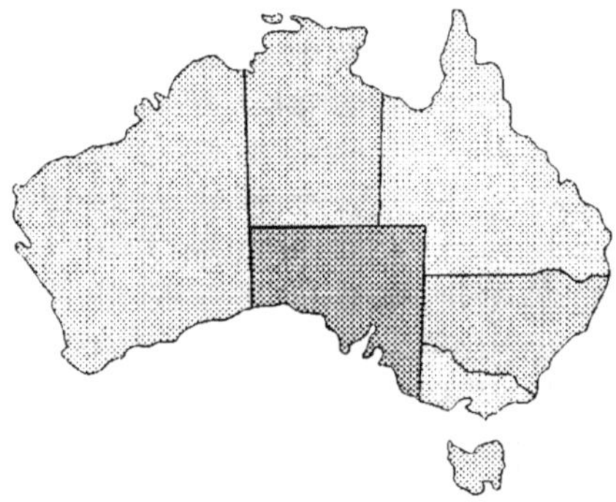

- Pair and share. Refine the results.
- Provide the areas of each state/territory.
- Using a calculator, check the estimate.

Volume and capacity

The learning activities associated with volume and capacity focus on the measurement related to three-dimensional shapes—properties of objects or the amount of space an object takes up, or the amount a container can hold.

Teaching strategies

- Begin with lots of activities involving packing and filling so as to develop a sense of volume and capacity. Fill containers with water (capacity), then immerse other objects to see displacement (Archimedes' Principle) (volume).
- In order to develop comparisons, use nesting boxes, babushka dolls and other containers that fit inside each other for direct comparisons of volume. Use items such as rice to fill one container, then pour the rice into another container and see whether there is leftover rice or leftover space to identify items with bigger and smaller capacities.

- Displacement of water can be done—items of greater volume displace more water than items with smaller volume.

TEACHING IDEA

Take an A4 sheet of paper and make the biggest container you can from it. Keep this task open-ended so that a range of shapes can be made (boxes, cylinders, etc.).

The results can create considerable discussion about what 'container' means—for example, does it need to have ends?

- Informal units of measure can include matchboxes fitted into other boxes (volume), or cups, jugs and spoons for liquid measures (capacity). Have students record their findings.
- Use of cubic centimetre blocks (base 10 blocks) leads into formal measuring units, since the unit/one block is 1 cm^3. Use the longs, flats or large cubes of the base 10 blocks to fill items.
- Benchmarks and referents should be developed through linking to common everyday objects—for example, 1 litre of milk, a 4-litre container of ice-cream, a 375 mL can of soft drink and a cubic centimetre. Once the litre has been introduced, fractions of the litre are introduced—500 mL, 250 mL, and so on.
- Larger units of measure can be difficult, but council or utility company account notices can be useful, since they report water usage in kilolitres.
- Organise activities so that students can discover the formula for calculating the volume of some objects (area of base × height), since it links to area (base of the shape). Centimetre cubes can be useful for this. A variety of objects should also be explored.
- Conversion between units of measure should be undertaken. Rather than relying on rote learning, provide experiences so that students can identify the relationships between length, area and volume and how these are intertwined. Ask students to determine how many cubic centimetres there are in a cubic metre: use a cubic metre and place base 10 (MAB) blocks along the bottom (10); students can then see that for the base, there are 10 × 10. This is then multiplied by another

10, since the cube will be 10 deep, making for a total of 1000 MAB unit blocks. There are 1000 cubic centimetres in one block; calculate how many there are in a cubic metre. How many millilitres are there in a cubic metre? Linking the visual, kinaesthetic and logical learning approaches enables students to make better sense of the relationships and numbers involved.

Mass

When students begin study in this area of the curriculum, the terms 'mass' and 'weight' tend to be used interchangeably, but as they progress through the primary-school years it is important that they come to recognise the difference between the two concepts (as discussed earlier in this chapter). This helps to prepare them for the more specific work of secondary-science experiences.

Young student weighing using 'beam balances'

Teaching strategies

- Practise free play with objects of varying masses. Activities can involve pushing, hefting (lifting) and balancing. Playground equipment such as a seesaw is useful because it enables students to link mass with the workings of an actual object.
- Use tools (such as balances) or skills (such as hefting) to identify heavy and light objects. To sort and order mass, begin by using objects that have quite different masses. As students become adept at identifying large differences in mass, move to smaller differences.
- Make your own balances with buckets, coat hangers, and the like and use them to identify different masses by comparison.
- When using equipment such as balances, use informal units such as blocks, pencils, marbles, and so on to measure the mass of objects. Students develop skills for using balances so that they will recognise that when the balance is even, the masses are the same. Use of informal units allows for indirect comparisons. Encourage students to record their measurements.

- The language of mass is common—most students will have some familiarity with kilograms and grams—so that its introduction to formal units follows the other sub-strands. Students need to gain experiences with recognising the mass of 1 kg, and activities need to be included to develop referent points—such as, 'I weigh 22 kg.' Weights of 1 g are difficult, since they have very little mass, but other multiples are useful, such as 30 g, 100 g, 500 g, and so on. Provide lots of experiences of weighing everyday items. Conversions between units of measure should be integrated into the work being undertaken.
- Explore the relationships between volume and mass. Explore the packaging of goods and the effect of transportation on foods.

Time

As introduced earlier, two aspects of time need to be considered: a point in time (day, time, event), and the passage of time. These learning experiences require different teaching approaches. Time is often the most difficult attribute for students to identify in a quantitative sense, since the qualitative experiences are so different. Waiting for a minute while your tooth is getting filled is very different from spending a minute at a birthday party. The everyday language used in families and other social contexts also results in confusion—for example, a parent who is talking to a friend at the supermarket might tell a child that they will 'only be a minute'.

Teaching activities for the passage of time include counting or clapping while an event occurs. When there are considerable differences in the amount of time elapsed, young students can identify this, but the closer events are in duration, the less efficient time estimation is. For a point in time, events such as birthdays are useful for months, and focusing on the day in an informal way helps to draw attention to the day of the week—'What do we do today [sport, music]?' 'So what day is today?' This strand often represents a point of confusion for some students. The language of duration of time links with the length sub-strand, and can be difficult to understand: how long is it from home to school—3 km or 10 minutes? How long did it take to cook dinner? Common social remarks can provide rich examples of this aspect of time: how far is it from Sydney to London? how long is a medium-sized book, or six movies?

Time is an integral component of contemporary life, and there are considerable opportunities for teachers to teach this aspect of the mathematics curriculum in an incidental manner so that it is constantly reinforced. For example, 'What is the time now?' 'How long before lunch?' 'We need to work on tidying up; we only have five minutes before play.' Accurate references to time should be made, thus avoiding confusion with everyday nomenclature, such as, 'We only have a couple of minutes left.'

Teaching strategies

Passage of time

Early experiences focus on informal units of measure, such as clapping or using egg-timers. Events with common starting points—'Who took longer to win the race?' (i.e. direct comparison)—are introduced first and followed by indirect comparisons, where the duration of an event is measured and compared. Informal units ('How many more sleeps to my birthday?') are used to create the need for formal units of measure (seconds, minutes, hours, days, weeks, etc.).

Once formal units of time have been introduced, students can measure the passage of time using clocks and stopwatches. Comparisons can be made to see which events take longer or shorter periods of time. At some stage—usually towards the upper years of schooling, once number and operations have been mastered—students begin calculating with time. This can pose problems, since there is often a tendency to work in base 10 rather than in the various units of time. When introducing this aspect of time, teaching episodes should be planned carefully, much like teaching operations. Begin by using whole units of time so that there is no need for regrouping (e.g. 'How many hours between 4 o'clock and 8 o'clock?'). Once this has been mastered, move to regrouping (e.g. 'If it is 4.30 p.m. and the train leaves at 6.10 p.m., how long do I have to wait?').

Student-constructed candle clocks

Points in time

- Students' experiences with points in time are developed in progressively smaller units. 'Morning', 'afternoon' and 'night' precede more specific moments in time. Students begin to move to times of the day that relate to their everyday lives. In the past, there has been an expectation of what would be appropriate times of the day for different events to occur, but teachers increasingly recognise that their students' lives can be substantially different from their own expectations. Pictures requiring minimal reading skills can be used to depict daily events for students to place in order (e.g. having breakfast, going to school, having dinner, having a bath, going to bed).

- Birthday graphs are a useful teaching tool for both the time and the data sub-strands.
- Use calendars, and relate these to longer periods of time (such as the seasons). Decades and centuries must also be learned, but these are often difficult for younger students, since when 'decade' is introduced, many students are not even that old!
- Long epochs of time are introduced, allowing the introduction of concepts of BC and AD.

Time as linear or cyclical

In Western thinking, time is considered a linear process. For many young students, this results in thinking that if their birthday is in, say, June, they have to wait longer for their birthday to arrive than their friend whose birthday is in, for example, March. In part, this is because the calendar is presented as a document from January to December.

Time-telling

- This skill takes considerable time for students to develop. It links closely with number knowledge (fives multiplication facts, fraction knowledge), so these concepts need to be developed in order to make the best gains in time-telling.

- Early experiences with clocks—analogue and digital—focus on features of clocks, which builds into time-telling of o'clocks and half-pasts. This builds into quarter-turns, before moving to minutes-past and minutes-to.
- Linking digital and analogue times is an important part of time-telling. Students' early experiences with digital time see them saying times such as '50 minutes past three'. Bingo games can be useful for this linkage, including the use of clock faces (analogue and digital) as well as written times in both numerals and words. Developing notions of a.m. and p.m. precedes learning 24-hour time. Once students are familiar with these aspects of time, international times can be introduced. Investigate time lines—Greenwich Mean Time, the International Date Line—and making international phone or Skype calls.

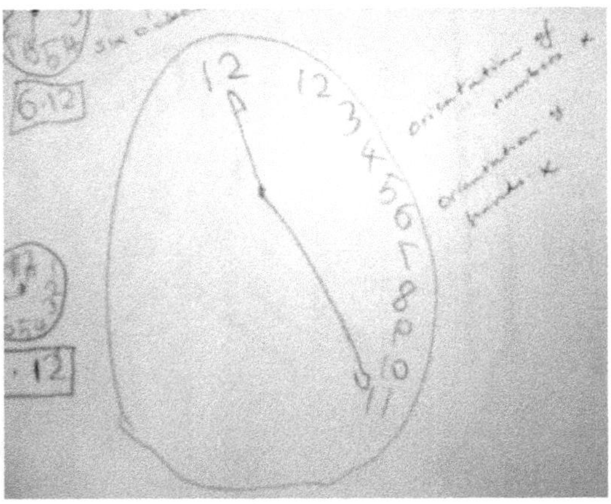
Student's early drawing of a clock

- Practical applications of timetables (e.g. train, bus, ferry) that are relevant to students' experiences are introduced. These experiences build towards calculating how long trips take.
- Using large clock faces has been shown to be a better strategy than using worksheets (Wieber et al. 2017).

TEACHING IDEA

Time bingo

Create a series of cards displaying different times in different formats. Play bingo.

3:00	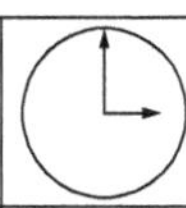
3 o'clock	Go home from school

There are several apps that are useful for reinforcing or practising time-telling. One that is supported by research is MathTappers: Clockmaster, available from: https://apps.apple.com/us/app/mathtappers-clockmaster/id336932114.

Temperature

There are a limited number of experiences that can be provided for students to identify this attribute. These usually consist of using sensory experiences of touch to feel whether things such as water, the asphalt on netball courts, and so on are hot or cold, as opposed to other attributes such as prickly, soft, fluffy, and so on. Students have a relative sense of temperature, so development of language is important in the early years—hot, cold, warm, freezing. This would be done before introducing thermometers.

Teaching strategies

- Provide practical experiences of feeling temperatures—cold, warm, hot—and ordering of the same temperatures.
- Use cut-outs of different climates (the North Pole, deserts, tropical areas) or the seasons (summer, winter, autumn, spring) to order.
- Keep a daily weather chart and plot the temperature—either measured by the class or found online or in a newspaper.
- Use thermometers to measure temperatures.

Money

The money sub-strand is potentially difficult for some students when discussing value. What young students value can be very different from what older students value. Value and worth can be two mutually exclusive constructs. Consider a home by the ocean—for someone who loves the sea, the house is worth more than it would be to someone who loves the outback. Unlike other areas of the mathematics curriculum, the concept of value is relative rather than absolute. For students, what is central is money-handling skills; thus, this topic is more commonly referred to as 'money' than 'value'. It is interesting to look at the teaching of number in another country and link this to the monetary system in that country. For example, in Portugal and Italy (pre-euro), the currencies had very high numerical denominations (which can be conceptualised as being akin to Australian currency if money was in cents only, rather than in dollars and cents). The units of values (escudos and lire, respectively) were very high (e.g. shoes could cost 3000 lire), so the decimal part of the currency was not in circulation. As such, decimal numbers did not feature in these countries' study of number. The introduction of the euro offered unique opportunities for teachers and researchers to study the transition and development of thinking in number.

Money-handling skills are becoming a popular topic in the curriculum. Teaching young people the importance of money handling from a very early age fosters an invaluable life skill (Sheldon & Duncan 2019). Early money activities can be difficult for Australian students, since the coins are not proportional to their value: while the silver coins increase in value as they become bigger in area, the gold coins reduce in area as they increase in value. In contrast, New Zealand's gold coins increase in size as they increase in value.

The money sub-strand is linked strongly with the study of number. Indeed, in some curriculum documents, to reinforce the link between number and money, and recognise that financial literacy is very important, money is located within the number strand. It has good application of decimals, since dollars can be considered whole units, and the cents as parts of the whole.

Teaching strategies

- Discuss items that have value to students, such as trading cards (e.g. Pokemon cards, football cards), and how some cards are worth more than others. Discuss items that have high or low value (e.g. cars and sweets).
- Create a class shop and have students put prices on items. Use money (plastic, paper copies) so that students can develop coin-recognition skills.
- Values and equivalence feature in the work undertaken, so that students will be expected to know the value of particular coins and notes as well as be able to recognise equivalence in values—for instance, that two 50-cent coins are the equivalent of one dollar. Bingo games are a useful tool for teaching this aspect of number.
- Operations with money link with number work. Initially, work will be in just one unit of measure (cents or dollars) until students

have become fluent with decimal fractions. This knowledge can then be applied to mixed units of measure. As number work develops, so too can money study. For example, as students work with multiplication of mixed numbers, they can do the same operations with money.

Making connections between sub-strands

Area and perimeter

Students often become confused when area and perimeter ideas are combined. In part, this can arise from poor understanding of area and length. Activities need to be given in which students' misconceptions can be challenged. The use of a geoboard can help students to see that the perimeter may be held constant (with a piece of string of a particular length), but the area can change depending on the shape—24 cm of string can make a shape of 6 × 6 cm, 3 × 9 cm and 4 × 8 cm, each of which has a different area.

Using grid paper, students can keep area constant using blocks, and then find out the different perimeters of the shapes they construct. In the following Teaching Idea, the area is four units; depending on how they are placed, a range of different perimeters can be found.

Digital tools can also be useful for exploring the relationship between perimeter and area (Bouck & Bouck, 2015). Such tools, including apps, can take a lot of the tedium out of pencil-and-paper constructions and allow students to explore the relationships more easily. One useful app

TEACHING IDEA

Grid shapes

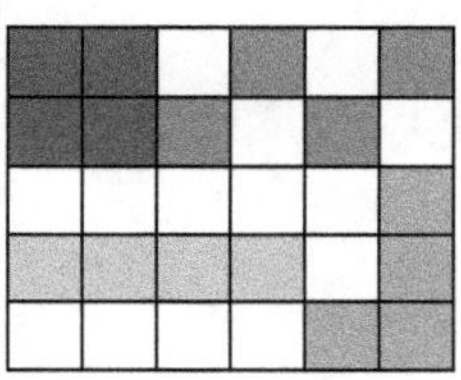

Using a sheet of grid paper, make as many different shapes as possible with four blocks. Colour the shapes different colours. What is the area and perimeter of each of the shapes? Record the results in a table. What do you notice?

is Area of Rectangles, available at: www.commonsensemedia.org/app-reviews/area-of-rectangles-interactive-lesson.

Practical, real-life contexts can provide a stimulating learning context. The relationship between a fence (perimeter) and a garden bed is one example that links to real contexts (Selmer et al. 2016). Farming contexts, which also feature paddocks, fencing and farming (grazing cattle or cropping), can extend understanding of the relationship between perimeter and area.

Volume and surface area

Just as many students get confused with perimeter and length, many also become confused with volume and area (Gough 2004). There is a misconception that if one quantity (e.g. area) is constant, then the other measure (volume) is also constant. Folding a piece of A4 paper into quarters in both directions can result in different volumes, as can folding A4 paper into a cylinder in both directions. Students can apply this knowledge to packaging and environmental issues so that they can find out how best to pack items with minimum surface area and maximum volume. This knowledge is also useful for understanding how food manufacturers can increase the price of goods without the public becoming aware of it. Current practices have seen a number of manufacturers reducing the content of containers but changing the shape (surface area) so that they look the same, while charging the same amount of money.

TEACHING IDEA

Volume and shape

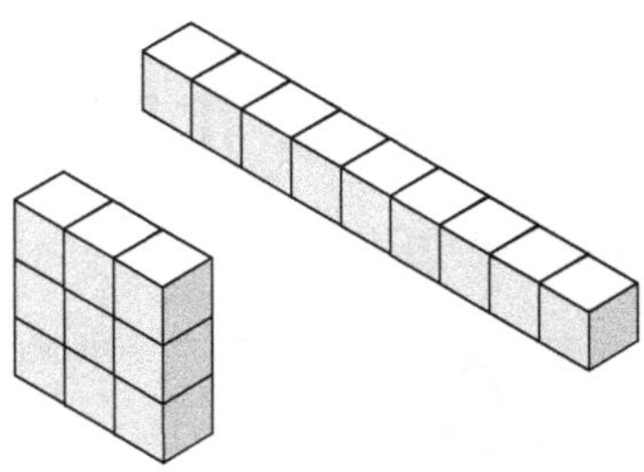

Using the same number of blocks, create different shapes. What is the surface area of each shape?

- Record results.

Volume and shape

Just as with the other examples, shape and volume can develop some interesting misconceptions. Allow students to construct various shapes using blocks. Keep the number of blocks constant (i.e. volume) and see what shapes they can make. This relates to the example above, but without formally acknowledging surface area.

Volume and capacity

The connection between liquid measures (mL) and solid measures (cm^3) often escapes students. This restricts their appreciation of the rich relationships in the metric system. Opportunities in which these relationships can be explored provide a rich basis for later work, particularly in science. They also help students to move away from rote memorisation of tables of equivalence, and allows them to recognise the connections between volume and capacity. Provide a range of items so that students can calculate (or count) the solid measures. Fill an item with water and then pour into a graduated container to see the corresponding liquid measure. Similarly, have particular measures of water in the graduated jug and then pour into the 1000s block container to see where it corresponds. Measures of half-litres, quarter-litres, and so on will be easy for students to see.

TEACHING IDEA

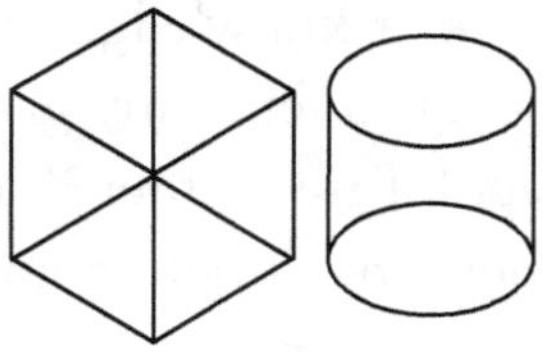

Volume and capacity

Pour water from a 1-litre jug into an empty 1000s cube. This allows students to see that 1 litre of water (1000 mL) takes up the same space as 1000 cm^3.

Volume and mass

The beauty of the metric system is its cross-linking. The activities undertaken in the previous section can be extended to measuring mass. Using water, have students weigh their containers (this is later subtracted, unless the scales have a zeroing mechanism). Find out how much different measures of water weigh. This does not hold true for other items, where the density will alter the weight. Students can fill the 1-litre or 1000-cm^3 container with different items (e.g. feathers, stones, metal, sticks) to see how much these weigh. This brings about the early concept of density, which is later explored in science and mathematics. This process can be used to reinforce the value of the metric system when the mass of 1 litre of water is weighed (1 L = 1 kg).

Teaching measurement in the middle years

In the middle years, students' understanding of measurement cannot be taken for granted. This means that teachers must continue to provide hands-on, active learning experiences to ensure that students' conceptual understanding of measurement topics becomes consolidated. Assessment studies have found that students in the middle years will make errors in measuring by not aligning the zero on the ruler with the edge of the object being measured. Further, students' knowledge of formulae for measurement calculations is tenuous, and students have difficulty understanding why formulae cannot be applied in non-standard contexts (Battista 2007). In summary, the research indicates that 'many students' measurement reasoning is superficial, with poorly understood procedures or formulas substituting for deep understanding' (Battista 2007, p. 893). In the middle years, the teacher's role is to provide opportunities for students to see the limitations of their formulaic measurement knowledge. The following scenario is a great way for students to question their knowledge of area and perimeter: 'If I have a fixed length of fencing to fence my garden, will I get the same area regardless of the shape of my garden?' Tasks such as these require students to think about how they might convince someone of their response. They will need to use concrete materials, and most likely some grid paper, so that they can argue their solution. Rather than completing pages and pages of exercises in which students have to find the area, the volume, the perimeter, and so on of various objects, look for activities that require them to think about the situation and their measurement knowledge. The modern classroom textbook is generally replete with such good investigative activities, and these are frequently overlooked. Finding and using these types of problems is the essence of teaching measurement in the middle years.

Review questions

13.1 List the general sequence for teaching any measurement sub-strand. Provide illustrations of the sequence.

13.2 List and explain three considerations when teaching time.

13.3 Current research indicates the importance of making connections in mathematics. How could you make connections between some of the topics in measurement?

13.4 The teaching of measurement is based on a Western worldview of quantifying. Many other cultures see the world differently. If you were working in an indigenous society that placed a greater emphasis on quality, what might this mean for teaching this aspect of mathematics?

Further reading

Aldridge, S. & White, A. (2002). What's the time, Ms White? *Australian Primary Mathematics Classroom,* 7(2), 7–12.

Friederwitzer, F. & Berman, B. (1999). The language of time. *Teaching Children Mathematics, 6*(4), 254–9.

Monroe, E.E., Orme, M.P. and Erickson, L.B. (2002). Working cotton: Toward an understanding of time. *Teaching Children Mathematics, 8*(8), 475–9.

Nitabach, E. & Lehrer, R. (1996). Developing spatial sense through area measurement. *Teaching Children Mathematics, 2*(8), 473–6.

Pelton, T., Milford, T. & Pelton, L.F. (2018). Developing mastery of time concepts by integrating lessons and apps. In N. Calder, K. Larkin, N. Sinclair (eds), *Using mobile technologies in the teaching and learning of mathematics* (pp. 153–66). Singapore: Springer.

Smith, T. & MacDonald, A. (2009). Time for talk: The drawing-telling process. *Australian Primary Mathematics Classroom, 14*(3), 21–6.

Walton, C. & Randolph, T. (2017). Alternative methods for understanding area formulas. *Illinois Mathematics Teacher, 64*(1) 40–5.

Weinberg, S. (2001). 'How big is your foot?' *Mathematics Teaching in the Middle School, 6*(8), 476–81.

Whitby, K. (2001). Teaching and learning about length. *Prime Number, 16*(4), 1–6.

Wright, S. (2006). *Children's multimodal meaning making: Giving voice to children through drawing and storytelling.* Rotterdam: Sense Publishers.

References

Aladé, F., Lauricella, A.R., Beaudoin-Ryan, L. & Wartella, E. (2016). Measuring with Murray: Touchscreen technology and preschoolers' STEM learning. *Computers in Human Behavior, 62*, 433–41. doi:10.1016/j.chb.2016.03.080.

Amagir, A., Groot, W., Maassen van den Brink, H. & Wilschut, A. (2018). A review of financial-literacy education programs for children and adolescents. *Citizenship, Social and Economics Education, 17*(1), 56–80. doi:10.1177/2047173417719555.

Battista, M. (2007). The development of geometric and spatial thinking. In F. Lester (ed.), *Second handbook of research on mathematics teaching and learning* (pp. 843–908). Charlotte, NC: Information Age.

Bouck, E., Flanagan, S. & Bouck, M. (2015). Learning area and perimeter with virtual manipulatives. *Journal of Computers in Mathematics and Science Teaching, 34*(4), 381–93.

Chigeza, P. & Sorin, R. (2016). Kindergarten children demonstrating numeracy concepts through drawings and explanations: Intentional teaching within play-based learning. *The Ausralian Journal of Teacher Education, 41*(5). doi:10.14221/ajte.2016v41n5.5.

Clarke, D.M., Sullivan, P., Cheeseman, J. & Clarke, B.A. (2000). The Early Numeracy Research Project: Developing a framework for describing early numeracy learning. In J. Bana and A. Chapman (eds), *Mathematics education beyond 2000* (pp. 180–7). Fremantle: MERGA.

Davies, P. (2015). Towards a framework for financial literacy in the context of democracy. *Journal of Curriculum Studies, 47*(2), 300–16. doi:10.1080/00220272.2014.934717.

Gough, J. (2004). Fixing misconceptions: Length, area and volume. *Prime number, 19*(3), 8–14.

Harris, P. (1991). *Mathematics in a cultural context: Aboriginal perspectives on space, time and money*. Geelong: Deakin University Press.

Huang, H.-M.E. (2017). Curriculum interventions for area measurement instruction to enhance children's conceptual understanding. *International Journal of Science and Mathematics Education, 15*(7), 1323–41. doi:10.1007/s10763-016-9745-7.

Kontorovich, I. & Zazkis, R. (2016). Turn vs. shape: teachers cope with incompatible perspectives on angle. *Educational Studies in Mathematics, 93*(2), 223–43. doi:10.1007/s10649-016-9699-2.

Lee, M.Y., & Francis, D.C. (2016). 5 ways to improve children's understanding of length measurement. *Teaching Children Mathematics, 23*(4), 218–24.

Lui, A.M., & Bonner, S.M. (2016). Preservice and inservice teachers' knowledge, beliefs, and instructional planning in primary school mathematics. *Teaching and Teacher Education, 56*, 1–13. doi:10.1016/j.tate.2016.01.015.

Mandell, L., & Schmid Klein, L. (2009). The impact of financial literacy education on subsequent financial behavior. *Journal of Financial Counseling and Planning, 20*(1), np.

McHugh, L., Kelly, A.M. & Burghardt, M.D. (2017). Teaching thermal energy concepts in a middle school mathematics-infused science curriculum. *Science Scope, 41*(1), 43–50.

Moore, D.A. (1999). Some like it hot: Promoting measurement and graphical thinking by using temperature. *Teaching Children Mathematics, 5*(9), 538–43.

National Council of Teachers of Mathematics. (1989). *Curriculum and evaluation standards for school mathematics*. Reston, VA: NCTM.

National Council of Teachers of Mathematics. (2000). *Principles and standards for school mathematics*. Reston, VA: NCTM.

Poole, R. (1998). Calendars, calibration and culture. *Mathematics in School, 27*(5), 14–16.

Selmer, S., Valentine, K., Luna, M., Rummel, S. & Rye, J. (2016). How can we best use our school garden space? Exploring the concepts of area and perimeter in an authentic learning context. *Australian Primary Mathematics Classroom, 21*(4), 3–10.

Sheldon, S. & Duncan, S. (2019). Teaching children money management skills. *Forever Families. 98*. https://scholarsarchive.byu.edu/foreverfamiles/98.

Tossavainen, T., Suomalainen, H. & Mäkäläinen, T. (2017). Student teachers' concept definitions of area and their understanding about two-dimensionality of area. *International Journal of Mathematical Education in Science and Technology, 48*(4), 520–32.

Wieber, A.E., Evoy, K., McLaughlin, T.F., Derby, K.M., Kellogg, E., Williams, R.L. & Rinaldi, L. (2017). The effects of a modified direct instruction procedure on time telling for a third grade student with learning disabilities with a brief comparison of interesting and boring formats. *Learning Disabilities: A Contemporary Journal, 15*(2), 239–48.

CHAPTER 14

STATISTICS AND PROBABILITY

What are probability and statistics?

Probability and statistics are encountered with regularity in everyday life. It is generally accepted that this is one of the key numeracies in the new millennium because of the data-saturated society in which we live, the rapid changes occurring in society and work, and the growing use of computer technology. Many educators and social commentators argue that the significant changes occurring in Western societies are akin to the changes that resulted from the move from agrarian communities to the organised workplaces of the Industrial Revolution. Contemporary theorists and educationalists argue that students need to enter this world able to make sense of the multifarious forms of information presented to them. For mathematics teachers, the task is one of preparing students for a society in which they need to be able to make informed decisions about whether or not a particular medical procedure or treatment is worth the risk, about their money and how to invest it, about which of many credit options to make purchases on, about risks associated with various types of purchases of cars or other items, and so on. To do this, they need to be able to evaluate the likelihood of events occurring (probability) and use information (data) to make the most appropriate decisions (Makar et al. 2011). Since much of the information they seek is now available online, internet-related activities should be included in teaching.

Statistical literacy

A renewed focus on statistics and probability in the Australian curriculum underscores how an understanding of chance and data is the linchpin of effective survival in the world beyond school. It is therefore vital that key skills and attitudes be developed. The emphasis of this strand goes beyond exploring probability and constructing graphs. It is far broader, and directly targets the development of statistical literacy. The characteristics that define this literacy include being able to identify the types of data that need to be collected; undertaking and subsequently organising the data collected; representing the data in ways that make sense and result in high levels of readability; and being able to interpret and critique data. Everyday literacy and numeracy are combined in this strand—today, even reading the newspaper involves interpretation of considerable statistical numeracy. Being able to interpret *and* evaluate the information being presented is a key life skill.

One of the more common experiences students will have of mathematics in their everyday lives is the interpretation of data—through the use of either 'averages' or graphs. Being able to read, interpret and analyse texts is a key literacy skill. Similarly, being able to read, interpret and analyse texts that involve mathematics is a key skill. This form of literacy frequently involves mathematics related to the chance and data strand. Statistical literacy involves the elements of text reading identified by Luke and Freebody (2001), as well as a numeracy element. Being 'literate' in these contexts demands that students have an understanding of, for example, how measures of central tendency are used in everyday contexts—for example, 'The average price of houses has increased by $40 000', 'The average salary of workers is $350 a week' or 'The average age of teachers is 48 years'. Knowing what these figures mean is an important component of being literate in today's data-drenched world. Students will learn the mathematical meanings of the central tendencies of mean, median and mode, and how using these measures can either inform or confuse.

Identifying the writer's intentions by analysing why one particular measure of central tendency (or graph) was used over others helps the reader identify the purpose of the text and any tools the writer used in order to convey particular meanings to the reader. Whereas the data strand once focused on the construction of graphs or measures of central tendency, it must now incorporate reading, interpreting and critically analysing information (or texts) in order to deconstruct meaning and any assumptions being built into texts. For example, different measures of 'average' are used to convey information to the public; being aware of what measures are used is important for understanding.

Much like the other strands in the mathematics curriculum, statistics and probability are arbitrary divisions of content constructed in order to make sense of an expansive and expanding body of knowledge. While they are presented as discrete sub-strands, there is considerable overlap between them. For example, when measuring experimentally the probability of an event occurring, such as throwing a six on a die, the throws would need to be collected and organised (usually as a table). In order to collect and collate the information, a knowledge of the data sub-strand is essential. The Australian Association of Mathematics Teachers has created a website that contains a wealth of resources to support the development of statistical literacy; see https://topdrawer.aamt.edu.au/Statistics. While visiting the site, also visit other 'drawers' for resources on patterns, reasoning and mental computation, to name a few.

Probability

Chance and probability refer to the same construct. Probability is concerned with measuring the chance that a particular event or outcome is likely to occur. It is measured in two quite different ways. One is experimental, where students actually undertake the process (such as throwing a die to see how many times it lands on each number) and then use this information to make predictions about future events. The other is theoretical, where students identify the number of possibilities in total and the number of possible times that a preferred outcome can occur. Often, chance is seen in the context of gambling options, and the activities undertaken in the teaching of the concept—using cards, dice, coins, and so on—frequently have strong gambling connotations. However, in daily life, such decisions as whether to drive a particular route, whether to undergo a medical procedure or take a particular medication, or whether to stay on a casual salary or move to a permanent salary are all influenced by chance. In order to make decisions that can influence quality of life in a positive way, understanding probability is an increasingly important life skill.

Statistics

This sub-strand refers to the collection, representation and interpretation of data. Considerable time is spent in this sub-strand on developing students' knowledge and skills in graph construction and interpretation. However, it has been recognised widely that, while students may be able to construct graphs, they may not transport this knowledge to their everyday lives and use it to interpret the graphs and other forms of data presentations with which they are confronted every day. More emphasis needs to be placed on reading, interpreting and critically appraising the information that is presented through graphs and tables.

This sub-strand also includes measures of central tendency (mean, median and mode). These measures of central tendency are another form of data representation, and students need to have this knowledge in order to interpret the information with which they are presented on a daily basis—for example, to be able to critically appraise reports in the daily newspaper and to assess the legitimacy of such claims as 'the average number of children per family is 1.7', 'the average Australian child is overweight' and 'the average number of hours of television that children watch per week is 35'. To do this, understanding the way in which 'average' has been measured is critical.

Teaching probability (chance)

The model used for planning in teaching measurement can be applied in the teaching of chance, since it is a process of measuring the probability of an outcome occurring. As stressed in the measurement chapter, organising for learning using this model is simply a planning process—and, while it has value, it is not meant to be the one true model for planning. It is, however, used widely in syllabus documents for organising the mathematical progression through school, and research suggests that students progress through particular and predictable patterns of chance understandings (Watson & Moritz 1998).

Identify the attribute

Students' everyday experiences of chance or probability can be used to discuss and identify the attribute. Using everyday terms such as likely, possible, maybe, unlikely and impossible, it is possible to identify the construct of probability. Using everyday experiences such as whether it is going to rain, snow or be hot, whether students will get certain presents for their birthdays or for Christmas, what will be in their lunchboxes, and so on can be useful for talking about the likelihood of events happening. Words that are part of students' everyday vocabulary can be listed so that students are able to build up a strong language of probability. A great deal of time during the early years of school is spent developing an informal understanding of probability through everyday experiences. Students should be given opportunities that allow them to predict the possible outcome of events and to discuss their level of confidence in their predictions (such as throwing a six on a die).

Comparing and ordering likelihood of outcomes

Using the informal language of probability, students categorise events according to the likelihood of their occurring. Events are categorised starting with terms such as possible and impossible, certain and uncertain, and likely and unlikely. This could be done through picture-sorting activities or using simple table formats. More complex categories can be constructed using terms such as 'more likely' or 'less likely'. Simple activities can be undertaken so that students can compare the likelihood of events happening when there are only two equal outcomes (e.g. tossing a coin, taking a red ball from a jar containing equal numbers of balls of two colours). Students should be

Student ordering the probability of events

encouraged to record their experiences on charts or tables so that they begin to see the emergence of the formal definition.

Activities can include the following:

- Toss a coin and record how many times it lands on heads and on tails. Compare which one occurred more often—heads, tails, or about the same.
- Have a bag containing balls of only two colours (e.g. red and blue). Students take a ball out, record the result and place the ball back. Repeat a number of times.
- Once two outcomes have been explored, move to experiences in which there are more equal outcomes (e.g. balls of three different colours in the bag).
- Throw a die and recording the outcomes on a table. Students often have the misconception that throwing a six, for example, is more difficult than any of the other numbers, based on their experiences of board games. Recording their outcomes can provide an opportunity for them to see that there is a relatively equal chance of each number occurring.

Measurement of probability

Prior to the formal measurement of probability, students should have a good understanding of number study in the areas of fractions and ratio, since probability is measured using these number skills and concepts. When an event is impossible, it is seen as zero (0) and when it is certain it is seen as one (1). Two processes are possible for understanding probability.

TEACHING IDEA

Toss a die 30 times. Record the outcomes in the table.

1	2	3	4	5	6

- Discuss individual outcomes. Combine small-group or whole-class outcomes. Discuss.

Theoretical probability

Theoretical probability is tied to understanding two aspects of probability. The first component is the number of times that the desired outcome can occur. For example, when considering throwing a die, the number of preferred outcomes is considered. A single outcome, such as throwing a 4 or a 1, is the usual starting activity. This can be increased to include other outcomes, such as throwing an even number (2, 4 or 6), so that there are three potential outcomes, or a multiple of three (3 or 6) so that there are two potential outcomes. The second consideration is the total number of outcomes that can occur. In throwing a die, these are throwing a 1, 2, 3, 4, 5 or 6, so that there are six potential outcomes in any one throw.

$$\text{Probability of an event} = \frac{\text{Number of individual outcomes}}{\text{Total number of possible outcomes}}$$

Thus, the probability of throwing a 6 can be recorded symbolically as Pr (6) = $\frac{1}{6}$. The relationship of probability fractions is seen from such symbolic notation.

TEACHING IDEA

When rolling two dice, what is the total number of potential outcomes? What is the probability of throwing a 6? Is it the same as that of throwing a 4? Is there a number that is more likely to come up than others? Why? Justify your answer.

Experimental probability

Actively involving students in explorations of probability provides them with the opportunity to link the components of the theoretical measures to their experiences. Throwing a die creates the opportunity for students to see the rationale for recognising the total number of possibilities, since they will need to construct some sort of table or recording tool for the numbers they throw. As they undertake the throws, the need for the table with six columns emerges, and with it the recognition of the number of possible outcomes. Using experimental situations can also prompt a more lateral thinking process. For example, earlier experiences with single die throwing can create a misconception about the probability of throwing numbers when using two dice. Students often think that there is an equal chance of throwing any number. They may also think that there is a particular range of outcomes. The two data tables below show the possible outcomes when two dice are thrown. The first table indicates the actual combinations possible (36 in total), and the second the number of times each total can occur. Thus, there is a greater chance of throwing a total of seven with two dice than a 12 or 2. Using an experimental approach gives students the opportunity to challenge their existing schema based on the simpler examples given when introducing probability. A very powerful digital resource for teaching experimental probability is found at www.random.org. At this website, students can perform large numbers of virtual trials with dice, coins, cards, and so on. This resource should be used after students have performed their own probability experiments.

Results of addition of two dice

+	*1*	*2*	*3*	*4*	*5*	*6*
1	2	3	4	5	6	7
2	3	4	5	6	7	8
3	4	5	6	7	8	9
4	5	6	7	8	9	10
5	6	7	8	9	10	11
6	7	8	9	10	11	12

Number of outcomes for each combination

outcome	2	3	4	5	6	7	8	9	10	11	12
number	1	2	3	4	5	6	5	4	3	2	1

Some situations cannot be measured theoretically, and hence are more open to experimental probability. Such situations can also lead to rich discussions about notions of fairness. Examples of these situations would include throwing a drawing pin and noting how often it comes up one way or the other. Similarly, the well-known question of 'which way up' a slice of buttered bread lands when it is dropped provides a rich discussion point. In this case, it would appear that there are only two outcomes, each equally likely, yet our everyday experience suggests that the bread almost always lands buttered side down! Asking questions such as 'Why?' or 'Is it really the case?' stimulates discussion.

Application to problem-solving contexts

The application of probability to problem-solving can involve two issues. One is the mathematical aspects of probability, whereby students can investigate the probability of horse racing or Lotto in order to decide whether or not they would gamble. While Lotto is, mathematically, a solid example, horse racing is fraught with other human interventions (such as cheating) that may unfairly skew the outcomes. However, horse racing is more complex than Lotto because it involves past performance, track conditions and other factors, and so presents considerably more variables. Less morally motivated applications of probability can involve Gallup polls and the lead-ups to elections, weather forecasting, insurance premiums (e.g. why it costs so much for young people to have full comprehensive insurance), familiar everyday situations (e.g. the time it will take to get home when leaving work at 4.00 p.m. vs. 4.30 p.m. due to peak-hour traffic), and why home insurance increases in some areas and not others after floods or fire.

The second aspect involves Critical Mathematics and the exploration of the social costs and issues of gambling for the community and individuals. Investigating why people would spend so much money on Lotto when there is such little potential for winning and, similarly, investigating social issues that arise from gambling offers issues that can be explored (and linked to other curriculum areas). Examples include embezzlement from family trusts or employers to supplement gambling, major sports being supported by gambling organisations, parents leaving young children in cars while they gamble in casinos, the breakdown of families due to gambling, and the support group Gamblers Anonymous. Such issues appear regularly in television and other media news reports.

Teaching points for probability

Within the teaching of probability, a number of teaching points need to be observed. Embedded in the activities outlined above are some key constructs that will be integrated into the teaching process.

Outcomes and sample space

Students need to gain an appreciation and understanding of the notion of outcomes and sample space. *Outcomes* refer to the number of outcomes for a particular event (e.g. an even number on a die). *Sample space* refers to the total range of potential outcomes. For example, when using a six-sided die, the number of outcomes is 6, and the sample space is 1, 2, 3, 4, 5, 6. If the preferred outcome is 4, because there is only one 4 on the die this would mean that the preferred outcome is 1. When throwing this die, it is reasonable, then, to expect that there is a probability of $\frac{1}{6}$ and that the outcome will be a 4. This is represented in the following way:

Pr (4) = $\frac{1}{6}$.

This can also be expressed as a decimal or percentage.

TEACHING IDEA

Fair go!

Students are asked to consider the following newspaper advertisement and then write to the consumer affairs minister justifying why the game should or should not be allowed at the local fair.

Tim Collins from Sideshow Alley asks all the kids in school to come and try their luck on his new game. The game is for you to throw two dice and win some money. If you throw a 1, 2, 3 or 4, you double your money. If you throw a 5, 6, 7 or 8, Tim keeps your money. If you throw a 9, 10, 11 or 12, your money goes to charity. Tim will not make any money on this game; the money he keeps on his scores is used to pay the double money students get on their scores. So come on down to the fair and double your pocket money!

Independence of events

Coming to appreciate that some events are independent of others can be difficult, since everyday misconceptions tend to sway judgements. For example, if a couple has had four sons, there is a perception that their chance of having a daughter next time must be much greater. Similarly, when tossing coins, if six heads have been thrown in a row, it is expected that the next toss will be much more likely to be tails. When students are confronted with this type of outcome, they tend to offer excuses such as, 'There are more holes [indentations] on the six dots [of the die], so it weighs less; that is why it comes to the top.' Falsely applying the law of averages logic can lead to an incorrect judgement. In both cases, each event is independent of the others—that is, the outcome of one event does not impact the others. Tossing a coin is not influenced by the throws from earlier tosses. Each toss has an equal probability of resulting in heads or tails, regardless of what has happened earlier.

In some cases, events are dependent on other outcomes (as in games such as Lotto), where the probability of the second event is influenced by the first event—for example, when a ball is removed from the pool in the first draw, leaving one less ball for the second draw. So the probability of the first ball is 1 in 45 (where there are 45 balls), and that of the second ball is 1 in 44, since there are only 44 balls left. In this case, the probability of the second event is influenced by the first event.

In order for students to gain an understanding of the independence of events, activities should be provided that allow them to explore the outcomes when sample sizes are increased. When tossing a coin, for example, the outcomes may be skewed (e.g. on ten tosses of the coin, there may be two heads and eight tails), but as the sample size increases, or as class results are pooled to increase the sample size, students can see that the actual outcome comes closer to the theoretical outcome (i.e. theoretically, the outcome for tossing a coin is 50/50).

Randomness

Randomness is a key aspect of probability. For an event to be random, it cannot be influenced by other factors. There is widespread confusion in both the conceptions and the perceptions of randomness (Batanero 2015) that need to be taken into consideration in teaching. The notion of randomness introduces considerations of fairness. If, at the end of a term, all class members' names were placed into a bag and one drawn out, with that person being given a chocolate bar, discussion about ensuring the event is random and fair could be developed.

Would students prefer to have everyone's name in the bag? Should the teacher be able to see the name before drawing it out? Should the girls be allowed to have their names in twice? By recognising biases in the sampling process, the notion of randomness becomes apparent.

Data collection, representation and interpretation

Current trends in teaching data are informed by the wider research community, where it is noted that, while students may be strong at constructing graphs, they do not appear to be as strong in interpreting the graphs and being able to use this information in decision-making. There is also a need for teachers to recognise that students may interpret problem contexts in ways that are unique to them (Busse 2011), so it is important for teachers to be aware of this and pose questions to ascertain students' thinking before making judgements. The emphasis in this sub-strand needs to be on a range of ways of collecting and representing data, along with interpreting the data itself. As with the other aspects of teaching mathematics, the role of the teacher is to guide learning in ways that are supported, so that students are able to achieve the desired outcomes. As has been noted, scaffolding of learning is a central part of curriculum planning and organising. Teachers need to be aware of what knowledge it is reasonable to expect from students, what knowledge students will be expected to construct, and where this will ultimately lead.

Students collecting traffic data

Collecting data

In order to collect data, a key decision has to be made regarding the purpose of collecting them. Knowing why the data should be collected will inform decisions about the types of data that are most appropriate. If data were required about modes of transport to school, information on the types of cars driven by teachers would not be the right data to collect; rather, the various forms of transport would need to be identified—cars, bikes, buses, walking, scooters, and so on. The sample from which the data are collected also needs to be considered—is it just the class or school, or the wider community? Collecting data for data's sake is a process that has been used in schools for some time, but the purpose of this needs to be questioned.

Data should be collected for realistic purposes. Why do we want to know about how we come to school (or work), other than to make a graph? What is now central is to consider the *purpose* of the graph. In this case, it may be part of a larger project on environmental issues for which modes of transport are a key issue; hence, the collection of data has a real purpose. In the early years, the purpose may link to the health curriculum, where notions of healthy eating are being considered, so collecting data on diets and lunches serves a real purpose. Ensuring that data collection (and subsequent representation) has real purpose means the activity will be seen as valuable, authentic and purposeful, rather than simply a time-filling exercise.

Organising data: teaching graphs

Rather than seeing graph work as a formal process of x-axis and y-axis, contemporary approaches adopt a scaffolding approach in which the need for various components emerges from problematising the process. When students are able to recognise the needs for the various components of a graph, they are likely to see these less as something demanded by the teacher and more as key aspects of the graph. Prior to constructing graphs, students should be given considerable experience in reading and interpreting them. The graphs with which students work—both interpreting and constructing—should be relevant to real-world experiences. There is a great deal of synergy between curriculum areas when using graphing (McHugh et al. 2017). Much of the work undertaken in number study is reinforced in this strand. For example, number study of large numbers (millions) will precede work with population graphs.

Simple picture graphs

Experiences with graphs in the early years of schooling are linked strongly to students' experiences. They often relate to when students' birthdays occur in the year, the types of pets they have, how many siblings are in their families, what they ate at a particular meal, and so on. The concreteness of these examples allows very concrete representations to be developed within the class. There is no real need for labels at this point, since the data and numbers are obvious. In these

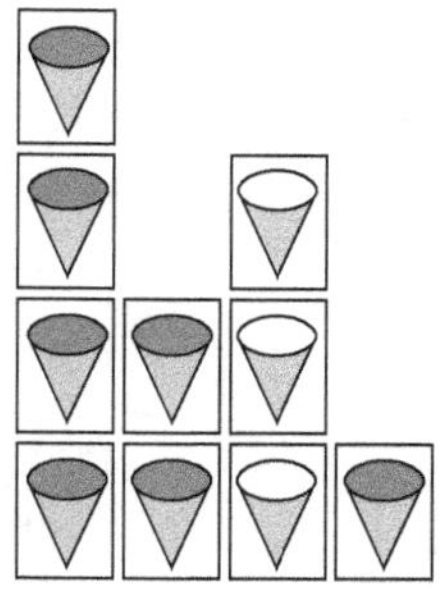

graphs, one image initially represents one count; as students become more experienced in representing data, one image can come to represent ten counts, 100 counts, and so on. Thus, once students have become familiar with organising data in this format, the task for the teacher is to progressively delete concrete evidence, thereby creating the need for labelling. The need for a common starting line and even spacing between the columns is central to these early experiences. For the teacher, when beginning work with pictographs, it is recommended that squares be used for the pictures so that the pictures' orientation will not matter. Square Post-it® Notes are a valuable resource for this activity, since they allow graphs to be made very easily—students draw on the data (e.g. their favourite ice-cream, what they had for lunch, what they do on holidays) and they can be stuck to a baseboard easily. This allows for the construction of many graphs quickly.

Early representations of graphing

It is advisable that early work with graphs focuses on two-column graphs, before building progressively to more columns as students demonstrate an understanding of the principles being developed.

Replace pictures with squares

In creating a need for labelling the *x*-axis, the teacher changes the pictures so that they do not visually cue the reader into seeing them as, for example, hair colour. This could mean using brown, black, red and yellow pieces of paper. Other resources, such as blocks and counters, can also be used. On seeing such a representation, the reader would be unaware of what the graph was representing, thus developing the need for the axis to be labelled.

Similarly, as the columns become larger and less obvious in number, the need for labelling the vertical axis (or *y*-axis) emerges. In the graph shown in the diagram, students have moved on from using picture graphs, but it is unclear just what the graph is representing. In order for it to be known to the reader, the axis must now be labelled. Since this is the object that is going to be measured, it is referred to, mathematically, as the *independent variable*. This might be hair colour, months of the year or foods eaten. Similarly, the other axis is used to represent the dependent variable, since this axis 'depends' on what is the object of the count. For example, the number of students represented in a particular column is dependent on what the focus of that count is—the number of students with brown or blonde hair, for example, or those who had pies or hot dogs for lunch. As with many mathematical ideas, it is important for teachers to know the mathematics but not to teach it to students explicitly. By knowing the concept, and hence by teaching it correctly, the teacher will ensure that students will have less chance of forming erroneous knowledge.

Transfer to grid paper

At some point in time, students need to record their graphs on paper. One of the easiest ways to achieve this is through the use of grid paper, which allows students to simply colour in the appropriate number of squares to represent the data. Initially, this might be done with different colours to represent different groups of data. After some time and experience, students will use one colour for all columns, thus creating a need to label the individual columns.

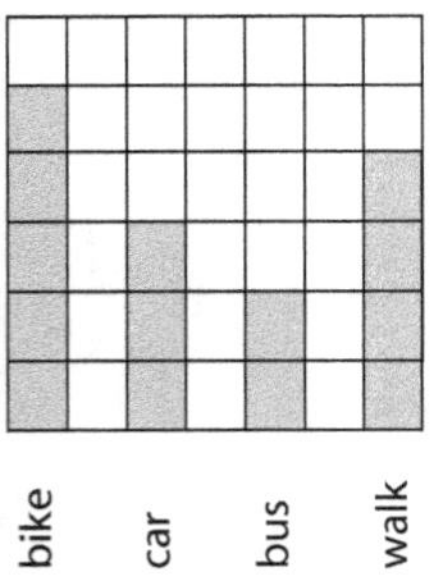

Grid representation of data

A need for labelling along the vertical or y-axis emerges as the data numbers increase. It is easy to recognise small groups, but as the data sets increase, a scale is needed to enable easy comparisons. In the early stages of graphing, the scale should be a one-to-one relationship. Once this relationship has been grasped, the scale can be increased so that students can have one square representing two, five or ten people, or whatever scaling is appropriate for the data being represented.

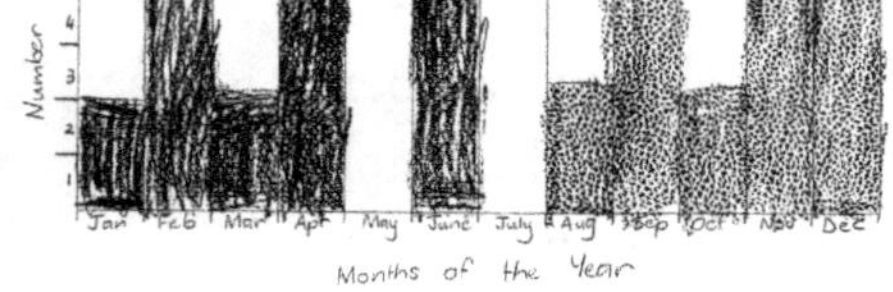

Student work sample: class birthday graph

Horizontal and vertical bar graphs should be used so that students will become familiar with both formats. The process of labelling the graphs and the axes, along with labelling any scales that might be used, is the focus of teaching.

Construct column graphs

The final phase in the construction of column graphs is the removal of grids and the drawing of columns that correspond with given amounts on the y-axis (when using vertical columns). Simple scales can be introduced. Students clearly label the y-axis and mark in scales and numbers on it. By progressively developing the need for the various components of graphs, students can recognise their purpose, and in so doing appreciate their appearance and role. When this understanding is developed, students are more likely to include the appropriate information and modes of construction when developing graphs than to memorise abstract features and hope that what is included is correct.

Types of graph

A common student misconception in the graphing sub-strand concerns the ways in which data can be displayed using column graphs. There is a range of column graphs, and their different purposes are the major reason for choosing a particular construction. One important aspect that students need to recognise is whether or not bars should touch. This is often understood poorly in the primary-school years, creating difficulties for students when they begin their formal study of statistics in high school. The difficulty lies in recognising the difference between continuous and discrete data, since these data types influence the type of graph constructed—histographs for continuous data, and bar graphs for discrete data. Other types of graphs with which students should become familiar are noted below.

Comparative bar graphs

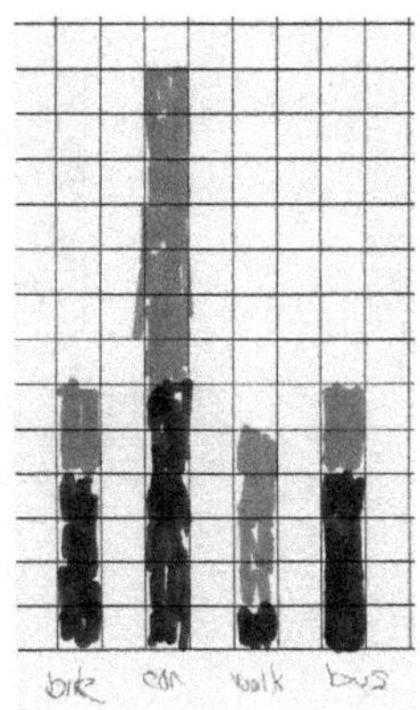

End-on-end comparative bar graph

Comparative bar graphs are used when there is a need to compare two sets of data. For ease of comparison, the two sets of data are placed on the same graph. For example, a teacher might want to consider how a group of students has performed on one assessment piece compared with another assessment piece. Students may want to compare the rainfall for each month over two years, how many pizzas a cohort of students may have eaten over two different time periods, or the pets owned by two cohorts of students.

Using the same data set, two different forms of comparative bar graph can be used. The data sets in the graphs pictured

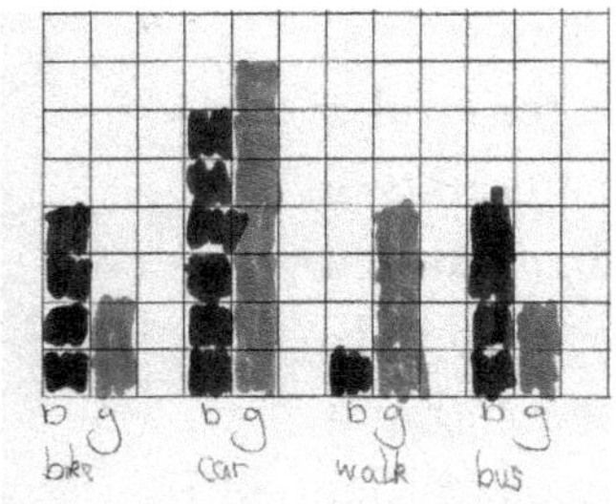

Side-by-side comparative bar graph

refer to students' end-on-end comparative bar graph preferences for four modes of transport to school. In the side-by-side comparative bar graph, columns are side by side so that comparisons between the gender and the transport can be made easily. The end-on-end graph allows the total transport modes across both genders to be compared.

It is critical, when using these forms of graph, that the data are of a form that can be compared. For example, it would not be possible to compare data that discuss income and expenditure, since these are not things that can be 'added'.

Grouped frequency histographs

Age in months of students in our class

Age in months	Tally	Frequency
85–96	|||	3
97–108	卌 卌 卌 |||	18
109–120	卌 卌 卌 卌 |||	23
121–132	卌 卌 卌 |	16
133–144	卌 ||	7

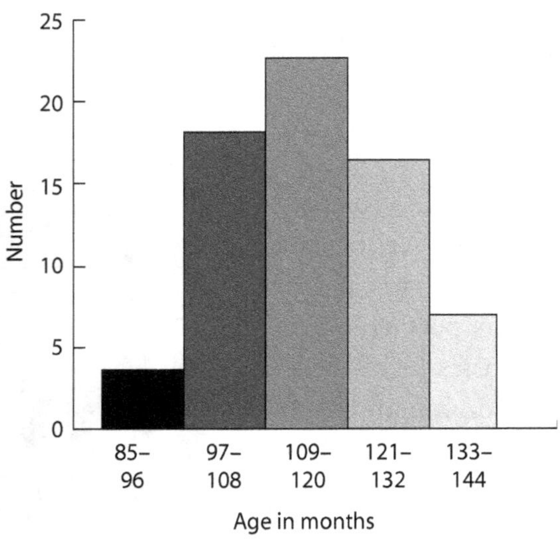

Age of students in our class

Continuous data are often 'clustered' so that small ranges can be used to organise them. This happens implicitly with all continuous data, although this is often not recognised or made explicit to those collecting the data. For example, when constructing an age graph, the data are collected based on the age of the student, so that, while the 'year' is seen as the category, it is how the category is constructed that determines how the data are grouped. Consider the ages of students in a multi-age Year 4/5/6 classroom. The expected age range is probably between seven and twelve years. If this age range is broken up into months, so that the ages are more accurate, it will run from 84 months to 144 months. Since a continuous range would not be used, the data are grouped and categories constructed. As discussed later in this chapter, in the section on discrete and continuous data, how these data are grouped depends on where cut-off points are made. In the data table, these are up to any given birthday. Once the student has had their birthday, they move into the next category. It could have been just as easy to have made categories that ranged from halfway between one birthday and the next—that is, 6½–7½ years.

The compiled data can be grouped into clusters of data that are useful for representing the final outcomes. The data can then be transferred from a table format into a graphical form. The data can be entered into a spreadsheet and then compiled into a graph using the functions available through the software.

Circle or pie graphs

Perhaps the most difficult graph for students to construct is the pie or circle graph. Considerable knowledge and skills—fraction knowledge, percentages, angles and operations—need to be brought together in order to construct this graph. When the graphs are introduced, simple data should be used so that the fractions to be interpreted (and, later, drawn) are halves, quarters and thirds. As students become more familiar with data representation, the teacher should ensure that the data can be transferred easily into a proportion of 360. For example, a sample of 18 will convert to an angle of 20°, which is an easy size for students to draw.

Informal introduction to circle graphs

By the time circle graphs are introduced in the primary school, students will have had substantial experiences with column graphs. To introduce the idea of circle graphs, two strategies can be used. One of the easiest ways to begin constructing the graph is to have the data constructed as a single long strip. Once the graph has been constructed on paper, the columns are cut out and glued end on end to form a long strip. Once the long strip has been constructed, it is then joined to form a circle. The centre of the circle is located, and the end of each section is then drawn to the centre.

A second method involves representing the columns through beads threaded onto a string. The items in a column are threaded onto the string. The string is then made into a circle. The centre is identified, and lines are drawn from the end of each coloured section into the middle. These simple methods can be used for introducing the idea of circle graphs. They are, however, inaccurate, since considerable guesswork is involved in identifying the centre of the circle and whether the shape made is circular. The process introduces the construction of a circle graph, but without the complex

Column graph to strip graph to create a circle graph

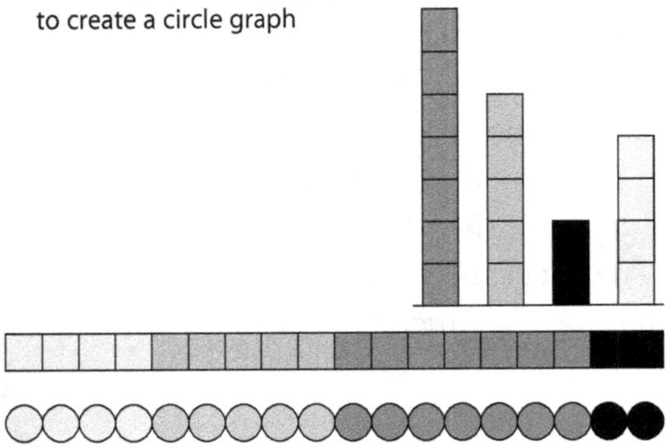

construction process. It allows students to appreciate the purpose of the graph and to develop an understanding of the graph itself without being confused by the mathematics involved in the process. The purpose, at this stage, is to introduce the new graphical format and an appreciation of its role in representing data as a whole set, and the individual data sets as parts of that whole. This is a new idea for students, so time needs to be spent developing it.

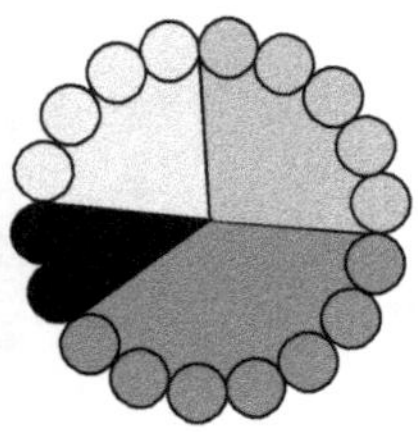

Using beads to construct pie graphs

Once students can construct circle graphs in this way, and have the requisite background knowledge in other areas of mathematics, they can construct graphs using the traditional methods. Initially, the examples should be ones in which students can divide the circle into recognisable sections such as halves, quarters, eighths, thirds, sixths or twelfths. The lines can be drawn from side to side (when using halves) or from the centre.

Other knowledges can be used to construct pie graphs. In particular, clock knowledge can be most useful. In the example here, the teacher has selected a sample with 60 items. Students can then apply their knowledge of minutes, quarters and halves (15, 30 and 45) to create appropriate divisions and then divide the circle graph into segments to appropriately represent the data.

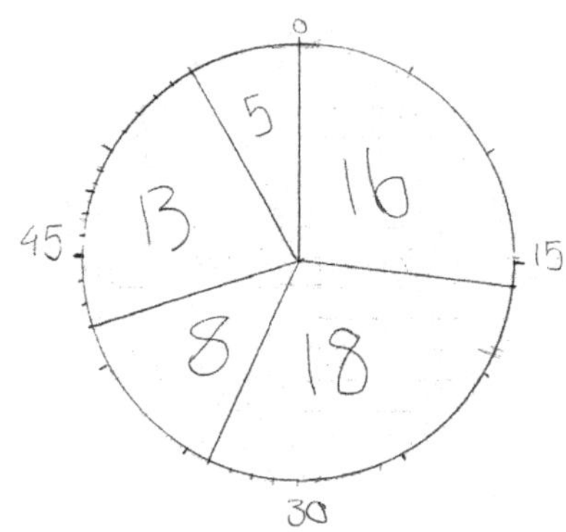

Student work sample: using knowledge of clocks to construct a pie graph

TEACHING TIP

Use students' Fitbit data to create data representations of their activities. This links with health and physical education. Ensure that there is discussion around the data and students' lives.

Formal construction of circle graphs

Constructing pie graphs, in which angles have to be calculated, is generally left for the upper years of primary school, since understandings of so many other aspects of mathematics are needed. Any examples developed by the teacher should be those that will compute easily into 360°. Early experiences need to focus on examples in which students can divide the total number of items into 360° to find out the size of individual segments, and then multiply this by the number of items in each data set.

As in other teaching strategies, the teacher's role is to move students to sounder methods of mathematising situations. Once students are confident with this simple method, a more general strategy can be developed so that they will be able to construct any circle graph, regardless of the numbers in the data set. Constructing tables such as those shown below helps students to work systematically through the long process. Rather than coming to rely on formulaic thinking, students should be encouraged to see whether there is any connection between the size of the fraction and the degrees, in order to see whether the task can be calculated more easily.

Early experiences need to focus on examples where students can divide the total number of items into 360° to find out the size of individual segments, and then multiply this by the number of items in each data set.

Constructing tables can help students to systematically work through this long process. Rather than coming to rely on formulaic thinking, students should be encouraged to see whether there is any connection between the size of the fraction and the degrees in order to see whether the task can be calculated more easily.

In the example shown, the students were constructing a pie graph to display the colour preference (blue, orange, red or pink) of 18 people surveyed. They constructed two tables in which to analyse the process of constructing a pie graph without relying on formulae. In the first table, the number of respondents who selected each colour was presented. Knowing that 18 divides neatly into 360, they knew that each person's preference would be equivalent to 20° on the pie graph. The third column of this table shows their workings to determine the size of the wedge on the pie graph required to represent the number of people who preferred each colour.

Colour	No. of people	Degress in pie graph
Blue	7	7 × 20 = 140°
Orange	5	5 × 20 = 100°
Red	2	2 × 20 = 40°
Pink	4	4 × 20 = 80°
Total	**18**	**18 3 20 5 360°**

Because pie graphs provide a visual representation of parts compared to the whole, the parts are often then converted to percentages. The second table shows the working for this in relation to the preferred colour activity.

Colour	No. of people	Fraction	Percent
Blue	7	7/18	7 ÷ 18 × 100 = 39%
Orange	5	5/18	5 ÷ 18 × 100 = 28%
Red	2	2/18	2 ÷ 18 × 100 = 11%
Pink	4	4/18	4 ÷ 18 × 100 = 22%
Total	**18**	**18/18**	**100%**

Stem and leaf plots

A stem and leaf plot is another way to represent data, and such a graph allows a greater appreciation of the scores or data set. A bar graph gives only the amount of scores or events within any particular column; no other information is apparent. Consider a scenario in which test scores are being plotted and have been grouped (as per a grouped frequency histograph). This type of graph would not tell us anything about the scores—for example, whether they were distributed equally across each score or were clustered around particular points. You might get results back from an external system-testing procedure and find that many of your students have scored in the 40–49 per cent range. Does this mean that they scored just below the pass mark, or were their scores much lower than this? A stem and leaf plot allows this information to be made apparent.

Much like the process that is undertaken for other graphs, the data must be collated and clustered. However, unlike the grouped frequency process, where only tallies are taken for any score, it is now necessary to keep track of the individual scores as they are clustered.

Consider the final scores for your class:

108	86	45	67	89	101
39	65	87	90	76	120
91	67	74	97	69	109
65	32	89	82	93	74

When these data are plotted on a column graph, it shows the spread of scores, but its usefulness is restricted because the values of the actual scores are not evident. For example, there is no indication of whether the scores in any of the categories were high, mid-range or low.

A stem and leaf plot preserves the scores. Finding the range of scores, then plotting the stem in an upward growth is the general format. However, some graphs use only the tens place value in the stem and the ones value in the leaves. First, the data must be grouped as in the table shown. Once this has been done, the data must then be organised in order so that the smaller numbers are closer to the stem and radiate outwards. What can be observed quite easily with this type of graph is that there are numerous high scores, and that students scored well within each scoring category.

120	0				
110					
100	8	1	9		
90	0	1	7	3	
80	6	9	7	9	2
70	6	4	4		
60	7	5	7	9	5
50					
40	5				
30	9	2			

Unsorted data showing scores

120	✓				
110					
100	✓	✓	✓		
90	✓	✓	✓	✓	
80	✓	✓	✓	✓	✓
70	✓	✓	✓		
60	✓	✓	✓	✓	✓
50					
40	✓				
30	✓	✓			

Tally marks do not show the actual scores

The graph is also useful if two data sets need to be compared. For example, you may want to compare boys' and girls' results for the maths tests and compare the different outcomes, or compare the results from two year levels that are in your class. In contexts beyond the classroom, the graph may be used for comparing scores for two teams over a season.

May		February
1	**12**	0
0	**11**	
9 3	**10**	1 8 9
9 4 3 1 0	**9**	0 1 3 7
9 7 3	**8**	2 6 7 9 9
8 5 4 0	**7**	4 4 6
8 7 7 6	**6**	5 5 7 7 9
1	**5**	
0	**4**	5
4	**3**	2 9

Sorted data using 2-sided representation

Using the same data set, the tests for a second test were produced, and it seemed that the students would have scored much better had a simple column graph been used. However, the stem and leaf provides the detail to allow a comparison.

Measures of central tendency

One of the more common statistical measures encountered by students in their everyday lives is that of *central tendency*. Rather than focus on the calculation of various averages, current teaching also requires students to know when particular measures of central tendency are used, and why; why they would report using such measures; the degree of accuracy needed; how different measures can be used to convince the reader; and so on.

The mean

Mean refers to the arithmetic average, where all scores are totalled and then divided by the number of addends. When teaching the mean, activities to support the development of the concept might include finding the mean height of the class. One way to do this is to measure lengths of streamer the same height as the students. These are taped end to end to form a single strip. This is then folded into the number of students in the group so that the mean height becomes apparent. An important understanding here is that there may be no students in the class who are the exact height of the folded piece of paper. Alternatively, using beads placed in an egg carton, students can represent the number of pieces of fruit they ate in one week. The corresponding number of beads is then placed in a hole representing each student, as shown in this example. The students then move the beads around until there are an equal number of beads in each hole, thus representing the mean.

Such experiences help students to understand the fundamental aspects of mean. Studies have shown that, while students may be able to calculate means, they often have little understanding of the concept, even by the later years of middle schooling (Years 8 or 9); yet the concept is critical for everyday knowledge and problem-solving.

TEACHING IDEA

Mean, median and mode

In what situations would it be best to use:

mean ....

median ....

mode ....?

The median

The *median* refers to the middle value in an ordered set of data. In the data set of 2, 4, 5, 9 and 9, the median is the value that falls midway in the set, so the point is the third value—5. Teachers often use the analogy of a highway where the mid-point of the highway is the median strip, thus illustrating the notion of 'middle point'. This measure of central tendency is less well known to primary-school students, but it is one with which they must become familiar, since it is used often in reporting. Teaching of this concept often takes place through the physical manipulation of data sets, whereby each score is written on an individual card and then organised in ascending or descending order. The end scores at each end are removed or marked off simultaneously so that there is only one (or two) cards left. The remaining card is the median. Where there are two scores, the mid-point between these scores is the median.

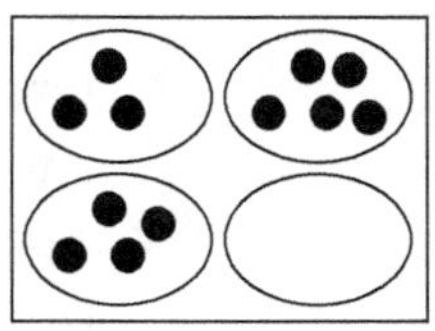

TEACHING IDEA

Consider the following range of salaries for staff within a new IT company:

$	$
35 000	42 000
31 500	76 000
28 000	120 000
37 000	41 000
39 400	46 500

- What is the mean? What is the mode? What is the median?
- Which is the best measure to report to the workers? Why?

The mode

The *mode* refers to the score that occurs most frequently within a data set. When considering this value in relation to a column graph, it is the value with the highest column. It is usually easy to find, and it is the score that is least affected by extreme values in a data set. The age of the students in a classroom is an excellent example of the application of mode, since in any one class there is likely to be a considerable number of students of the same age. A commonly reported mode in recent times is that of the ageing teacher population, where there are considerable numbers of teachers in their mid- to late forties. Calculating the mean age would not provide authorities with a true picture of the depth of the issue, but when the mode is used it becomes clear that a large number of teachers are in this upper age bracket and that they are likely to retire within a few years, thereby creating staffing problems.

Using the different measures of central tendency provides a useful forum for students to explore why different measures are used in different contexts, and why writers might prefer to use one measure over another, depending on the task at hand. This is an important life skill, since the wider public needs to develop critical reading skills in relation to descriptive statistics in order to be able to analyse the ways in which texts are used (rightly or wrongly) to persuade the reader. Much of the literature on critical numeracy focuses on the use and interpretation of statistics. Knowing how statistics can be used to empower and disempower readers is an important skill to be learned.

TEACHING IDEA

Book club

There are 7 students in the Ned Kelly reading group. If the mean number of books read by the group was 5 and the mode was 4, how many books could the students have read individually?

Note: This is a complex question and would be suitable for challenge work.

Statistical literacy: interpretation of data representations

While considerable time is spent teaching students how to construct graphs, less time is spent on teaching them how to interpret graphs—perhaps one of the most neglected aspects of the graphing sub-strand.

Where interpretation is taught, it tends to be very simplistic and not prepare students adequately for the new forms of statistical literacy of the world beyond school. As noted earlier in this chapter, statistical literacy is a key life skill for contemporary times (Steen 1999), and students need to leave school with this skill well developed. Much like our expectation that students will exit school with appropriate levels of traditional literacy, so too must they now exit with the capacity to read the many types of data that saturate contemporary society. This means that they must have significant experiences in reading, interpreting, analysing and being able to critique the multiple forms of data they may experience. Data will be displayed through various media—including newspapers, the internet, television, brochures and radio. Understanding what is being said is critical to the ability to make informed decisions.

Critical numeracy

Critical numeracy is a key skill related to statistical literacy. Interpretation of data is often embedded within superficial learning environments. The deforestation of the Amazon, the emission of greenhouse gases from various countries, the amount of energy consumed by different countries, the water quality of the local creek, or the amount of government money spent on education in comparison with defence are all interesting data sets with which to work in the classroom. However, students need to work further with such data sets to explore questions of why such outcomes occur, and to look at what can be done to challenge some of the important issues that have traditionally been masked through statistics. For example, students may collect statistics about the water quality in the local creek and use spreadsheets to record and display their findings. Within a critical numeracy approach, questions are posed about the effects of the findings on local ecosystems; depending on the findings, students might decide to write a letter to their local member of parliament or take other social action to draw attention to the water quality. Critical numeracy seeks to empower students by allowing them to gain appropriate statistical knowledge, and then enabling them to use that knowledge beyond the classroom setting.

Teaching notes for graphing

Types of data

The two main types of data that can be collected—discrete and continuous—have a significant impact on what can be constructed. While this aspect of data is not taught to students in primary school, teachers need to be very cognisant of it so that they will not create misconceptions among students, who may then have to relearn it in their later years of schooling.

Discrete data

Discrete data is that which falls into one category or another. Examples include types of cars (Toyota or BMW), flavours of ice-cream (strawberry or chocolate), grades of enrolment (Year 4 or Year 5), and so on. Data that belong to one category or the other are presented in *bar graphs*—graphs where the columns do not touch.

In the early years of schooling, data are often collected around themes that appear to be discrete, such as the season or the colour of hair or eyes. These categories may *appear* to be discrete, since the season is either autumn or winter, or the hair is either black or brown, but this is not entirely accurate. A category such as a season may commence on the first of the month in some countries but on the equinox in other countries. Thus, an arbitrary decision is being implemented in regard to category.

Continuous data

The second category is *continuous data*. This type of data relates to information that exists along a continuum, with arbitrary categories being made at some point. The data are usually something that can be measured—time, length, temperature, and so on. When constructing graphs of ages of students in classes, for example, categories of 5, 6 and 7 years might be used. This may appear to be discrete—you are either 5, 6 or 7. However, on closer examination, we can see that the ways in which the construct of 5 is made can vary. Is it when someone turns 5, or just before they turn 6? Is it somewhere between 4½ and 5½? Is it how old they are on a particular day? From 4 years 10 months to just before 5 years 10 months? Each of these categories means that the data concerning the students in a class can be organised in different ways. The reason here is that the construct of time is continuous: to make sense of the 5 years category, we need to make decisions, and cut-offs along a continuum, to establish how 5 will be defined. Since the data are continuous and the categories are 'slippery'—depending on how the construct is defined—the corresponding column graph is a histograph (or histogram), where the bars or columns touch.

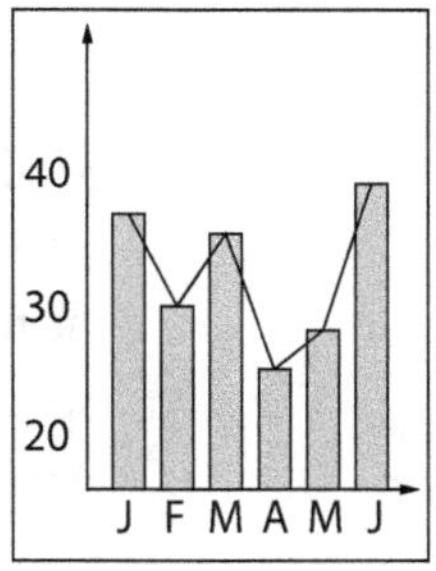

Rainfall in semester

The unique feature of histographs is that the bars are related to one another, and hence can be compared over a period. This means that this column graph can be converted to a line graph should the data need to be considered over time. This conversion is not possible with a bar graph.

Purposes of graphs

In and of themselves, graphs are not entirely 'free choice' in nature. The type of graph possible is influenced by the data that have been collected and the purposes for which it was collected. When using graphs to represent data, the purpose of the graph is an important consideration, and different graphs serve different purposes. Graphs are used to display information quickly and easily, so using the right type of graph for the purposes of the display is important. For example, suppose a class collected data on lunches because the unit of work being undertaken was on healthy lifestyles, with cross-curricular links. If the purpose of the task was to compare the different types of food items eaten, a column graph would be the most appropriate, since it would allow for quick reckoning of which lunches were eaten most frequently, least frequently,

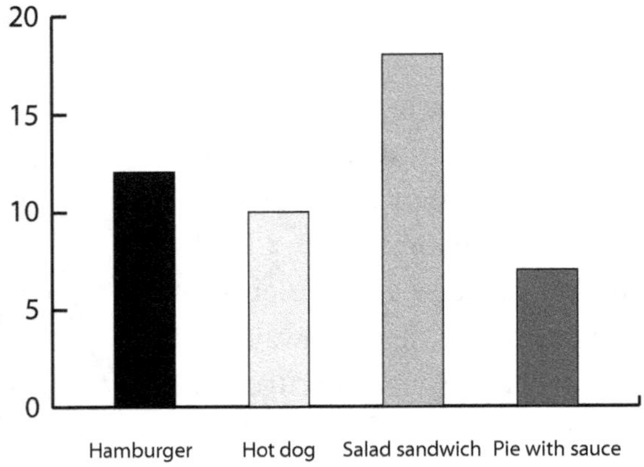

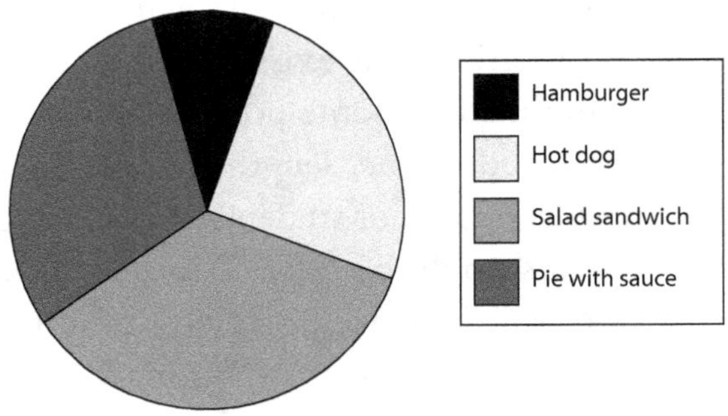

Lunches eaten in Ms W's class

Using spreadsheets to construct graphs

and so on. If the task was to consider what the class ate overall, a column graph would not allow this interpretation easily, and a pie or circle graph would be far more useful in conveying the information.

Digital technology

Just as the calculator has a unique role to play in developing students' number and computational sense, the use of computers (and particularly spreadsheets) has the potential to radically impact the teaching of statistics. The construction of graphs has often been an arduous task, akin to the task of calculation in the number strand. The use of technology can support and enhance statistical and graphical understandings. Once students have developed a sense of the types of graphs, and their construction, interpretation and purpose, the task of constructing them can be replaced through the use of technological tools such as the spreadsheet. This frees up class time for more complex and meaningful activities.

Entering data into the spreadsheet, then making key decisions regarding the type of graph most useful for displaying the data, requires high levels of mathematical reasoning. This, in turn, allows a greater amount of teaching time to be spent on developing the appropriate thinking skills associated with construction and interpretation of data. Being able to estimate and evaluate whether the graph constructed is correct demands that students are familiar with the processes associated with data representation. Decision-making processes are informed by students' knowledge, and hence demand the appropriate levels of understanding. The NCTM has created Data Grapher (www.nctm.org/Classroom-Resources/Illuminations/Interactives/Data-Grapher/), a tool that can assist statistical understanding while removing some of the tediousness of creating data representations by hand.

As in the debates associated with the use of calculators in number work, some sectors of the community may argue that the use of digital technology 'de-skills' students; however, there is a substantive body of research that counters this position (Asp et al. 1993; King 1993; Vincent 2000).

Teaching probability and statistics in the middle years

In this chapter, the importance of extensive experiences for promoting probabilistic reasoning and statistical literacy have been emphasised. The early-years learning experiences must extend students beyond the 'birthday graph' to thinking about reasons why data are collected and what the data tell us, as well as engaging students in discussing chance outcomes. In the middle years, learning experiences must be opportunities for capitalising on promoting meaning for key probabilistic terms, such as randomness, variation and independence, and extending the critical numeracy skills associated with questioning data.

When students come to the middle years, they often carry with them misconceptions or naïve conceptions about chance events. They need to be immersed in a range of chance activities so that they can make conjectures and test them out. Common errors associated with the language of chance are highlighted when students are asked to describe the likelihood of an outcome when there are more than two possible outcomes and they respond 50/50 (Tarr 2002; Watson 2005). Quantifying events as 50/50 indicates students' awareness of measuring chance, but not a movement to the mathematical analysis required.

Another common misconception relates to the randomness of sequences of outcomes and independence of events. When a coin is tossed three times and it has come up tails all three times, many students will state that the next one will be heads because 'it is the heads' turn'. Because calculation of mathematical probabilities (theoretical probability) rarely matches actual outcomes (experimental probability), students need to reconcile experimental and theoretical probability outcomes by building understanding of the notion of 'in the long run'. This is where digital resources such as www.random.org are very useful. Real-world examples can be drawn on to stimulate discussion about theoretical and experimental probability outcomes in relation to, for example, insurance premium calculations, football betting, horse racing, health outcomes, and so on. Making probability real and relevant promotes students' appreciation of probability and its direct application to our daily lives.

A further common misconception is in relation to populations and sampling. Research shows that many middle-years students are unsure about representative sampling. When asked about the best method to survey students in a school by either (a) putting all 600 students' names in a hat and drawing out 60, (b) surveying 60 friends, or (c) asking volunteers to participate in the survey as they approach a specific location in the school (tuckshop), only 12 per cent of Year 7–12 students recognised choice (c) as a potential source of bias, and one-third of students thought (b) was a good method (Shaughnessy 2007). To promote statistical thinking in the middle years, population and sampling are important. Other key teaching points in relation to statistics, as suggested by Shaughnessy (2007), are as follows:

- *Emphasise variability*—Rather than just describing sets of data in relation to mean (median and mode) scores, the outliers are also important to consider the variability in the data.
- *Compare data sets*—Students should be immersed in a range of opportunities to collect data that they analyse 'by eye' (rather than determining measures of central tendency—mean, median, mode and other data-analysis methods).

In relation to probabilistic reasoning, the big ideas are 'chance variation, randomness, independence (and their complementary elements stability, regularity and co-occurrence)' (Jones et al. 2007).

Review questions

14.1 List the general sequence for teaching about the creation of graphs. Provide illustrations of the sequence.

14.2 Discuss the difference between theoretical and experimental probability.

14.3 When is it more useful to use theoretical probability as opposed to experimental probability in the teaching of probability? Provide illustrative examples to justify your argument.

14.4 Students often develop misconceptions about the independence of events. Select a situation in which the events are independent and discuss how you would go about teaching this concept so as to overcome misconceptions.

14.5 Identify particular instances of when different measures of central tendency are more appropriate than others.

14.6 Discuss the value of using digital technology to support learning about probability and statistics.

Further reading

Aspinwall, L. & Shaw, K. (2000). Enriching students' mathematical intuitions with probability games and tree diagrams. *Mathematics Teaching in the Middle School*, *6*(4), 214–20.

Callingham, R. (2000). Come in spinner! Innovative assessment in the primary classroom. *Australian Primary Mathematics Classroom*, *5*(1), 9–14.

Feicht, L. (1999). Making charts: Do your students really understand the data? *Mathematics Teaching in the Middle School*, *5*(1), 16–18.

Mulligan, J. & Bobis, J. (1998). Investigating data exploration. *Reflections*, *23*(1), 19–22.

Perry, B. (1998). Frameworks for statistical thinking in primary schools. *Reflections*, *23*(1), 85–8.

Quinn, R. (2001). Using attribute blocks to develop a conceptual understanding of probability. *Mathematics Teaching in the Middle School*, *6*(5), 290–4.

Ritson, R. (2000). A question of choice. *Australian Primary Mathematics Classroom*, *5*(3), 10-14.

Stephens, M. (1998). Using data representations to ask hard questions. *Australian Primary Mathematics Classroom*, *3*(4), 23–7.

Tarr, J. (2002). Providing opportunities to learn probability concepts. *Teaching Children Mathematics*, *8*(8), 482–7.

Watson, J. (2007). The foundations of chance and data. *Australian Primary Mathematics Classroom*, *12*(1), 4–7.

Wiest, L.R. (1998). In the face of uncertainty. *Australian Primary Mathematics Classroom*, *3*(4), 19–22.

References

Asp, G., Dowsey, J. & Stacey, K. (1993). Teaching mathematics with technology: Computer spreadsheets and graphing applications. In J. Mousley and M. Rice (eds), *Mathematics: Of primary importance* (pp. 86–92). Melbourne: Mathematics Association of Victoria.

Batanero, C. (2015). Understanding randomness: Challenges for research and teaching. In K. Krainer and N. Vondroga (eds), *CERME 9—Ninth Congress of the European Society for Research in Mathematics Education* (pp. 34–49). Prague: Congress of the European Society for Research in Mathematics Education.

Busse, A. (2011). Upper secondary students' handling of real-world contexts. In G. Kaiser, W. Blum, R. Borromeo Ferri and G. Stillman (eds), *Trends in teaching and learning of mathematical modelling* (pp. 37–46). Dordrecht: Springer.

Jones, G., Langrall, C. & Mooney, E. (2007). Research in probability: Responding to classroom realities. In F. Lester (ed.), *Second handbook of research on mathematics teaching and learning* (pp. 909–56). Charlotte, NC: Information Age.

King, M. (1993). Data handling using spreadsheets in primary schools. In J. Mousley and M. Rice (eds), *Mathematics: Of primary importance* (pp. 80–5). Melbourne: Mathematics Association of Victoria.

Luke, A. & Freebody, P. (2001). *Report of the literacy review for Queensland state schools.* Brisbane: Education Queensland.

Makar, K., Bakker, A. & Ben-Zvi, D. (2011). The reasoning behind informal statistical inference. *Mathematical Thinking and Learning, 13*(1–2), 152–73. doi:10.1080/10986065.2011.538301.

McHugh, L., Kelly, A.M. & Burghardt, M.D. (2017). Teaching thermal energy concepts in a middle-school mathematics-infused science curriculum. *Science Scope, 41*(1), 43–50.

Shaughnessy, J.M. (2007). Research on statistics learning and reasoning. In F. Lester (ed.), *Second handbook of research on mathematics teaching and learning* (pp. 957–1010). Charlotte, NC: Information Age Publishing.

Steen, L.A. (1999). Numeracy: The new literacy for a data-drenched society. *Educational Leadership*, October, 8–13.

Tarr, J. (2002). The confounding effects of '50–50 chance' in making conditional probability judgments. *Focus on Learning Problems in Mathematics, 24*, 35–53.

Vincent, J. (2000). *Computer enriched mathematics for Years 5 and 6.* Melbourne: Mathematics Association of Victoria.

Watson, J.M. (2005). Variation and expectation as foundations for the chance and data curriculum. In P. Clarkson, A. Downton, A. McDonough, R. Pierce, A. Roche, M. Horne, D. Gronn (eds), *Building Connections: Theory, Research and Practice: Proceedings of the 28th Annual Conference of the Mathematics Education Research Group of Australasia* (pp. 35–42). Sydney: MERGA.

Watson, J.M. & Moritz, J.I. (1998). Longitudinal development of chance measurement. *Mathematics Education Research Journal, 10*(2), 103–27.

CHAPTER 15

GEOMETRY

The topic of geometry includes developing students' knowledge about aspects such as two-dimensional shapes and three-dimensional objects, symmetry, tessellations, transformation and geometric reasoning. Geometry also includes learning about position, location and arrangement in relation to maps, plans, scaled drawings and grid systems. Spatial knowledge extends from learning the simple language of position and arrangement in the early years—inside, outside, near, far, and so on—to using a map to determine distances in the later years. Also important in the topic of geometry is developing students' spatial visualisation and spatial reasoning abilities, as well as making links to natural and built environments. Learning activities in this topic can involve exploration and investigation of buildings, bridges, consumer products, shapes and symmetry in plants and animals, and shapes and functions of natural and manufactured (constructed) objects. Everyday tasks such as map reading and furniture arrangement, and the construction of items from plans, involve spatial reasoning. The geometry topic also links to many occupations—architecture, landscape gardening, golf-course design, civil engineering, surveying, town planning, painting, crafts, art, building and dress designing. It is important to make connections from space to measurement (Battista 2010; Trafton & Hartman 1997) and number topics, and to other curriculum areas such as art, craft, physical education, science and social studies.

A large body of research continues to highlight the link between early spatial reasoning and later performance in mathematics (Verdine, Golinkoff, Hirsh-Pasek & Newcombe, 2017). This is an important consideration given the emphasis—perhaps overemphasis—on number in many school curricula. Spatial reasoning is considered critical for everyday tasks, and helps us to understand, appreciate and interpret our geometric world (NCTM 2000). Mix and Cheng (2012) suggest that the link between spatial and mathematics ability is so well established that it no longer makes sense to ask whether developing spatial reasoning is related to mathematical learning. Given this, it is encouraging to note that school mathematical curricula are becoming increasingly spatial and graphic in nature, and are moving away from predominantly word-based tasks to include a range of quantitative information (Lowrie, Logan, & Ramful, 2017). In Australia, spatial reasoning is identified as a core component in the Numeracy general capability within the Australian Curriculum, and thus is to be incorporated into all subjects across the curriculum. The good news for teachers is that a large body of evidence suggests that most spatial skills are malleable and that these skills can improve as a result of spatial experiences, both more generally and as a result of specific, intensive training (Stieff & Uttal 2015).

The geometry strand is one of the richest in terms of language. Developing a range of strategies to support learning of the spatial language is important. Creating posters, books (including big books, for early-years learners), word walls, flash cards, dictionaries, word games, and so on are useful and practical strategies for supporting learners to better understand the nuanced language of this strand. Words in this strand have specific meanings in relation to spatial concepts that are different

TEACHING TIP

As spatial language is so important, create a word bank for the topic area you are covering. These should be created to suit students' needs.

- e.g. simple—triangle, square, rectangle, circle, sides, angles, line, slope
- e.g. more complex—quadrilateral, rhombus, obtuse, acute, right angle, parallel, converging, gradient, convex, concave, regular, irregular.

to other strands of the mathematics curriculum, as well as having non-mathematical meanings. A good example of this is the term 'base', which in the geometry strand may refer to the bottom edge of a triangle or the bottom face of a prism, whereas in the number strand 'base 10' refers to a particular counting system. In the non-mathematics register, the term 'base' can have many meanings, such as 'base camp' or someone being 'based' (located) at a particular site.

There are many dynamic geometric programs available on computers and handheld devices. The potential of these tools to create new understandings appears to be more significant than the traditional pencil-and-paper approaches of the past (Chorney & Sinclair 2018). Use of dynamic geometry software has become an accepted part of contemporary mathematics education, and so this needs to be incorporated in teaching (Jones & Tzekaki 2016). However, as educators (Larkin & Milford 2018) caution in their work with apps, not all software is a panacea for good teaching, so a critical analysis of the packages available is essential.

Topics in the geometry strand

Topics within the geometry strand include location (position, direction, coordinates), networks (maps, scale drawing, topology), three-dimensional objects (surface, corner, edge), two-dimensional shapes (sides, angles, polygons), symmetry (line symmetry, rotational symmetry), tessellations (tiling, design, art applications), dissections (puzzles), similarity (shape, perspective) and transformations (flips, slides, turns). Many of these topics are interrelated, and the geometry strand can thus be organised in different ways. In this chapter, we have loosely organised the topics into the categories of shape and structure, transformation and symmetry, and location and arrangement.

Visualisation

Being able to visualise images is an important foundation for problem-solving in mathematics (Lowrie et al. 2017). Developing a strong facility for visualisation will result in an increased capacity to solve problems. Visualisation is fostered through the study of geometry, so developing spatial sense is very important.

The study of geometry assists the development of visual processing—the capacity to create, manipulate and transform spatial images in the mind. The exercise in the diagram requires students to be able to visualise the rectangular image and rotate it in order to respond to the question. Visualisation also assists in interpretation of information represented in diagrams, graphs and maps. Lowrie and colleagues (2017) have found that graphical literacy—the capacity to read diagrams, maps and graphs—is an important component of the geometry strand, but that it is a skill that needs to be developed, particularly from around the middle years of schooling (i.e. from Year 4 onwards). Their research has indicated marked gains in mathematics achievement following an intervention that developed students' abilities in mental rotation, spatial orientation and spatial visualisation (Lowrie et al. 2017). Lowrie and Diezmann (2012) report that in the middle years, there are gender differences in graphical literacy of which teachers also need to be cognisant, and it is important to develop strategies to enhance these skills and redress any differences.

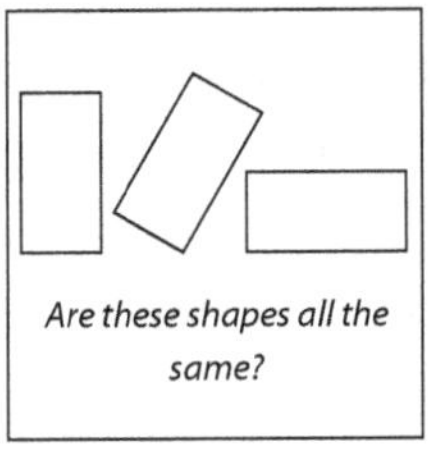

Creating, manipulating, transforming images

Visual processing is a key skill in mathematical problem-solving across the strands. It is recognised that the teaching of spatial concepts and processes is the key to students developing their visualisation and spatial sense (Pegg & Davey 1989). The most effective problem-solvers have a strong visual sense, as well as verbal, symbolic and abstract thinking. Throughout their study of geometry, students should continually be challenged so that they do not form inflexible images. An example of an inflexible image is when students correctly identify a square when it is aligned horizontally but insist that it is a diamond when it is tilted on one corner. Students holding inflexible images often cannot identify as triangles forms other than an equilateral triangle. In a similar vein, triangles with their bases uppermost may be regarded as 'wrong' or 'upside down'. Thus, in order to enhance visualisation skills, early shape recognition activities must provide students with a range of forms and a variety of orientations for each particular shape. Teachers need to be able to recognise and articulate the properties of shapes and solids, including examples and non-examples (Tsamir et al. 2015).

The van Hiele levels of geometric thinking

The most cited and used theory underpinning the teaching of geometry is that proposed by Dutch researchers van Hiele and van Hiele, who note that students' geometric thinking can typically be represented by key characteristic styles, which they refer to as 'levels'. According to this theory, students' progression across the levels occurs in response to instruction, emphasising the impact of teaching on students' thinking. (The levels are not intended in the sense of broad developmental or biological progression.) The theory provides a useful model for planning teaching (Howse & Howse 2015). The model has five levels, but for the primary-school teacher, the first three are most relevant. When teaching geometric thinking, it is useful, particularly in the primary-school years, for students to have access to manipulatives such as attribute blocks (Howse & Howse 2015).

Level 1: Visualisation

At level 1, students' thinking about shapes relates solely to what they see. Students react to figures as wholes—a square is a square because it is a square. Similarly, an angle is an angle because it is an angle. Students do not think about the components of a shape or its properties. They might regard a square that has been rotated through 45° as a diamond—that is, they no longer see it as a square. Students at this level justify a circle being a circle because 'it looks like a circle' (Ferguson 2017).

Level 2: Analysis

At level 2, students begin to focus on the properties of shapes and to informally analyse the components of shapes—a square is a square because it has four equal sides and four right angles. However, they do not focus on relationships between properties.

Level 3: Abstraction or informal deduction

At level 3, students order properties logically, form abstract definitions and identify necessary and sufficient conditions associated with properties. For example, a square has some properties related to other shapes—a square is both a special rectangle and a special rhombus; if the opposite angles of a quadrilateral are equal, then the opposite sides are parallel; if I know that a parallelogram has one right angle, then it must be a rectangle; a triangle with three equal sides must have three equal angles.

Level 4: Deduction

Level 4 involves formal reasoning, where axioms, terms, underlying logical systems, definitions and theorems become the focus. This level is commonly associated with high-school geometry, and includes developing formal arguments and proving theorems.

Level 5: Rigour

Level 5 involves comparing systems based on different axioms—for example, Euclidean geometry versus spherical geometry. Typically, the study of geometry associated with this level of thinking is undertaken beyond the high-school years.

The importance of the van Hiele model

Central to the van Hiele model is that progression in geometric thinking is due to effective teaching, and is not necessarily associated with increasing age or maturation—effective teaching is the essential element. By focusing on the different ways in which students think geometrically, the van Hiele model allows teachers to identify how students are conceptualising geometric ideas, and how best to then plan and organise effective learning so that students can progress to more advanced thinking (Alex & Mammen 2016). For example, a student working at level 2 may not be able to see that a square is a special kind of a rectangle. By recognising that the student is not seeing interconnections between properties, the teacher is able to organise appropriate learning experiences that have the goal of enabling the student to see connections between the properties of rectangles and of squares. In this way, students can become aware that a square is a special kind of a rectangle—a rectangle with the additional property that its adjoining sides (as well as its opposite sides) are equal.

For students in the primary years, the first three levels of the van Hiele model are generally very applicable. Using the model to carefully plan teaching can lead to enhanced understandings of geometry by students. However, as with any model, it should not be used to constrain learning and learners. If students are working beyond the first three levels, then teaching should scaffold students beyond these levels.

Teaching notes

The research of van Hiele and van Hiele has informed much of the current teaching of geometry. Central to their work is the important role of teaching of geometric ideas. They argue that the complexity in students' geometric thinking arises from the experiences provided by teachers. Geometry activities should link to and draw from the everyday world and build on students' experiences with spatial aspects of their environments. Geometry activities should be multisensory, informal before formal, and provide ample opportunities for the development of visualisation.

Appropriate geometric activities should use and develop important skills such as:

- *visualising*—manipulating and transforming spatial images in the mind; interpreting figural information; interpreting and reading diagrams or maps
- *communicating*—developing accurate and precise geometric language, both spoken and written (start with children's natural language and gradually develop more formal terminology)
- *drawing and modelling*—sketching and modelling with appropriate accuracy
- *thinking and reasoning*—asking students to give reasons for their answers; 'Why did you say that?' 'How do you know that?' 'What evidence is there for that?'
- *applying geometric concepts and knowledge*—drawing attention to real-life applications; deriving problems from real-life situations; considering relationships between shape and function.

Spatial concepts: early activities

Play activities in the early years are important for developing spatial concepts. Rather than being random, play activities should be planned carefully and their purpose should be clear. Early activities include:

- *sorting*—according to students' criteria or criteria suggested by the teacher (e.g. rolling or sliding; height, pointiness; games of identifying which one is different or which one is the same)

- *building*—Lego, blocks, junk; making the tallest or widest structure; making a symmetrical building; making an enclosure for this object
- *packing*—placing materials neatly into storage boxes
- *modelling*—with plasticine, clay, straws and string, cardboard and tape or rubber bands, toothpicks
- *touching*—using touch to identify a shape (feely bag); describing the object to others
- *matching*—shapes to diagrams; matching a picture and a model; building a three-dimensional structure from its picture; matching containers with models
- *drawing*—solids from different viewpoints; identifying and drawing shapes after walking through the park, playground or garden; discussing reasons particular shapes have been used for various structures
- *visualising*—replicating structures behind a screen that have been shown for a few seconds.

Shape and structure

Two-dimensional shapes

The study of two-dimensional shapes relates to topics such as polygons, lines and angles. Polygons are closed, two-dimensional shapes with three or more straight sides. The word 'polygon' comes from two Greek words: *poly*, meaning many, and *gon*, meaning sides. Hence, a polygon is a many-sided, closed figure. Polygons are a type of plane shape, since they lie on the two-dimensional plane. The name of a specific polygon includes a prefix determined by its number of sides (see the following table).

Polygons are classified as regular or irregular. Regular polygons have all sides equal, and therefore all angles equal. An irregular polygon has either some or all of its sides of different lengths, or some or all of its angles of different sizes, or both sides and angles differing. In some cases, students might be presented with polygons that are almost exclusively regular. Limiting students' experiences in this way can result in their having an impoverished notion of polygons. They might have difficulty in naming certain polygons (Ozkan & Bal 2017). Thus, it is important that students engage in activities with both regular and irregular polygons.

In the early years, the focus is on assisting students to identify common shapes such as squares, triangles and circles. (Note that a circle is not a polygon, since a polygon must have only straight sides.) Richer conceptual

Names of some polygons

No. of sides	Prefix	Name
3	tri-	triangle
4	quad-	quadrilateral
5	penta-	pentagon
6	hexa-	hexagon
7	hepta-	heptagon
8	octa-	octagon
9	nona-	nonagon
10	deca-	decagon
...	...	...
many	poly	polygon

understanding is promoted through activities that direct students to consider properties of shapes, and by providing a variety of examples of particular polygons—both regular and irregular—in various spatial orientations (Seah 2018). Sorting activities are useful for preventing the development of inflexible conceptualisations of particular shapes.

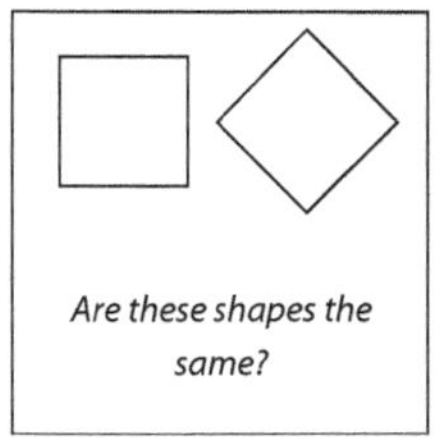

Visualising shapes

When leaning about basic geometric shapes, students need to become familiar with shapes presented in a range of orientations. If students' experience of two-dimensional shapes is mainly limited to shapes in an upright orientation (i.e. where the base line is on the horizontal orientation), this can result in misconceptions. For example, students may regard only one of the two squares in the diagram as a square. Therefore, it is important that students are presented with examples of basic two-dimensional shapes in a variety of orientations.

7-piece tangram

Tangrams are inexpensive and useful in the teaching of two-dimensional shape. Tangram activities not only support the learning of the names of shapes but also help students to develop spatial visualisation. Various tangram puzzles are available, but a seven-piece tangram is most commonly used (van Hiele 1999). The tangram allows students to make many shapes, both prescribed and free-form. It also allows them to explore aspects of transformational geometry as they create or copy shapes.

TEACHING IDEA

Geoboard activities

Exploring lines

- Make the shortest line segment possible on your geoboard.
- Make the longest line segment possible.
- Make three line segments of different lengths.

Exploring angles

- Make a narrow angle.
- Make a wide angle.
- Make an angle like the corner of a square.
- Make an angle with 2 nails between the rays.
- Make an angle with 0 nails between the rays.

Exploring regions

- Construct a region that has just one nail inside it.
- Construct a region that has 0 nails inside it.
- Construct a region that has 1 nail outside it.
- Construct a region that has 1 nail outside and 3 nails inside it.
- Construct a region that has 3 nails on the boundary.

Line and angle

Learning about line and angle requires students to be engaged in constructing a variety of lines and angles and discussing them (Host et al. 2014). Geoboards are a valuable resource for investigating lines and angle. Geoboards are wooden (or plastic) boards with nails protruding in a grid pattern. They are frequently in a square array, but can be in circular patterns as well. A variety of two-dimensional shapes can be readily created on a geoboard using rubber bands, and language associated with shapes, lines and angles can be fostered (Britton & Stump 2001).

The geoboard is also a valuable resource for constructing and learning about angles. Many students develop misconceptions about angles (Mitchelmore 2000), particularly in relation to the size of the arms of an angle versus the size of an angle. The notion of angle relates to measuring an 'amount of turn'. A simple way to introduce students to the idea of angle is to ask them to stand with their arms outstretched in front of them and then to open one arm to the side through an angle of 90°. This can be linked to turning all the way around, which is a turn through 360°, and turning halfway around, a turn of 180°. Outstretched arms can be dropped to the side of the body and turning can continue in order to emphasise that the notion of angle relates to an *amount of turn* rather than the length of the arms or the length of the lines making an angle.

TEACHING IDEA

Sorting shapes

- Which of these shapes are triangles? Put a cross on the shapes that are triangles.
- Cut out the shapes and sort the triangles into one group. Explain why you sorted the shapes in that way.

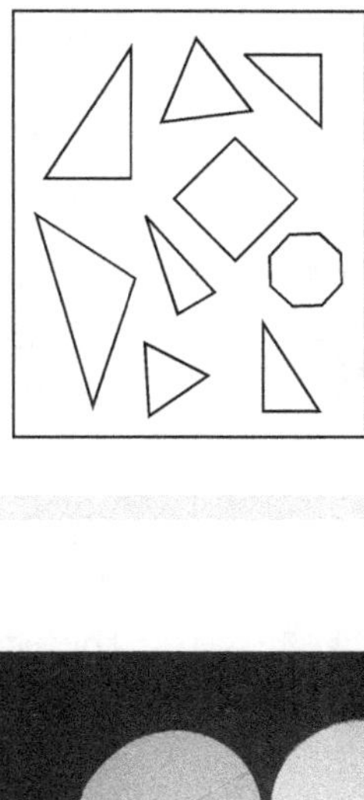

To focus further on angle (e.g. the angles made by the hands of an analogue clock), an angle tester can be created. This is simply a circle of paper folded into four quarters. The corner created is a right angle and can be used to find angles that are exactly 90°, less than 90° and greater than 90°. An angle wheel (see the diagram) can be constructed using two circles of equal size cut from different coloured pieces of paper. By cutting a split to the mid-point of both circles, fitting the circles together and

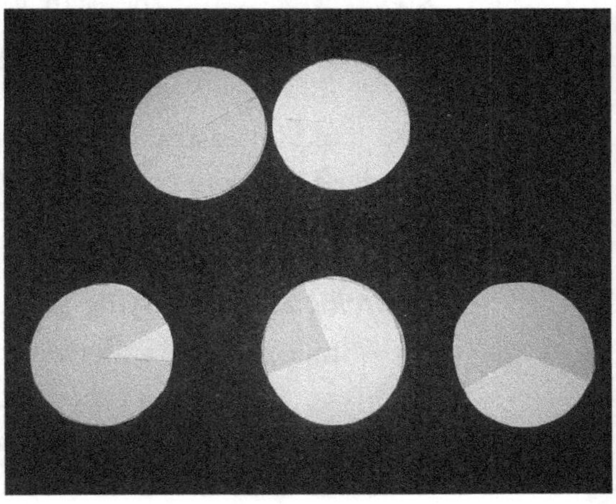

Angle wheels

turning them, parts of each circle can be seen, from either side, forming angles of different sizes. Angles can initially be classified as acute (sharp), right-angled (like the corner of a square), obtuse (blunt), and straight (flat line). A website for developing two-dimensional understanding of shapes and angles can be found here: www.mathopenref.com/planegeometry.html. The site is interactive; for example, students can create various angles, or create various polygons using a drag and drop action.

Three-dimensional objects

The topic of three-dimensional (3D) objects focuses on learning names and properties of solids (3D objects), relationships among solids and visualisation skills associated with this topic. Learning about solids can include learning about 3D objects that are important in geometry (e.g. cubes, spheres, pyramids, prisms, cylinders and cones), and linking this knowledge to objects that occur in the natural and built environments. For example, a block of wood might have the shape of a prism, while a basketball has the shape of a sphere and a funnel the shape of a cone. Activities involving the construction of solids can support students' development of spatial knowledge (Ambrose & Falkner 2002).

Many syllabi list the topic of 3D objects before the topic of two-dimensional (2D) shapes. This is because young students can think more easily about solids than they can about planar (two-dimensional) shapes. Solids can be regarded as more experientially real for young students because they are essentially objects that can be handled easily, whereas the notion of a 2D shape is more abstract. Indeed, students are often confused when teachers pick up an object and call it a shape.

Naming and classifying solids

It is important to understand the nomenclature of solids. There are parallels between the nomenclature of solids and that of plane figures, and teachers should be aware of the links between these two naming systems. The term 'solid' is used in an inclusive sense—any 3D object, whether it is 'hollow' or not, is a solid. Objects in the sub-group of solid figures that have flat surfaces (faces) only are referred to as *polyhedra* (singular:

'polyhedron'); the word 'polyhedra' comes from two Greek words: *poly*, meaning many, and *hedra*, meaning faces. Just as circles and semi-circles are 2D shapes that are not polygons, so spheres, cylinders and cones are examples of solids that are not polyhedra (because of their curved surfaces). Examples of polyhedra are cubes, rectangular prisms and pyramids.

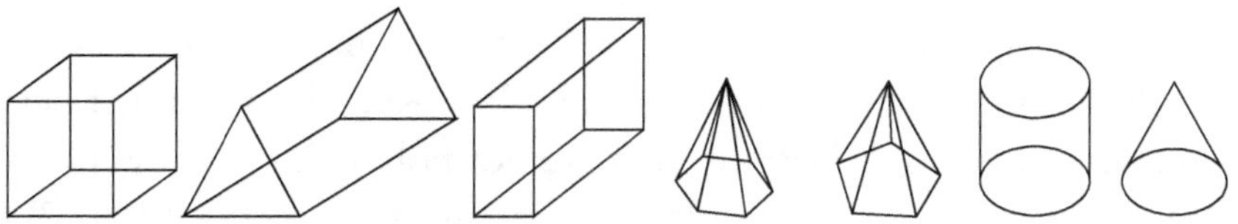

Polyhedra (prisms and pyramids) and solids (cones and cylinders)

Prisms and pyramids

Look at the diagram of solids above. Compare the prisms with the pyramids. What do you notice? The first thing you will likely notice is that the pyramids form a point. Pyramids are named by the two-dimensional shape of their base. All other sides of the shape are triangles. Prisms have at least one pair of opposite faces of particular polygons (e.g. rectangle, triangle, pentagon, hexagon), and the remaining four faces are rectangles.

TEACHING IDEA

Sorting solids

Which of these solids roll and which of these solids do not roll?

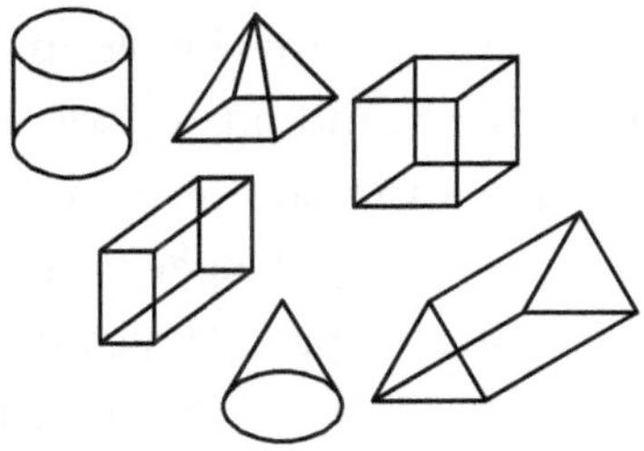

Nets of solids

The notion of a 'net' is the basis of an important activity in geometry that helps to connect ideas in 2D and 3D space. The net of a solid is a plane shape that can be folded up to make a hollow representation of the solid. Early learning activities for nets commence when students take a cardboard box (such as a biscuit box) and unfold it. They can see how the 3D biscuit box was constructed from a flat 2D shape. The specific shapes that were used to construct the 3D box can be seen readily. Students can then be challenged to create their own net to create a new box. More advanced activities can involve students drawing nets for a range of solids, then using them to construct the solids. Keep in mind that nets can be made for some solid figures that are not polyhedra (e.g. cylinders and cones). It is important for students to realise that a net is a flat, 2D shape that is folded to create the 3D object. Even though the net can be seen to consist of several 2D shapes, they are connected together in one flat shape ready for folding. A very useful, research-based app that can support understanding of nets of a cube is Click the Cube (https://apps.apple.com/au/app/click-the-cube/id1157365733). When using the app, students can manipulate the cube to 'see' the net being formed.

Nets for pyramids and prisms

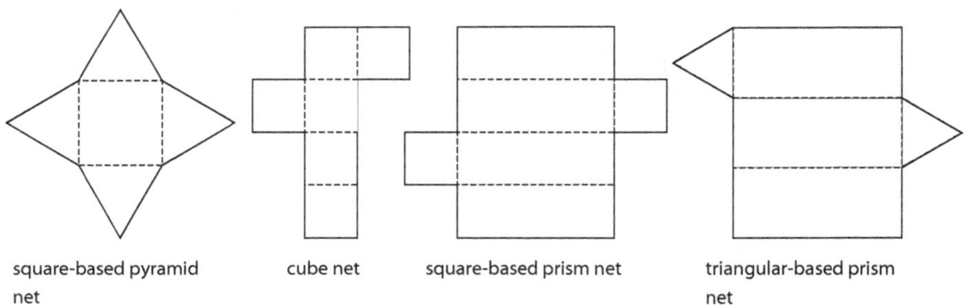

Solid, hollow and skeletal forms

Use of a wide array of resources is invaluable in creating opportunities for students to construct understandings of solids and their properties. Seeing the same solid in three different forms (solid, hollow and skeletal) is a key teaching strategy to develop deeper understandings of 3D shape properties. Teaching aids for 3D space typically include a set of solids made of wood. Hollow objects—that is, objects made from nets or from connecting plane figures—enable students to focus on the number of flat surfaces of the shape (i.e. its faces). Skeletal figures can be made easily using straws or toothpicks connected with marshmallows or Blu Tack®. These shapes promote familiarity with the language used for 3D objects, such as vertices, faces and edges. When making hollow objects, students are working with each separate flat surface of the object—its faces. When making skeletal models, the 'skeleton bones' clearly highlight the edges of the object, and the connectors (marshmallows or Blu Tack®) emphasise the vertices.

Students making skeletons with straws and Blu Tack®

Euler's formula

To expand students' understanding of 3D solids, an exploration of Euler's formula provides an interesting investigation. Euler (pronounced 'Oiler') was a mathematician of the eighteenth century. Provide students with a range of 3D polyhedra (remember: 3D solids with curved surfaces are not polyhedra, so don't include cones and cylinders) and have students create a data table of the number of vertices, faces and edges of each of the shapes. After they have completed their table, ask them whether they can see a relationship between the number of vertices, faces and edges. They will see that, for each shape, the number of vertices plus the number of faces is equal to the number of edges minus 2 (V + F = E – 2).

	Vertices	Faces	Edges
Cube	8	6	12
Rectangular prism	8	6	12
Square-based pyramid	5	5	8
Triangular-based pyramid	4	4	6
Triangular-based prism	6	5	9

Transformation and symmetry

Transformation and symmetry are interrelated topics. *Transformation*, as the name implies, is about altering a shape in some way. This could be changing its position in space (congruent transformation), its size (projective transformation) or its features (topological transformation).

The terms 'object' and 'image' are often used in association with transformation. The figure prior to the transformation is referred to as the *object*; after the transformation, it is referred to as the *image*. *Symmetry* includes line symmetry and rotational symmetry in 2D space, and plane symmetry and rotational symmetry in 3D space. The topic of tessellations (or tiling) is an aspect of 2D space that draws on notions of both transformation and symmetry, and can also reinforce conceptual understanding of this area.

Congruent transformations

A shape transformation that leaves the original shape unchanged but alters its position is called a rigid or congruent transformation. Congruency is often a source of confusion for learners due to the way in which it has been taught (Plaksin & Patkin 2016). Providing experiences with various shapes (triangles and quadrilaterals, in the first instance) assists in building understandings of congruence. A rigid transformation occurs when a shape is reflected, translated or rotated from its original position; in some curricula, these transformations are referred to, respectively, as *flips*, *slides* and *turns*. In congruent transformations, the image is always congruent with the object. The transformations of flip, slide and turn relate closely to the topics of symmetry and tessellations (see below).

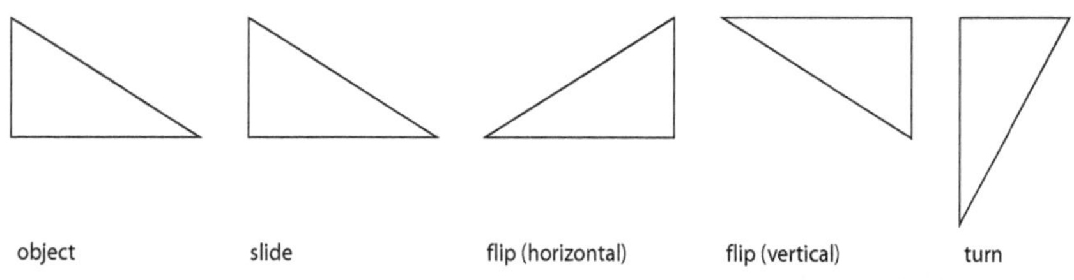

Transformations of flip, slide and turn

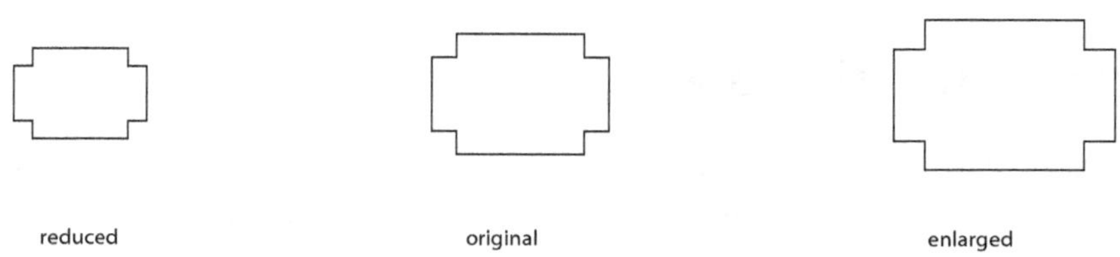

Transformations of enlargement and reduction

Projective transformations

Projective transformations are enlargements (dilations) or reductions of the original object. An enlargement (dilation) is a transformation whereby the object is made larger while maintaining the original shape. A reduction is a transformation whereby the object is made smaller while maintaining the same shape. Thus, enlargements and reductions preserve shape, but not size. They also preserve directions of lines and orientation. Projecting shapes onto a wall and tracing the image, and then comparing this with the original object, provides a great way to introduce students to this type of shape transformation. Shadows can be linked to explorations of projective transformations.

Topological transformations

Topological transformations involve stretching and bending without breaking or tearing. Properties such as lengths of lines, sizes of angles and straightness are ignored. An object and its image under a topological transformation are said to be topologically equivalent. Thus, a quadrilateral,

TEACHING IDEA

Topological shapes

Shapes can be classified topologically by the number of holes in their surface:

- Start with a lump of playdough with no holes in its surface. Make it into a ball shape. Now make it into a cube shape.
  1. Did the amount of playdough change from what you started with?
  2. Are the shapes the same topologically?
- Start with a lump of playdough and put a hole through it (use a pencil). Make a doughnut, and then the number 9 out of the playdough.
  1. Does each shape have just one hole?
  2. Did the amount of playdough you started with change?
  3. Are the shapes topologically the same?
- With your playdough, make the digits 0 to 9. Classify each one topologically.
- Find out if all the letters in your name belong to one particular topological category.
- Is it possible to have a given name in which all the letters are from one topological category?

for example, is topologically equivalent to any other polygon. A soccer ball that is not inflated is close to a flat shape. When inflated, it takes on the (spherical) shape of a soccer ball. The flattish shape and the air-filled shape are said to be topologically equivalent.

A lump of plasticine is an ideal starting point for exploring topological geometry. Rolling the plasticine into a sphere, then changing its shape into a cube demonstrates that these two shapes are topologically equivalent. Other shapes made by manipulating the plasticine (e.g. a bowl, a plate) are also equivalent. However, if a hole is put in the plasticine, the shape is no longer topologically equivalent. Thus, for example, a sphere and a doughnut shape are not topologically equivalent (see the Teaching Idea on p. 389).

The Möbius strip can be the basis of a rich topological investigation. Constructing a Möbius strip involves taking a long strip of paper, turning one end through one half-turn, and taping the two ends together. If a line is drawn down the middle of the Möbius strip, it eventually meets its beginning point. Thus, a Möbius strip is considered to have one side only. Interesting activities can begin with making a longitudinal cut about one-third in from the edge of the strip (see the following Teaching Idea). Making cuts at other distances in from the edge of the strip yields different results. These activities can lead to conjecture, hypotheses, discussion and questioning.

TEACHING IDEA

The Möbius strip

1. Make a Möbius strip by cutting out a flat strip of paper about 50 cm long. Twist the paper (once), then join the two ends to make a closed ring.
2. Try colouring one side of the strip red and the other side green (or any two different colours). What do you notice?
3. Try to draw a line along the centre of the strip, continuing until you come back to the same point from which you started. Are you convinced that this strip has only one side?
4. Predict what you think will happen if you cut along the middle of the strip of paper. Try it. Were you surprised by the result?

5. Make another Möbius strip. This time, draw a line along the strip one-third in from the edge. Continue to draw the line the same distance from the edge until you return to the point from which you started. Cut along the line you have just drawn. What happened? Did you expect that this would occur?
6. Make a Möbius strip that is twisted twice before the ends are joined. Repeat activities 3, 4 and 5.
7. Make Möbius strips with three or four twists, and repeat activities 3, 4 and 5. Can you discover a pattern emerging?

Symmetry

Symmetry is a geometrical property that can be identified in everyday objects in both the natural and built environments as well as in some geometric figures (2D shapes and 3D objects).

Line or lateral symmetry

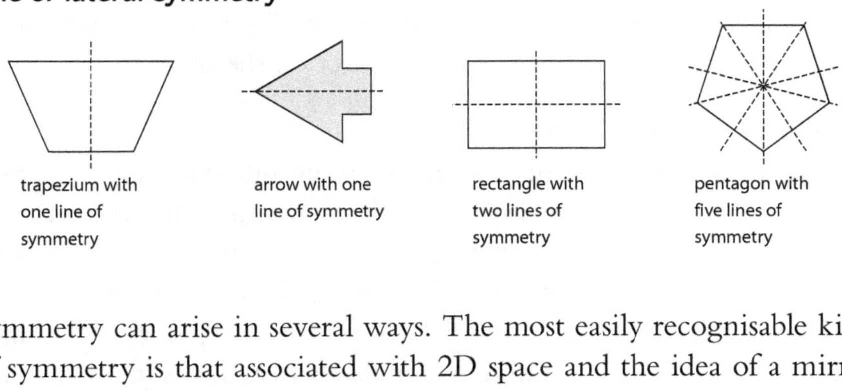

trapezium with one line of symmetry

arrow with one line of symmetry

rectangle with two lines of symmetry

pentagon with five lines of symmetry

Symmetry can arise in several ways. The most easily recognisable kind of symmetry is that associated with 2D space and the idea of a mirror image. This kind of symmetry can be seen in natural objects such as leaves and, in geometry, plane figures (2D figures) (Raisor & Hudson 2018). As seen in the examples above, symmetry of this kind is associated with a *line of symmetry*. This kind of symmetry is referred to as line symmetry, lateral symmetry or mirror-image symmetry. An important activity for students is to use a mirror to reflect images of objects back on themselves in order to determine lines of symmetry. Another excellent resource is the Mira (a hard piece of translucent plastic), which can be placed perpendicular to a shape so that students can see both the original shape and the mirror shape. Activities using the Mira can be found here: www.mrlondonsclass.com/mira.html. Activities using mirrors or Miras can be applied to everyday flat objects (e.g. leaves), to pictures (or a person's face), to symbols (such as many upper-case letters) and to 2D shapes. Place the mirror on the centre line of an object and see whether the reflected half of the image fits exactly on the actual object. Line symmetry can also be investigated through paper-folding along a line. To test whether a line of symmetry exists, cut the shape out of paper and fold it in half back on itself along the predicted line of symmetry. If it folds back exactly, that is the line of symmetry.

Try to categorise the 26 upper-case letters according to the number of lines of symmetry. Students have opportunities to determine all the possible lines of symmetry for a range of 2D shapes—various triangles and quadrilaterals (square, rhombus, kite, etc.), and various pentagons, hexagons, and so on. There is also a correspondence between line symmetry and the congruent transformation of a shape by flip or reflection (see above). When a shape is flipped, it becomes a mirror image of the original object. A line is drawn to indicate the exact point where the flip occurred to result in the mirror image. This is the line of symmetry for an object under a flip transformation. The figure on one side of the line of symmetry is the image of the object on the other side.

Rotational symmetry in 2D space

Symmetry in 2D space can arise in two ways: line symmetry and rotational symmetry. Rotational symmetry is associated with turning the 2D shape through a unit fraction (½, ⅓, ¼, etc.) of a full circle. To test for rotational symmetry, draw an outline of the shape on another piece of paper. Place the shape on its outline. Mark the 'top point' of the shape with a dot. Turn the shape until it completely matches its outline. Count the number of times it fits its outline until it reaches the original position. An object or figure is said to have rotational symmetry of order two if, when turned through one-half of a circle (180°), the image corresponds exactly to the object in terms of the position it occupies. Rotational symmetry of order two is the simplest case of rotational symmetry, and is referred to as *point symmetry*. In a similar vein, an object or figure is said to have rotational symmetry of order three if, when turned through one-third of a circle (120°), the image corresponds exactly to the object in terms of the position it occupies. Rotational symmetries of orders four, five, six, and so on are defined similarly.

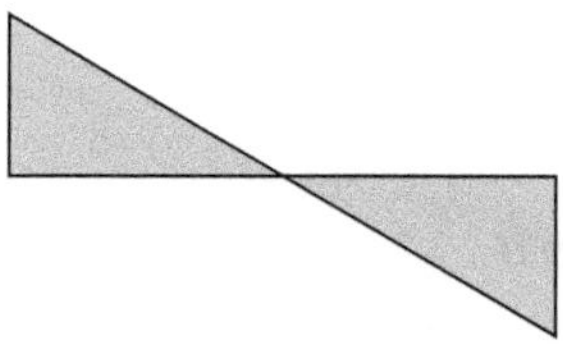

order two rotational symmetry

order four rotational symmetry

order three rotational symmetry

As in the case of line symmetry, activities relating to rotational symmetry can involve flat objects, symbols (upper-case letters) and 2D shapes. Try to categorise the 26 upper-case letters in terms of whether or not they have rotational symmetry and, if so, of what order. A final point to keep in mind is the general separateness of line symmetry and rotational symmetry. Although an equilateral triangle has three lines of symmetry and rotational symmetry of order three, and a square has four lines of symmetry and rotational symmetry of order four, line symmetry and rotational symmetry are not always associated in this way. Try to draw figures that have rotational symmetry of various orders and no lines of symmetry, or figures with one or more lines of symmetry but no rotational symmetry.

Symmetry in 3D space

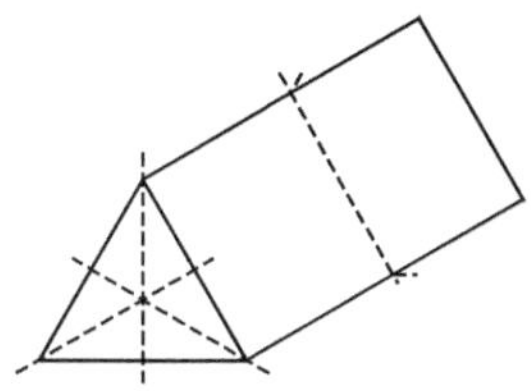

Symmetry can also be considered in 3D objects. Imagine cutting an object in half. Do you have two identical objects? If yes, then the cut that yielded the two objects that mirror each other is called a *plane of symmetry*. Consider a solid such as a cylinder—how many planes of symmetry does it have? Since circles have an infinite number of lines of symmetry, cylinders have an infinite number of planes of symmetry. Rotational symmetry can also be considered in 3D space. Rotational symmetry about a point in 2D space corresponds with rotational symmetry about an axis in 3D space. As in the case of planes of symmetry, students could investigate axes of symmetry, and the order corresponding to each axis, which is associated with 3D mathematical objects and other everyday solids.

TEACHING IDEA

Symmetry in 3D space

- Find the planes of symmetry in the five regular solids.
- Identify some everyday objects (e.g. boats, cars, caravans, etc.) that may have planes of symmetry.
- Classify the objects as symmetrical and non-symmetrical

Tessellations

The topic of tessellations is associated with 2D shape, and links to the everyday notion of tiling and area. A 2D shape is said to tessellate if it can cover an area without gaps or overlaps. Below are some examples of tessellating shapes. They extend in all directions without gaps or overlaps.

Tessellating shapes

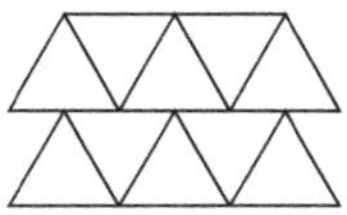

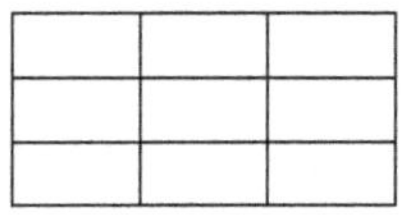

The following examples do not tessellate, since there are gaps when the shapes are placed together.

Non-tessellating shapes

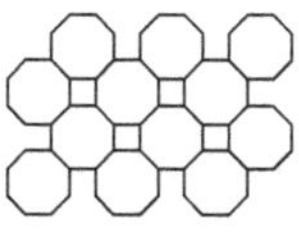

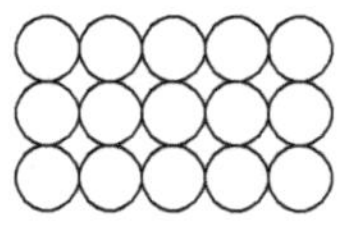

TEACHING IDEA

Tessellations

- Which of the regular polygons tessellate? Think about each of these in turn—equilateral triangle, square, regular pentagon, regular hexagon, and so on.

- Consider a scalene triangle—for example, with sides of 6 cm, 8 cm and 13 cm. Will this triangle tessellate?
- Consider a quadrilateral with sides of 4 cm, 7 cm, 9 cm and 12 cm. Will this figure tessellate?
- Try to find a pentagonal shape that will tessellate.

One of the appealing aspects of the topic of tessellation is its potential for integrating 2D geometry and art. Use of tessellation in artistic and decorative contexts has a long history. This topic lends itself well to investigations of library sources and websites for artistic examples of tessellations as they occur in a range of cultural and historical contexts (Paznokas 2003). Investigations could include finding information about, and examples of, the work of Mauritz (M.C.) Escher, who embraced both art and geometry. His work included unusual and picturesque examples of tessellation (see the following Teaching Idea). Students can use tessellation as a basis for their own artwork.

TEACHING IDEA

Escher tessellations

Take a regular tessellating shape, such as a rectangle.

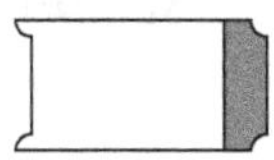

Make a cut from one end and then slide this to the opposite end and glue. Do the same to the corresponding edges.

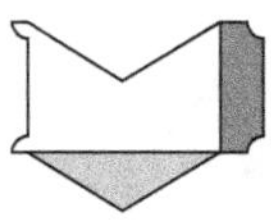

This odd shape may not appear as if it will tessellate. Students should be allowed to predict whether it will tessellate or not. Provide opportunities for students to create their own tessellations and share these with the class.

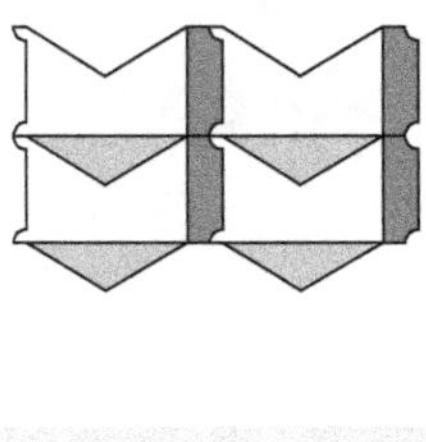

To create Escher-type artworks, begin with a shape that tessellates. Change the shape of one or more sides—for example, by introducing curves and using congruence transformations (flips, slides or turns) as the basis for additional and compensating changes that preserve the tiling aspect of the shape. In this way, the topic of tessellation is seen to link closely to the topics of transformations and symmetry (Whitaker 2001).

Technology and geometry

Several software programs have been developed to support teaching and learning in geometry. LOGO®, for example, was developed more than 30 years ago, and, together with its recent derivations (e.g. Microworlds®) supports learning of geometry, and in particular the sub-topic of location. LOGO® has useful applications across the primary-school years and is also useful for teaching coding. More recently developed programs for geometry are Cabri®, Geometer's Sketchpad®, Cinderella® and GeoGebra®—commercial programs suitable for the upper years of primary schooling and into early secondary schooling. The usefulness of these programs relates to the ways in which they enable students to engage in dynamic geometry activities (Battista 2002).

Location and arrangement

The topic of location focuses on ways to characterise position within a space. In curriculum documents, this sub-strand has a range of labels—location, arrangement or position—and includes learning about coordinate systems. At more advanced levels, this includes developing notions associated with map reading—for example, coordinates and scales—and using a globe to find locations and learn about topics such as latitude and longitude.

The language of location and movement

Locational or positional language is a key component of early location work. Students come to school with varying experiences of language relating to the notion of location. Some students will have a well-established location language, whereas others may not. Teachers need to spend time developing this aspect of location. The language of location includes words to describe objects in relation to others—the chair is 'behind' the table; the vase is 'above' the cabinet. Other language can be related to movement—move 'forward' one step; jump 'backward' three squares.

Following is a list of terms associated with the language of location:

above	top
below	bottom
right	forward
left	backward
in front of	beside
behind	around
	next to

Early location experiences

Students can begin to learn about coordinate systems in the early years (Dobler & Klein 2002). Early experiences focus on students learning to read the grid references corresponding to a region where items appear. Typically, these involve an alpha-numerical pairing—that is, letters are used on one axis and numerals are used on the other. Each square on the map is characterised by a distinct alpha-numerical pair. When using a map such as this, by convention, students would work with letters on the *x*-axis (horizontal) and numerals on the *y*-axis (vertical). Using the alpha-numerical scale supports students' learning to read the *x*-axis first and *y*-axis second. This is important for both the topic of location and further work with graphs and functions. Students need to experience two aspects of location. First, the coordinates are given and the student identifies the site—'What is located at B2?' Second, the site is given and the student determines the coordinates—'What are the coordinates for the bag of money?'

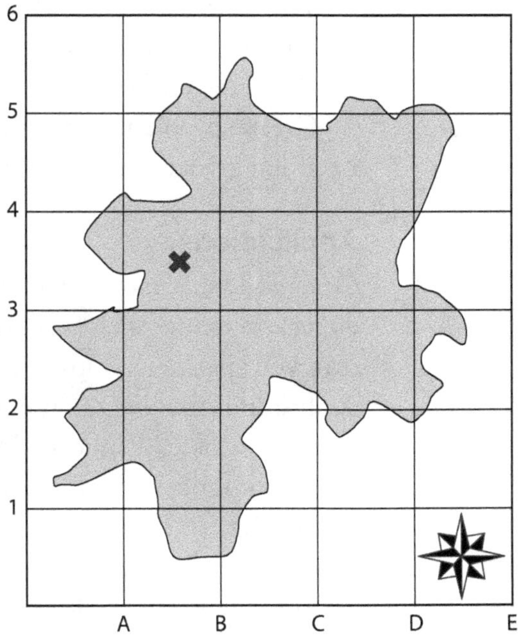

Simple map with alpha-numeric coordinates

As students develop knowledge of alpha-numerical scales, they can be introduced to the use of ordered numerical pairs. Thus, activities involving reading the value on the *x*-axis, then reading the value on the *y*-axis, serve to develop knowledge of the more abstract notion of an ordered pair. The term 'ordered' is used in the sense that, by mathematical convention, the first value of the pair corresponds to the value on the *x*-axis, and the second corresponds to the value on the *y*-axis. Notation such as (3, 5) is used for an ordered pair.

Later experiences of location

Once students have learned to read ordered pairs, teaching can focus on associating each ordered pair with a point rather than a region. As part of the progression to associating each ordered pair with a point, each numeral on the axes corresponds with a grid line rather than an interval on the axis. This can be seen on the map on p. 397, where the numerals are aligned with the grid lines.

Initial learning experiences may involve the items being placed on intersecting grid lines. As students' fraction knowledge develops, items to be located can be placed at points where students can readily identify the point—this can initially involve simple fractions such as halves and quarters (1.5, 2–2.5), and later tenths (1.3, 1.8). This approach enables students to see the usefulness of finer gradations on the axes—points can be characterised more precisely.

Create problems for students whereby they come to 'discover' the need for gradations in the pairings.

Applications

Once students have developed knowledge of ordered pairs, they can engage in games such as 'Battleships', where the objective is to locate and sink your partner's ships (see the following Teaching Idea). More realistic applications involve the use of maps (analogue and digital) to locate sites. Initially, this can involve local street directories, which typically use alpha-numerical scales and regions. Later experiences can focus on using maps in an atlas or online. An important point to keep in mind when working with maps in atlases is that coordinates are specified and read as follows: first latitude (north or south), then longitude (east or west). Thus, the order of reading these map coordinates is at odds with the order of reading coordinates on a grid system—first horizontal (*x*-axis), then vertical (y-axis). Students may struggle with this at first, in part due to the order of reading the pairs, and also because of the use of four quadrants (NE, NW, SE, SW), but it is a life skill that is important for students to understand.

TEACHING IDEA

Battleships

Using grid paper, have students put a number of ships on the paper. In pairs, students list off coordinates to identify and sink their partner's ships. The winner is the one who sinks all the ships first.

Moving through space

A common teaching practice is for students to be provided with a map and then given a series of directions—such as move forward two units, turn right, then move forward x units. Such activities often use street names to incorporate an element of reality into the task. Commonly, this activity may involve starting at a certain point, such as a railway station, and then navigating through streets, turns, distances, and so on to arrive at a particular place, which the student must name. This requires the coordination of a number of variables and is quite complex. It is a useful activity, in that it resembles out-of-school experiences of using maps to find sites or moving through new territories. Diezmann and Lowrie (2010) report that in this activity, students often make errors by not coordinating the various elements of the task, such as overlooking one element—ordinal language ('the third street'), positional language ('at the school' or 'into Lambert Road') or positional language ('turn right'). They report that the development of this skill appears to plateau in the middle years around Year 6 or 7.

Location and technology

Research is beginning to emerge about students' map-reading skills. We have shown the typical teaching process, whereby elements are built into the curriculum progressively so that students learn the required mapping skills. This approach seems to limit what students can do when they use digital environments. Typically, experiences in school mathematics have been contained to two-dimensional space—such as maps on paper. However, in the ICT and games environments, young children are exploring 3D representations in which objects are moved along the two dimensions of the screen but also appear to 'fall' into three-dimensional spaces. Lowrie (2003) has shown how young children (eight-year-olds) have been found to construct complex images of the maps beyond immediate visual cues and to construct three-dimensional images. In his work in this area, Lowrie has shown how young children not only create three-dimensional mental maps of the worlds in which they play (such as Mario Brothers, Pokemon or Minecraft) but are also capable of creating mental images of the maps outside the screen area. The impact of this has not yet been recognised formally in curriculum documents, but it challenges the static current views of map reading. Furthermore, middle-school students are often involved in complex computer games in which they are required to navigate through worlds to solve problems. The challenges they confront in these worlds are not only useful in terms of problem-solving but also help to develop visualisation, as students create mental maps of the worlds they will encounter as they move from one screen to the next. This information is not available to students, but must be created before they are able to move successfully into the next screen. The impact of these games will have a profound influence on how map reading is to be conceptualised in the emerging theorising of curriculum and knowing. Clearly, these technologies will create new learning opportunities for spatial thinking and visualisation that were inconceivable in the traditional pencil-and-paper format of previous times.

A further advance in modern society is the increasing availability of global-positioning systems (GPS). Various tools are now available that allow students to undertake location activities. GPS tools allow for activities such as orienteering or linking to other curriculum areas (such as science, to plot the location of species or to track animal migrations). While some of these tools may be expensive, others—such as car-navigation tools—are more readily accessible, and offer new representations of space and movement through space. Similarly, these tools are available widely on mobile phones and can readily be used to support learning activities. As mobile devices become increasingly available, access to such resources will enable new modes of teaching and learning to be made available to students and teachers.

Direction

Direction is another aspect of the geometry topic of location, and involves learning about compass points. Initially, students learn about the four compass points: north, south, east and west. Depending on syllabus requirements, this is extended to eight points in all by the inclusion of north-east, north-west, south-east and south-west, or to 16 points by also including NNE, ENE, NNW, WNW, SSE, ESE, SSW and WSW.

In students' initial learning of the four major compass points, north and south are usually learned relatively easily, but some confusion can arise when learning east and west. Mnemonic devices—such as 'Never Eat Soggy Weetbix'—can be used for learning these as students rotate clockwise through the points. A common instructional activity involves students finding a particular point by following directions on a grid map—'two paces north, three paces west', and so on. Orienteering activities involving the use of a compass are also useful for this aspect of location.

The cultural relativity of direction is an interesting area of application. For most students growing up in Western societies, the concept of direction is relative—that is, it is relative to where you are located. Terms such as left and right are commonly used for finding or giving directions. In some cultures, however, different applications are used. Some Indigenous Australian cultural groups, for example, use compass points in their everyday discussions (Harris 1990). In other contexts, different methods are developed. For example, in Hawaii, relative (but significant) points are used as the terms of reference ('Diamond Head', the mountains), so that compass points are not used when giving directions.

Teaching geometry in the middle years

The van Hiele levels of geometric thinking, summarised at the beginning of this chapter, are the foundation for instruction in the primary curriculum. Levels 4 and 5 would be the focus of activities in the primary years only informally. However, the middle years should be laying the foundation for students to achieve at level 4, in particular, where students develop the capacity to argue the case for why certain geometric situations are mathematically sound. This is the notion of proof, which is something that is not frequently used in the discourse of the middle-years mathematics curriculum. However, calls have been made for proof to be an integral part of the middle-years curriculum, by providing convincing arguments in mathematics and showing that 'what is offered as a convincing argument by one person must be accepted by others' (Harel & Sowder 2007, p. 808). Harel and Sowder discuss the notion of a proof scheme, rather than proof in the formal sense, whereby members of a (classroom) community determine the parameters for ascertaining and persuading the community about the 'truth' of an assertion. In such a community, students would be given tasks whereby they must persuade others of the legitimacy of their thinking. Their argument is based on examples and generalisations rather than reference to an authoritative source, such as the teacher, the internet or a textbook. To emphasise the importance of engaging students in discussions about geometric relationships and notions, Harel and Sowder provide the following assessment item, used to determine Year 11 students' understanding of mathematical argumentation and proof: 'Jim says, "If a 4-sided figure has all equal sides, it is a square". Which of the following figures might be used to prove that Jim is wrong?'

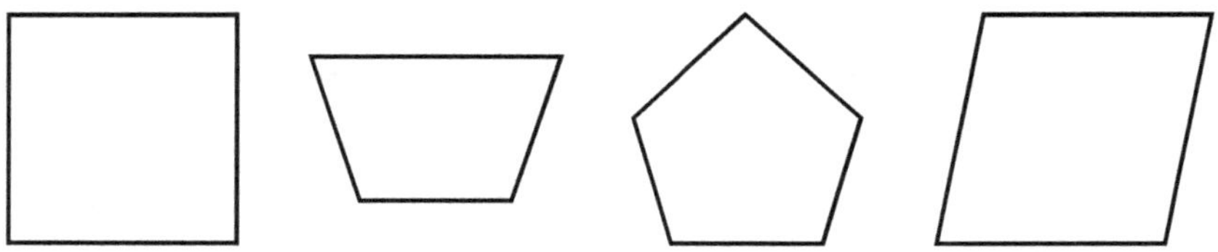

The number of students selecting each figure was 15, 31, 24 and 31 per cent, respectively. The point is that middle-years teachers need to move students beyond level 3 of the van Hiele model. Proof schemes require students to build mathematical arguments to convince others of their solutions, necessitating revisiting knowledge and understanding of properties of shapes and relationships.

To support students' movement between the van Hiele levels, extensive opportunities to explore and experience spatial activities are required. In the primary years, hands-on activities including building, designing, constructing and representing are advocated for space. This must continue in the middle years, with tasks becoming increasingly sophisticated. For example, constructing shapes from 2 cm cubes, then drawing these designs on isometric paper, promotes visualisation and arrangement, which support reading and interpretation of the architectural drawings that students will most likely encounter in their lives in the future.

Review questions

15.1 What common observations might you expect from your students when teaching basic geometric plane (2D) shapes that would confirm the van Hiele theory?

15.2 Explain the difference between nets and skeletal forms, using examples to illustrate the difference(s).

15.3 Describe three introductory activities that would be appropriate for the topic of tessellation.

15.4 Students need to have considerable experiences in language when working in the geometry strand. Give examples of early language that should be developed, and of how you would go about providing learning experiences to develop this language.

15.5 Outline learning activities suitable for the topic of symmetry in the middle-primary years.

15.6 In a multi-age classroom, there is considerable diversity in student learning. How would you use one stimulus map to cater for a diverse range of learning when teaching mapping skills to students?

15.7 Find out how other cultural groups use directions in their everyday lives. Compare and evaluate this with standard curriculum expectations.

Further reading

Battista, M. & Clements, D.H. (1998). Finding the number of cubes in rectangular cube buildings. *Teaching Children Mathematics*, *4*(5), 258–64.

Britton, B. & Stump, S. (2001). Unexpected riches from a geoboard quadrilateral activity. *Mathematics Teaching in the Middle School*, *6*(8), 490–3.

Clements, D. & Sarama, J. (2000). The earliest geometry. *Teaching Children Mathematics*, 7(2), 82–6.

Cockcroft, W.C. & Marshall, J. (1999). Educating Hannah: It's a what? *Teaching Children Mathematics*, *5*(6), 326–9.

Fox, T. (2000). Implications of research on young children's understanding of geometry. *Teaching Children Mathematics*, *6*(9), 572–6.

Furner, J.M. (2017). Using fairy tales and children's literature in the math classroom: Helping all students become Einstein's in a STEM world. *Journal of Advances in Education Research*, *2*(2), 103–12.

Koester, B.A. (2003). Prisms and pyramids: Constructing three-dimensional models to build understanding. *Teaching Children Mathematics*, *9*(8), 436–42.

Liedtke, W.W. (1995). Developing spatial abilities in the early grades. *Teaching Children Mathematics*, *2*(1), 12–18.
Lindquist, M.M. & Clements, D.H. (2001). Geometry must be vital. *Teaching Children Mathematics*, 7(7), 409–15.
Mitchelmore, M. (2000). Teaching angle measurement without turning. *Australian Primary Mathematics Classroom*, *5*(2), 4–8.
Simpson, D. (2001). Shapes, shapes, everywhere! *Prime Number*, *16*(3), 4–7.
Southwell, B. (1998). Give me space. *Reflections*, *23*(1), 52–6.
Trafton, P.A. & Hartman, C. (1997). Exploring area with geoboards. *Teaching Children Mathematics*, *4*(2), 72–5.
van Hiele, P.M. (1999). Developing geometric thinking through activities that begin with play. *Teaching Children Mathematics*, *5*(6), 310–16.
van Hiele, P.M. (2002). Similarities and differences between the theory of learning and teaching of Skemp and the van Hiele levels of thinking. In D. Tall and M. Thomas (eds), *Intelligence, learning and understanding in mathematics: A tribute to Richard Skemp* (pp. 2–47). Brisbane: Post Pressed.
Wheatley, G. (1990). Spatial sense and mathematics learning. *Arithmetic Teacher*, *37*(6), 10-15.
White, P. & Mitchelmore, M. (1997). Sharpening up on angles. *Australian Primary Mathematics Classroom*, *3*(1), 19–21.
Yackel, E. & Wheatley, G.H. (1990). Promoting visual imagery in young pupils. *Arithmetic Teacher*, *37*(6), 52–8.

References

Alex, J.K. & Mammen, J.K. (2016). Geometrical sense making: Findings of analysis based on the characteristics of the van Hiele theory among a sample of South African Grade 10 learners. *Eurasia Journal of Mathematics, Science and Technology Education*, *12*(2), 173–88.
Ambrose, R.C. & Falkner, K. (2002). Developing spatial understanding through building polyhedrons. *Teaching Children Mathematics*, *8*(8), 442–7.
Battista, M. (2002). Learning geometry in a dynamic computer environment. *Teaching Children Mathematics*, *8*(6), 333–9.
Battista, M.T. (2010). Thoughts on elementary students' reasoning about 3D arrays of cubes and polyhedra. In *Future curricular trends in school algebra and geometry: Proceedings of a conference* (pp. 183–99). Charlotte, NC: Information Age.
Britton, B. & Stump, S. (2001). Unexpected riches from a geoboard quadrilateral activity. *Mathematics Teaching in the Middle School*, *6*(8), 490–3.
Chorney, S. & Sinclair, N. (2018). Fingers-on geometry: The emergence of symmetry in a primary school classroom with multi-touch dynamic geometry. In N. Calder, K. Larkin, and N. Sinclair (eds), *Using mobile technologies in the teaching and learning of mathematics* (pp. 213–30). Cham: Springer.
Crowley, M. (1987). The van Hiele model of the development of geometric thought. In M.M. Lindquist (ed.), *Learning and teaching geometry, K–12: 1987 yearbook* (pp. 1–16). Reston, VA: National Council of Teachers of Mathematics.

Diezmann, C. & Lowrie, T. (2010). Students as decoders of graphics in mathematics. In L. Sparrow, B. Kissane and C. Hurst (eds), *Shaping the future of mathematics education: Proceedings of the 33rd Annual Conference of the Mathematics Education Research Group of Australasia* (pp. 161–8). Fremantle: MERGA.

Dobler, C.P. & Klein, J.M. (2002). First graders, flies, and a Frenchman's fascination: Introducing the Cartesian coordinate system. *Teaching Children Mathematics*, *8*(9), 540–5.

Ferguson, S. (2017). Teaching shape in primary school: Much more than 'name that shape'. *Prime Number, 32*(3), 22–3.

Harel, G. & Sowder, J. (2007). Towards comprehensive perspectives on the learning and teaching of proof. In F. Lester (ed.), *Second handbook of research on mathematics teaching and learning* (pp. 805–42). Charlotte, NC: Information Age.

Harris, P. (1990). *Mathematics in a cultural context: Aboriginal perspectives on space, time and money.* Geelong: Deakin University Press.

Host, E., Baynham, E. & McMaster, H. (2014). Using digital technology to see angles from different angles: Part 1: Corners. *Australian Primary Mathematics Classroom*, *19*(2), 18–22.

Howse, T.D. & Howse, M.E. (2015). Linking the van Hiele theory to instruction. *Teaching Children Mathematics, 21*(5), 304–13.

Jones, K. & Tzekaki, M. (2016). Research on the teaching and learning of geometry. In Á. Gutiérrez, G.C. Leder and P. Boero (eds), *The second handbook of research on the psychology of mathematics education: The journey continues* (pp. 109–49). Rotterdam: Sense Publishers.

Larkin, K. & Milford, T. (2018). Using cluster analysis to enhance student learning when using geometry mathematics apps. In L. Ball, P. Drijvers, S. Ladel, H.-S. Siller, M. Tabach and C. Vale (eds), *Uses of technology in primary and secondary mathematics education: Tools, topics and trends* (pp. 101–18). Cham: Springer.

Lowrie, T. (2003). Posing problems in ICT-based contexts. In L. Bragg, C. Campbell, G. Herbert and J. Mousley (eds), *Mathematics education research: Innovation, networking and opportunity. Proceedings of the 26th Annual Conference of the Mathematics Education Research Group of Australasia* (vol. 1, pp. 499–506). Geelong: MERGA.

Lowrie, T. & Diezmann, C.M. (2011). Solving graphics tasks: Gender differences in middle-school students. *Learning and Instruction*, *21*(1), 109–25.

Lowrie, T., Logan, T. & Ramful, A. (2017). Visuospatial training improves elementary students' mathematics performance. *British Journal of Educational Psychology, 87*(2), 170–86. doi:10.1111/bjep.12142.

Mitchelmore, M. (2000). Teaching angle measurement without turning. *Australian Primary Mathematics Classroom*, *5*(2), 4–8.

Mix, K.S. & Cheng, Y.-L. (2012). Space and math: The developmental and educational implications. In J. Benson (ed.), *Advances in child development and behavior* (pp. 179–243). New York, NY: Elsevier.

National Council for Teachers of Mathematics (2000). *Principles and standards for school mathematics*. Reston, VA: NCTM.

Ozkan, M. & Bal, A.P. (2017). Analysis of the misconceptions of 7th grade students on polygons and specific quadrilaterals. *Eurasian Journal of Educational Research (67)*, 161–82.

Paznokas, L. (2003). Teaching mathematics through cultural quilting. *Teaching Children Mathematics*, *9*(5), 250–6.

Pegg, J. & Davey, G. (1989). Clarifying level descriptors for children's understanding of some basic 2-D geometric shapes. *Mathematics Education Research Journal*, *1*(1), 16–27.

Plaksin, O. & Patkin, D. (2016). Exploring the minimum conditions for congruency of polygons. *Learning and Teaching Mathematics, (21)*, 42–6.

Raisor, J.M. & Hudson, R.A. (2018). Bottle caps as prekindergarten mathematical tools. *Teaching Children Mathematics, 24*(6), 370–7. doi:10.5951/teacchilmath.24.6.0370.

Seah, R. (2018). Choosing the right resources to support the learning of polygons. *Australian Primary Mathematics Classroom, 23*(1), 3–8.

Stieff, M. & Uttal, D. (2015). How much can spatial training improve STEM achievement? *Educational Psychology Review,* 27(4), 607–15. doi:10.1007/s10648-015-9304-8.

Trafton, P.A. & Hartman, C. (1997). Exploring area with geoboards. *Teaching Children Mathematics*, *4*(2), 72–5.

Tsamir, P., Tirosh, D., Levenson, E., Barkai, R. & Tabach, M. (2015). Early-years teachers' concept images and concept definitions: triangles, circles, and cylinders. *ZDM: The International Journal on Mathematics Education, 47*(3), 497–509.

van Hiele, P.M. (1999). Developing geometric thinking through activities that begin with play. *Teaching Children Mathematics*, *5*(6), 310–16.

Verdine, B.N., Golinkoff, R.M., Hirsh-Pasek, K. & Newcombe, N.S. (2017). Links between spatial and mathematical skills across the preschool years. *Monographs of the Society for Research in Child Development, 82*(1), 1–184.

Whitaker, R. (2001). Tessellating can be fun. *Vinculum*, *38*(2), 10–12

INDEX

For Product Safety Concerns and Information please contact our EU
representative GPSR@taylorandfrancis.com
Taylor & Francis Verlag GmbH, Kaufingerstraße 24, 80331 München, Germany

www.ingramcontent.com/pod-product-compliance
Lightning Source LLC
LaVergne TN
LVHW080629011026
849160LV00014B/1158
* 9 7 8 1 7 6 0 5 2 9 7 3 4 *